THE MECHANICAL DESIGN PROCESS

Schaum's Outline Series

Schaum's Outline of Acoustics by William W. Seto
Schaum's Outline of Continuum Mechanics by George E. Mase
Schaum's Outline of Engineering Mechanics, 4/e William G. McLean and F.W. Nelson
Schaum's Outline of Fluid Dynamics, 2/e William F. Hughes and J.A. Brighton
Schaum's Outline of Heat Transfer Donald Pitts and Leighton e. Sissom
Schaum's Outline of Machine Design A.S. Hall and HG Laughlin
Schaum's Outline of Mechanical Vibrations Graham Kelly
Schaum's Outline of Elementary Statics and Strength of Materials John J. Jackson & Harold G. Wirtz
Schaum's Outline of Thermodynamics for Engineers Merle C. Potter & Craig Somerton

Schaum's Software Series

Schaum's Electronic Tutor for Mechanical Vibrations
Schaum's Interactive Fluid Mechanics & Hydraulics

Classic McGraw-Hill Handbooks

Marks' Standard Handbook for Mechanical Engineers, 10/e
The Marks' Electronic Handbook for Mechanical Engineers
Shock and Vibration Handbook, 4/e
Mechanical Design Handbook

THE MECHANICAL DESIGN PROCESS

Second Edition

David G. Ullman
Oregon State University

The McGraw-Hill Companies, Inc.

New York St. Louis San Francisco Auckland Bogotá Caracas Lisbon London Madrid Mexico City Milan Montreal New Delhi San Juan Singapore Sydney Tokyo Toronto

McGraw-Hill

A Division of The ***McGraw-Hill*** *Companies*

THE MECHANICAL DESIGN PROCESS
International Editions 1997

1 2 3 4 5 6 7 8 9 0 CWP PMP 9 8 7

This book was set in Times Roman by Publication Services, Inc.
The editors were Debra Riegert and John M. Morriss;
the production supervisor was Denise L. Puryear.
The cover was design by Farenga Design Group.

Library of Congress Cataloging-in-Publication Data

Ullman, David G.,
The mechanical design process / David G. Ullman. – 2nd ed.
p. cm.
Includes index.
ISBN 0-07-065756-4
1. Machine design. I. Title.
TJ230.U54 1997
621.8'15–dc20 96-30987

http://www.mhcollege.com

When ordering this title, use ISBN 0-07-115576-7

Printed in Singapore

ABOUT THE AUTHOR

David G. Ullman is an active product designer who has taught, researched, and written about design for over twenty years. He is Professor of Mechanical Design at Oregon State University and Vice President of New Product Development for BikeE Corporation. He has professionally designed fluid/thermal, control, and transportation systems. He has published over twenty papers focused on understanding the mechanical product design process and the development of tools to support it. He is on the editorial board of four journals in this field and is founder of the Design Theory and Methodology Committee of the American Society of Mechanical Engineers.

CONTENTS

Preface xiii

1 Why Study the Design Process? 1

1.1 Introduction 1
1.2 Measuring the Design Process with Product Cost, Quality, and Time to Production 2
1.3 The History of the Design Process 6
1.4 The Life of a Product 9
1.5 The Many Solutions of Design Problems 12
1.6 The Basic Actions of Problem Solving 14
1.7 Knowledge and Learning During Design 15
1.8 Summary 16
1.9 Sources 16
1.10 Exercises 17

2 Describing Mechanical Design Problems and Process 18

2.1 Introduction 18
2.2 Decomposition of Mechanical Systems 18
2.3 Importance of Product Function, Behavior, and Performance 21
2.4 Different Types of Mechanical Design Problems 23
2.4.1 Selection Design 23
2.4.2 Configuration Design 24
2.4.3 Parametric Design 26
2.4.4 Original Design 27
2.4.5 Redesign 27

2.5 Languages of Mechanical Design 30
2.6 Constraints, Goals, and Design Decisions 31
2.7 The Value of Information 32
2.8 Design as Refinement of Abstract Representations 33
2.9 Summary 33
2.10 Sources 36
2.11 Exercises 36

3 Designers and Design Teams 37
3.1 Introduction 37
3.2 A Model of Human Information Processing 38
3.2.1 Short-Term Memory 40
3.2.2 Long-Term Memory 42
3.2.3 Control of the Information Processing System 43
3.2.4 External Environment 44
3.2.5 Implications of the Model 45
3.3 Mental Processes that Occur During Design 45
3.3.1 Understanding the Problem 45
3.3.2 Generating a Solution 46
3.3.3 Evaluating the Solution 47
3.3.4 Controlling the Design Process 47
3.3.5 Problem-Solving Behavior 47
3.4 Characteristics of a Creative Designer 51
3.5 Engineering Design Teams 54
3.5.1 Team Goals 55
3.5.2 Team Roles 55
3.5.3 Building Team Performance 56
3.6 Summary 57
3.7 Sources 57
3.8 Exercises 58

4 The Design Process 60
4.1 Introduction 60
4.2 Overview of the Design Process 60
4.2.1 Plan for the Design Process 64
4.2.2 Develop Engineering Specifications 64
4.2.3 Develop Concepts 65
4.2.4 Develop Product 65
4.3 The Design Process: An Organization of Techniques 66
4.4 Simple Design Process Examples 68
4.5 A More Complex Example: Design Failure in the Space Shuttle *Challenger* 70
4.6 Communication During the Design Process 72
4.6.1 Design Records 73
4.6.2 Documents Communicating with Management 73
4.6.3 Documents Communicating the Final Design 74
4.7 Introduction of a Sample Design Problem 74
4.8 Summary 75
4.9 Sources 76
4.10 Exercises 76

5 Planning for the Design Process 77
5.1 Introduction 77
5.2 Background for Developing a Design Project Plan 78
5.2.1 Types of Design Projects 78
5.2.2 Members of Design Teams 79
5.2.3 Structure of Design Teams 81
5.3 Planning for Deliverables 83
5.4 The Five Steps in Planning 85
5.5 A Design Plan for Low-Volume Products 89
5.6 A Design Plan for Mass-Produced Products: Xerox Corporation 90
5.7 A Design Plan for Mass-Produced Products: The Splashgard 92
5.8 Summary 96
5.9 Sources 96
5.10 Exercises 96

6 Understanding the Problem and the Development of Engineering Specifications 98
6.1 Introduction 98
6.2 Step 1: Identify the Customers: Who Are They? 102
6.3 Step 2: Determine the Customers' Requirements: What Do the Customers Want? 104
6.3.1 The Kano Model of Customer Satisfaction 104
6.3.2 Collection Methods for Customers' Requirements 105
6.3.3 The Types of Customers' Requirements 108
6.3.4 The Splashgard Requirements 111
6.4 Step 3: Determine Relative Importance of the Requirements: Who versus What 112
6.5 Step 4: Identify and Evaluate the Competition: How Satisfied Is the Customer Now? 113
6.6 Step 5: Generate Engineering Specifications: How Will the Customers' Requirements Be Met? 115
6.7 Step 6: Relate Customers' Requirements to Engineering Specifications: Hows Measure Whats? 116
6.8 Step 7: Identify Relationships Between Engineering Requirements: How Are the Hows Dependent on Each Other? 116
6.9 Step 8: Set Engineering Targets: How Much Is Good Enough? 116
6.10 Further Comments on QFD 117
6.11 Summary 118
6.12 Sources 119
6.13 Exercises 119

7 Concept Generation 120
7.1 Introduction 120
7.2 A Technique for Functional Decomposition 122
7.2.1 Step 1: Find the Overall Function That Needs to Be Accomplished 123
7.2.2 Step 2: Create Subfunction Descriptions 125
7.2.3 Step 3: Order the Subfunctions 129
7.2.4 Step 4: Refine Subfunctions 131

7.3 A Technique for Generating Concepts from Functions 133
7.3.1 Step 1: Developing Concepts for Each Function 133
7.3.2 Step 2: Combining Concepts 134
7.4 Sources for Concept Ideas: Techniques to Prime Creativity 140
7.4.1 Using Patents as an Idea Source 141
7.4.2 Finding Ideas in Reference Books and Trade Journals 145
7.4.3 Using Experts to Help Generate Concepts 145
7.4.4 Brainstorming as a Source of Ideas 145
7.4.5 Using the 6-3-5 Method as a Source of Ideas 147
7.4.6 Using Existing Products and Concepts as Idea Sources 148
7.5 Communication During Concept Generation 149
7.6 Summary 149
7.7 Sources 149
7.8 Exercises 150

8 Concept Evaluation 152
8.1 Introduction 152
8.2 Information Representation in Concept Evaluation 154
8.3 Evaluation Based on Feasibility Judgment 155
8.4 Evaluation Based on Technology-Readiness Assessment 157
8.5 Evaluation Based on Go/No-Go Screening 158
8.6 Evaluation Based on a Decision Matrix 160
8.7 Product Safety and Liability 165
8.7.1 Product Safety 165
8.7.2 Products Liability 167
8.8 Communication During Concept Evaluation 169
8.9 Summary 171
8.10 Sources 171
8.11 Exercises 171

9 The Product Design Phase 173
9.1 Introduction 173
9.2 The Importance of Drawings 175
9.3 Drawings Produced During Product Design 175
9.3.1 Layout Drawings 176
9.3.2 Detail Drawings 177
9.3.3 Assembly Drawings 177
9.4 The Information on Drawings: Dimensioning and Tolerancing 179
9.5 Bills of Materials 181
9.6 Product Data Management 181
9.7 Summary 182
9.8 Sources 183
9.9 Exercises 183

10 Product Generation 184
10.1 Introduction 184
10.2 Form Generation 186
10.2.1 Collect Spatial Constraints 186
10.2.2 Configure Components 187
10.2.3 Develop Connections: Create and Refine Interfaces for Functions 188

10.2.4 Develop Components 189
10.2.5 Refine and Patch 195
10.3 Materials and Process Selection 199
10.4 Vendor Development 200
10.5 Generating Product Design for the Splashgard 201
10.6 Summary 210
10.7 Sources 210
10.8 Exercises 210

11 Product Evaluation for Performance 212

11.1 Introduction 212
11.2 The Importance of Functional Evaluation 212
11.3 The Goals of Performance Evaluation 213
11.4 Accuracy, Variation, and Noise in Modeling Products 215
11.5 Modeling for Performance Evaluation 218
11.6 Robust Design 221
11.7 Sensitivity Analysis: A Window on Quality 225
11.8 Parameter and Tolerance Design for Quality 232
11.9 Robust Design Through Testing 234
11.10 Evaluation of the Splashgard 238
11.11 Summary 240
11.12 Sources 241
11.13 Exercises 241

12 Product Evaluation for Cost, Manufacture, Assembly, and Other Measures 243

12.1 Introduction 243
12.2 Cost Estimating in Design 243
12.2.1 Determining the Cost of a Product 244
12.2.2 Making a Cost Estimate 248
12.2.3 The Cost of Machined Components 248
12.2.4 The Cost of Injection-Molded Components 251
12.3 Value Engineering 253
12.4 Design for Manufacture 255
12.5 Design-for-Assembly Evaluation 255
12.5.1 Evaluation of the Overall Assembly 259
12.5.2 Evaluation of Component Retrieval 267
12.5.3 Evaluation of Component Handling 269
12.5.4 Evaluation of Component Mating 273
12.6 Design for Reliability (DFR) 274
12.6.1 Failure-Potential Analysis 276
12.6.2 Reliability 277
12.7 Design for Test and Maintenance (DFTM) 279
12.8 Design for the Environment 279
12.9 Summary 282
12.10 Sources 282
12.11 Exercises 282

13 Launching the Product 284

13.1 Introduction 284
13.2 Design Records 284

13.3 Patent Applications 286
13.4 Assembly, Quality Control, and Quality Assurance Documentation 288
13.5 Installation, Operation, Maintenance, and Retirement Instructions 288
13.6 Engineering Changes 289
13.7 Vendor Relations 289
13.8 Sources 291

Appendix A Properties of 25 Materials Most Commonly Used in Mechanical Design 292

A.1 Introduction 292
A.2 Properties of the Most Commonly Used Materials 293
A.3 Materials Used in Common Items 306
A.4 Sources 307

Appendix B Normal Probability 309

B.1 Introduction 309
B.2 Other Measures 313
B.3 Sources 313

Appendix C The Factor of Safety as a Design Variable 314

C.1 Introduction 314
C.2 The Classical Rule-of-Thumb Factor of Safety 315
C.3 The Statistical, Reliability-Based, Factor of Safety 317
C.3.1 Introduction 317
C.3.2 The Allowable Strength Coefficient of Variation 320
C.3.3 The Applied Stress Coefficient of Variation 321
C.3.4 Steps for Finding the Reliability-Based Factor of Safety 324
C.4 Sources 324

Appendix D Human Factors in Design 325

D.1 Introduction 325
D.2 The Human in the Workspace 326
D.3 The Human as Source of Power 330
D.4 The Human as Sensor and Controller 330
D.5 Sources 335

Index 337

PREFACE

I have been a designer all my life. I have designed vehicles, medical equipment, furniture, and sculpture, both static and dynamic. Designing objects has come easy for me. I have been fortunate in having whatever talents are necessary to be a successful designer. However, after a number of years of teaching mechanical design courses, I came to the realization that I didn't know how to teach what I knew so well. I could show students examples of good-quality design and poor-quality design. I could give them case histories of designers in action. I could suggest design ideas. But I could not tell them what to do to solve a design problem. Additionally, I realized from talking with other mechanical design teachers that I was not alone.

This situation reminded me of an experience I had once had on ice skates. As a novice skater I could stand up and go forward, lamely. A friend (a teacher by trade) could easily skate forward and backward as well. He had been skating since he was a young boy, and it was second nature to him. One day while we were skating together, I asked him to teach me how to skate backward. He said it was easy, told me to watch, and skated off backward. But when I tried to do what he did, I immediately fell down. As he helped me up, I asked him to *tell* me exactly what to do, not just show me. After a moment's thought, he concluded that he couldn't actually describe the feat to me. I still can't skate backward, and I suppose he still can't explain the skills involved in skating backward. The frustration that I felt falling down as my friend skated with ease must have been the same emotion felt by my design students when I failed to tell them exactly what to do to solve a design problem.

This realization led me to study the design process, and it eventually led to this book. Part has been original research, part studying U.S. industry, part studying foreign design techniques, and part trying different teaching approaches on design classes. I came to four basic conclusions about mechanical design as a result of these studies:

1. The only way to learn about design is to do design.
2. In engineering design the designer uses three types of knowledge: knowledge to generate ideas, knowledge to evaluate ideas, and knowledge to structure the design process. Idea generation comes from experience and natural ability. Idea evaluation comes partially from experience and partially from formal training, and is the focus of most engineering education. Generative and evaluative knowledge are forms of domain-specific knowledge. Knowledge about the design process is largely independent of domain-specific knowledge.
3. A design process that results in a quality product can be learned, provided there is enough ability and experience to generate ideas and enough experience and training to evaluate them.
4. A design process should be learned in a dual setting—in an academic environment and, at the same time, in an environment that simulates industrial realities.

I have incorporated these concepts into this book, which is organized so that readers can learn about the design process at the same time they are developing a product. The first few chapters present background on mechanical design, define the terms that are basic to the study of the design process, and discuss the human element of product design. Chapters 4–13, the body of the book, present a step-by-step development of a design method that leads the reader from the realization that there is a design problem to a solution ready for manufacture and assembly. This material is presented in a manner independent of the exact problem being solved. The techniques discussed are used in industry, and their names of have become buzzwords in mechanical design: quality function deployment, Pugh's method, concurrent engineering, design for assembly, and Taguchi's method for robust design. These techniques have all been brought together in this book. Although they are presented sequentially as step-by-step methods, the overall process is highly iterative, and the steps are merely a guide to be used when needed.

As mentioned earlier, domain knowledge is somewhat distinct from process knowledge. Because of this independence, a successful product can result from the design process regardless of the knowledge of the designer or the type of design problem. Even students at the freshman level could take a course using this text and learn most of the process. However, to produce any reasonably realistic design, substantial domain knowledge is required, and it is assumed throughout the book that the reader has a background in basic engineering science, material science, manufacturing processes, and engineering economics. Thus, this book is intended for upper-level undergraduate students, graduate students, and professional engineers who have never had a formal course in the mechanical design process.

ADDITIONS TO THE SECOND EDITION

Knowledge about the design process is increasing rapidly. A goal in writing the second edition was to incorporate this knowledge into the unified structure that was one of the strong points of the first edition. Throughout the new edition topics have been updated and integrated with other best practices in the book. Some specific additions to the new edition include the following:

1. Nine key features of concurrent engineering are introduced in the first chapter and reflected on through the material in the remaining chapters.
2. Chapter 3, which focuses on the cognitive aspects of design, has been extended to include more information on team problem solving.
3. The example that flows through Chap. 4–11 has been improved.
4. A new chapter on planning for design projects (Chap. 5) has been added.
5. The coverage of the quality function deployment method in Chap. 6 has been extended.
6. The function modeling methods in Chap. 7 have been refined and improved.
7. Chap. 11 presents a greatly enhanced coverage of product performance evaluation. This includes an introduction to robust design and the design of experiments.
8. Exercises for students have been added at the end of each chapter.
9. An instructor's manual is now available from the publisher, which contains methods for teaching the material in the book.
10. A software program is availabe on the McGraw-Hill Web site (http://www.-mhcollege.com/engineering/mecheng.html). This software supports the estimation of costs for injection molded, machined and forged parts.

Beyond these, many small changes have been made to keep the book current and useful.

ACKNOWLEDGMENTS

Special appreciation is extended to Kris Wood of the University of Texas for his help with improving the chapter on function modeling. Also, I would like to express special appreciation to David Gobeli of the Business College at Oregon State University for his help with the new chapter on planning.

I would like to thank the following reviewers for their helpful comments:

Rudy Eggert (Boise State University)
Mark Costello (U.S. Military Academy)
John Starkey (Purdue University)
Young Chun (Villanova University),

Larry Stauffer (University of Idaho)
Michael Brady (U.S. Air-Force Academy)
Christian Burger (Texas A&M University)
Adan Akay (Wayne State University)
C. Wesley Allen (emeritus, University of Cincinnati)
Gale Nevell (University of Florida)
F. C. Appl (Kansas State University)
Donald R. Flugrad (Iowa State University)
David C. Jansson
R. T. Johnson (University of Missouri–Rolla)
Geza Kardos (emeritus, Carleton University)
Ray Murphy (Seattle University)
Drew Nelsson (Stanford University)
Erol Sancaktar (Clarkson University)
Bryan Wilson (Colorado State University)

Additionally, I would like to thank Debra Riegert of McGraw-Hill for her interest and encouragement in this project and Gordon Reistad, my chairman at Oregon State University, for his giving me time and supportive teaching assignments. Last and most important my thanks to my family, Adele and Eric, for their never-questioning confidence that I could finish this project.

The preface for the first edition ended with a caricature of me standing next to a fanciful bicycle as shown below. At that time I had a strong interest in nontraditional bicycles.

May everywhere you ride
be downhill !
The Author

(Figure with permission of Chuck Meitle)

This preface ends with a photograph of me (without a beard) standing next to a production bicycle I designed in 1992 after the release of the first edition. It bears a striking and purely coincidental similarity to the caricature bike above.

Photo compliments of BikeE Corporation, Corvallis, Oregon

THE MECHANICAL DESIGN PROCESS

CHAPTER 1

WHY STUDY THE DESIGN PROCESS?

1.1 INTRODUCTION

Beginning with the simple potter's wheel and evolving to complex consumer products and transportation systems, humans have been designing mechanical objects for nearly five thousand years. Each of these objects is the end result of a long and often difficult design process. This book is about that process. Regardless of whether we are designing gearboxes, heat exchangers, satellites, or doorknobs, there are certain techniques that can be used during the design process to help ensure successful results. Since this book is about the *process* of mechanical design, it focuses not on the design of any one type of object but on techniques that apply to the design of all types of mechanical objects.

If people have been designing for five thousand years and there are literally millions of mechanical objects that work and work well, why study the design process? The answer, simply put, is that there is a continuous need for new, cost-effective, high-quality products. Today's products have become so complex that most require a team of people from diverse areas of expertise to develop an idea into hardware. The more people involved in a project, the greater is the need for assistance with communication and for structure to ensure that nothing important is overlooked. In addition, the global marketplace has fostered the need to develop new products at a very rapid and accelerating pace. To compete in this market, a company must be very efficient in the design of its products. It is the process that we will study here that determines the efficiency of new product development. Finally, it has been estimated that 85 percent of the problems with new products not working as they should, taking too long to bring to market, or costing too much are the result of a poor design process.

The goal of this book is to give you the tools to develop an efficient design process regardless of the product being developed. In this chapter the important features of design problems and the processes for solving them will be introduced. These features apply to any type of design problem, whether for mechanical, electrical, software, or construction projects. Subsequent chapters will focus more on mechanical design, but even these can be applied to a broader range of problems.

1.2 MEASURING THE DESIGN PROCESS WITH PRODUCT COST, QUALITY, AND TIME TO PRODUCTION

The three measures of the effectiveness of the design process are cost, quality, and time. Regardless of the product being designed—whether it is an entire system or some small subpart of a larger product—the customer and management always want it cheaper, better, and faster.

The actual cost of design is usually a small part of the manufacturing cost of a product, as can be seen in Fig. 1.1, which is based on data from Ford Motor Company. The data shows that only 5 percent of the manufacturing cost of a car (the cost to produce the car but not to distribute or sell it) is for design activities that were needed to develop it. This number varies with industry and product, but for most products the cost of design is a small part of the manufacturing cost.

The effect of the quality of the design on the manufacturing cost is much greater than 5 percent. This is most accurately shown from the results of a detailed study of 18 different automatic coffeemakers. The results of this study are shown in Table 1.1. Here the effects of changes in manufacturing efficiency, such as material cost, labor wages, and cost of equipment, have been separated from the effects of design quality—the results of the design process. Note that manufacturing efficiency and design quality have the same influence on the cost of manufacturing a product. The results of the design process can change the cost of manufacturing a product by 50 percent (±25 percent for an average manufacturing process) or more. Xerox also attributes 50 percent of the final cost to the results of the design process. In some industries this effect is as high as 75 percent. Thus it can be concluded that *the decisions made during the design process have a great effect on the cost of a product but cost very little.* Design decisions directly determine the materials used, the goods

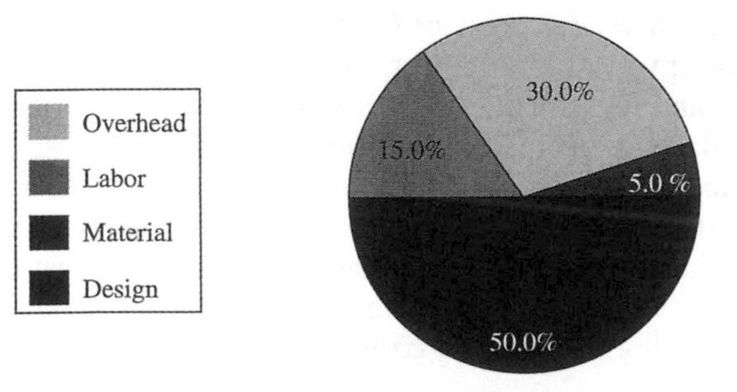

FIGURE 1.1
Design cost as fraction of manufacturing cost.

TABLE 1.1
The effect of design quality on manufacturing cost

	Efficient manufacturing	Average manufacturing	Poor manufacturing
A good design	$.50	$ .75	$ .94
An average design	$.75	$1.00	$1.25
A poor design	$.94	$1.25	$1.50

Data reduced from "Does Product Design Really Determine 80% of Manufacturing Cost?" by K. T. Ulrich and S. A. Pearson, working paper 3601-93, Sloan School of Management, 1993.

purchased, the parts, the shape of those parts, the product sold, and, in the end, the scope of management.

Another example of the relationship of the design process to cost comes from Xerox. In the 1960s and early 1970s, Xerox controlled the copier market. However, by 1980 there were over 40 different manufacturers of copiers in the marketplace and Xerox's share of the market had fallen significantly. Part of the problem was the cost of Xerox's products. In fact, in 1980 Xerox realized that some producers were able to sell a copier for less than Xerox was able to manufacture one of similar functionality. In one study of the problem, Xerox focused on the cost of individual parts. Comparing plastic parts from their machines and ones that performed a similar function in Japanese and European machines, they found that Japanese firms could produce a part for 50 percent less than American or European firms. Xerox attributed the cost difference to three factors: materials costs were 10 percent less in Japan, tooling and processing costs were 15 percent less, and the remaining 25 percent (half of the difference) was attributable to how the parts were designed.

Not only is much of the product cost committed during the design process, it is committed early in the design process. As shown in Fig. 1.2, 75 percent of the

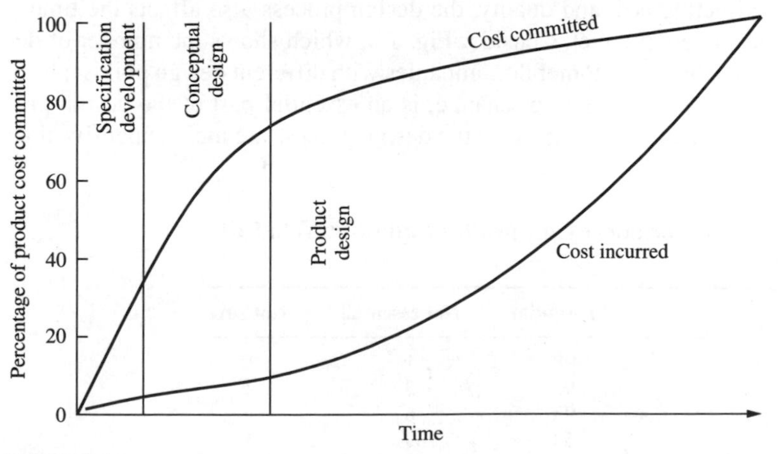

FIGURE 1.2
Manufacturing cost commitment during design.

manufacturing cost of a typical product is committed by the end of the conceptual phase process. This means that decisions made after this time can influence only 25 percent of the product's manufacturing cost. Also shown in the figure is the amount of cost incurred, which is the amount of money spent on the design of the product. It is not until money is committed for production that large amounts of capital are spent.

The results of the design process also have a great effect on product quality. It is clear that *quality cannot be built into a product unless it is designed into it.* In a survey taken in 1989, American consumers were asked, "What determines quality?" Their responses, shown in Table 1.2, indicate that "quality" is a composite of factors that are the responsibility of the design engineer. Thus the decisions made during the design process determine the product's quality as perceived by the customers.

Another indicator of quality comes from another Xerox study. This one focused on "line fallout," a measure of the number of components that do not fit together during assembly—the components that literally "fall out" of the assembly line. If it is assumed that a company uses quality-control techniques (for example, measuring a component's geometry and/or other properties during production) to ensure that the components meet the production specifications, and it is further assumed that components that reach the assembly line are within the design specifications, then components that do not fit are poorly designed. These design failures must be either scrapped or reworked; either choice adds costs in the manufacturing process.

The results of Xerox's line fallout study are shown in Fig. 1.3. In the first year it took data, 1981, Xerox had 30 times the line fallout of its Japanese competitors, who had 1 component per 1000. It restructured its design process (in ways similar to those that will be presented in this book), and by 1989 it had reduced its line fallout by a factor of 30. Xerox's ultimate goal, 125 parts per million (0.125 per thousand) was reached in the early 1990s. Xerox now uses a measure that, besides line fallout, includes defective parts identified in the supplier's preshipping inspection, in Xerox's incoming inspection, and in warehouse inspection.

Besides affecting cost and quality, the design process also affects the time it takes to produce a new product. Consider Fig. 1.4, which shows the number of design changes made by two automobile companies with different design philosophies. As shown in this book, iteration, or change, is an essential part of the design process. However, changes occurring late in the design process are more expensive than

TABLE 1.2
Results of a consumer survey on product quality: What determines quality?

	Essential	Not essential	Not sure
Works as it should	98	1	1
Lasts a long time	95	3	2
Is easy to maintain	93	6	1
Looks attractive	58	39	3
Incorporates latest technology	57	39	4
Has many features	48	47	5

Based on a survey published in *Time,* Nov. 13, 1989.

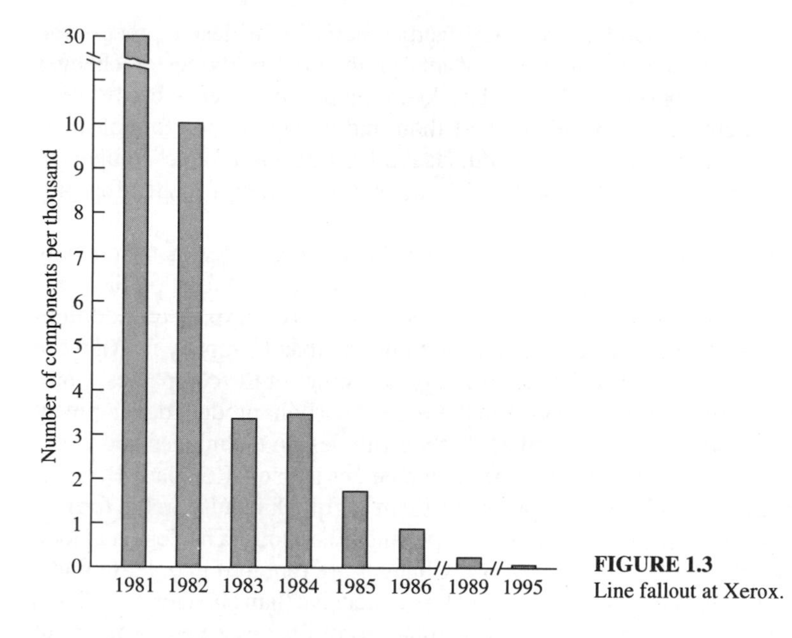

FIGURE 1.3
Line fallout at Xerox.

those occurring earlier. The curve for Company B indicates that the company was still making changes after the design had been released for production. In essence, Company B was still designing the automobile as it was being sold as a product. This causes tooling and assembly-line changes during production and the possibility of recalling cars for retrofit, both of which would necessitate significant expense.

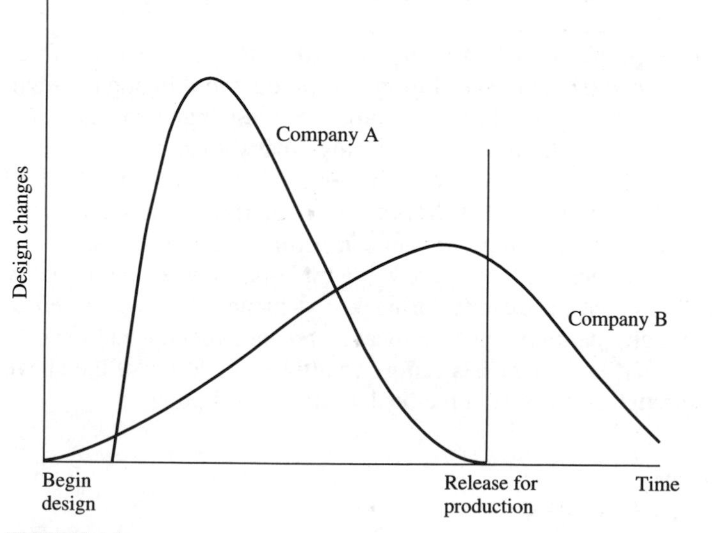

FIGURE 1.4
Engineering changes during automobile development.

Company A, on the other hand, made many changes early in the design process and finished the design of the car before it went into production. Early design changes require more engineering time and effort but do not require changes in hardware or documentation. A change that would cost $1 thousand in engineering time if made early in the design process may cost $10 thousand later during product refinement and $1 million or more in tooling, sales, and goodwill expenses if made after production has begun.

Figure 1.4 also indicates that Company A made more changes than Company B. This implies that it explored more design alternatives, which at least partially explains why modifications were not still being made at the end of the project. Additionally, Company A took less time to design the automobile than Company B. All these differences are due to differences in the design philosophies of the companies. Company A assigns a large engineering staff to the project early in product development and encourages these engineers to utilize the latest in design techniques and to explore all the options early to preclude the need for changes later on. Company B, on the other hand, assigns a small staff and pressures them for quick results, in the form of hardware, discouraging the engineers from exploring all options. The design axiom, *fail early, fail often,* applies to this example. Changes are required in order to find a good design, and early changes are easier and less expensive than changes made later.

The curves of Fig. 1.4 are actual representations of the design philosophies of a Japanese company (A) and an American company (B) in the early 1980s. During this period the time to design a car in the United States was a little over five years from the presentation of the initial problem to the production of the final product. For the Japanese the same activities took three and one-half years, and the product was perceived to be so superior to the American that the United States imposed import quotas on Japanese cars in the 1980s. However, American car manufacturers responded to the challenge like Xerox. They instituted better design practices and improved the quality of their products.

Here is one last example from Xerox. In the 1970s it took Xerox about three years to progress from establishing the need for a new product to bringing that item to production. By 1990 Xerox had reduced that process to less than two years, and by 1995, to less than 30 weeks. Xerox's goal is to halve that yet again.

Caterpillar, the manufacturer of heavy equipment, reduced its 50-month 1980 development cycle to 20 months in 1993. Xerox's and Caterpillar's reductions are typical of the performance of most companies in competitive environments.

For many years it was believed that there was a trade-off between high-quality products and low development and manufacturing costs—namely, that it cost more to develop and produce high-quality products. However, recent experience has shown that increasing quality and lowering costs can go hand in hand. Some of the above examples and ones throughout the rest of the book reinforce this point.

1.3 THE HISTORY OF THE DESIGN PROCESS

During design activities ideas are developed into hardware that is usable as a product. Whether this piece of hardware is a bookshelf or a space station, it is the result of a process that combines people and their knowledge, tools, and skills to develop a new creation. This task requires their time and costs money, and if the people are good at what they do and the environment they work in is well-structured, they can do it efficiently. Further, if they are skilled, the final product will be well-liked by those who use it and work with it—the customers will see it as a quality product. *The design process, then, is the organization and management of people and the information they develop in the evolution of a product.*

In simpler times, one person could design and manufacture an entire product. Even for a large project such as the design of a ship or a bridge, one person had sufficient knowledge of the physics, materials, and manufacturing processes to manage all aspects of the design and construction of the project.

By the middle of the twentieth century, products and manufacturing processes had become so complex that one person no longer had sufficient knowledge or time to focus on all the aspects of the evolving product. Different groups of people became responsible for marketing, design, manufacturing, and overall management. This evolution led to what is commonly known as the "over-the-wall" design process (Fig. 1.5).

In the structure shown in Fig 1.5, the engineering design process is walled off from the other product development functions. Basically, people in marketing communicate a perceived market need to engineering either as a simple, written request or, in many instances, orally. This is effectively a one-way communication and is thus represented as information that is "thrown over the wall." Engineering interprets the request, develops concepts, and refines the best concept into manufacturing specifications (i.e., drawings, bills of materials, and assembly instructions). These manufacturing specifications are thrown over the wall to be produced. Manufacturing then interprets the information passed to it and builds what it thinks engineering wanted.

Unfortunately, often what is manufactured by a company using the over-the-wall process is not what the customer had in mind. This is because of the many weaknesses in this product development process. First, marketing may not be able to communicate to engineering a clear picture of what the customers want. Since the design engineers have no contact with the customers and limited communication with marketing, there is much room for poor understanding of the design problem. Second, design engineers do not know as much about the manufacturing processes as manufacturing specialists, and therefore some parts may not be able to be manufactured as drawn or manufactured on existing equipment. Further, manufacturing

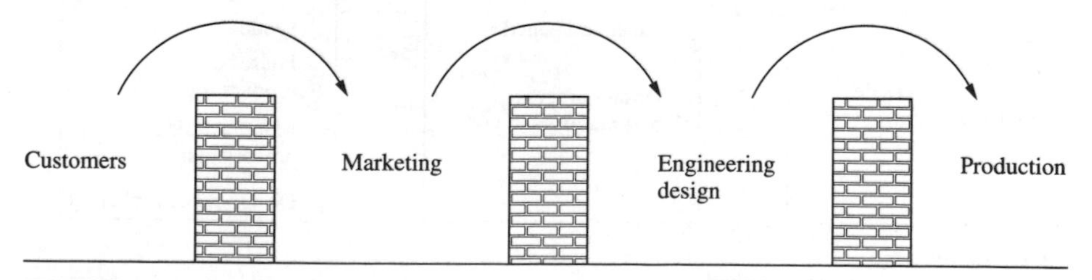

FIGURE 1.5
The over-the-wall design method.

TABLE 1.3
The nine key features of concurrent engineering

Focus on the entire product life (Chap. 1)
Use and support of design teams (Chaps. 3 and 5)
Realization that the processes are as important as the product (Chaps. 4 and 5)
Attention to planning for information-centered tasks (Chap. 5)
Careful product requirements development (Chap. 6)
Encouragement of multiple concept generation and evaluation (Chaps. 7 and 8)
Attention to designing in quality during every phase of the design process (throughout)
Concurrent development of product and manufacturing process (Chaps. 9–13)
Emphasis on communication of the right information to the right people at the right time (throughout)

experts may know less expensive methods to produce the product. Thus, this single-direction over-the-wall approach is inefficient and costly and may result in poor-quality products. Although many companies still use this method, most are realizing its weaknesses and are moving away from its use.

In the late 1970s and early 1980s, the concept of *simultaneous engineering* began to break down the walls. This philosophy emphasized the simultaneous development of the manufacturing process with the evolution of the product. Simultaneous engineering was accomplished by assigning manufacturing representatives to be members of design teams so that they could interact with the design engineers throughout the design process. The goal was the simultaneous development of the product and the manufacturing process.

In the 1980s the simultaneous design philosophy was broadened and called *concurrent engineering* or *integrated product and process design (IPPD).* Although the terms *simultaneous, concurrent,* and *integrated* are basically synonymous, the change in terms implies a greater refinement in thought about what it takes to efficiently develop a product. Throughout the rest of this text, the term *concurrent engineering* will be used to express this refinement.

Concurrent engineering is built around a concern for the nine key features listed in Table 1.3. These nine are covered in the chapters shown and are integrated into the concurrent engineering method as shown in Fig. 1.6. This figure looks much

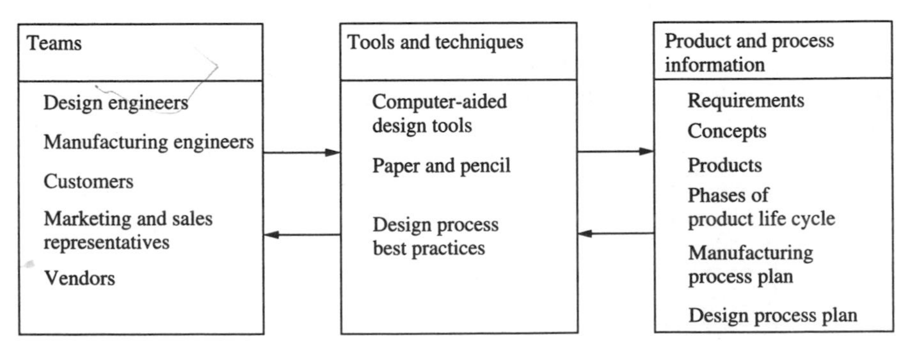

FIGURE 1.6
The concurrent engineering method.

different from the over-the-wall method in Fig. 1.5. In concurrent engineering the primary focus is on the integration of teams of people having a stake in the product, design tools and techniques, and information about the product and the processes used to develop and manufacture it.

The use of teams, including all the "stakeholders" (people who have a concern for the product), eliminates many of the problems with the over-the-wall method. During each phase in the development of a product, different people will be important and will be included in the product development team. This mix of people with different views will also help the team address the entire life cycle of the product.

A key point in the concurrent engineering method is a concern for information. Drawings, plans, concept sketches, and meeting notes all provide information that must be shared with the right people at the right time. This concern is not only about the development of this information but also about its distribution. Traditionally, the greatest emphasis during the design process has been on the development of drawings. In concurrent engineering this has been broadened to include concern for information about requirements, concepts, and process plans—items that cannot be represented as formal drawings.

Tools and techniques connect the teams with the information. Although many of the tools are computer-based, much design work is still done with pencil and paper. In fact, concurrent engineering is 80 percent company culture and only 20 percent computer support. Thus, the emphasis in this book is not on computer-aided design but on the techniques that affect the culture of design and the tools used to support them.

An important aspect of concurrent engineering is concern for both the development of the product and the associated processes. Two processes listed among the key features are the product development process—the focus of this book—and the manufacturing process. Concurrent engineering forces engineers to put effort into these processes while designing the product.

Finally, during the 1980s and 1990s many design process techniques were introduced and became popular. They are essential to the concurrent engineering philosophy and are introduced throughout the book.

1.4 THE LIFE OF A PRODUCT

Whether the over-the-wall or the concurrent design philosophy is being used, every product has a life history as described in Fig. 1.7. Here, each box represents a phase in the product's life. These phases are grouped into four broad areas. The first area concerns the development of the product, the focus of this book. The second group of phases includes the production and delivery of the product. The third group contains all the considerations important to the product's use. And the final group focuses on what happens to the product after it is no longer useful. Each phase will be introduced in this section, and all are detailed later in the book. Note that the designers, who are involved with the first five phases, must fully understand all the subsequent phases if they are to develop a quality product. The product life phases are discussed in the following.

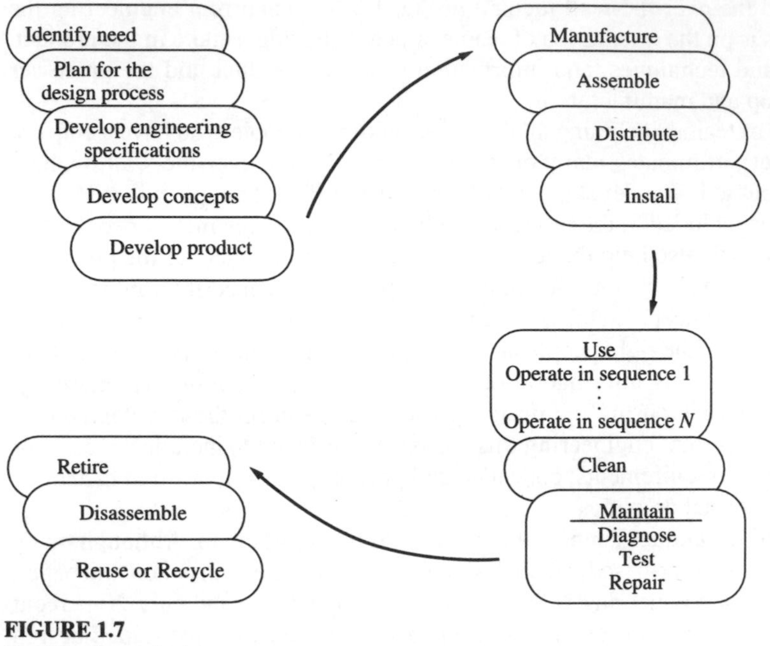

FIGURE 1.7
The life of a product.

IDENTIFY NEED. Design projects are initiated by either a market requirement, the development of a new technology, or the desire to improve an existing product.

PLAN FOR THE DESIGN PROCESS. Efficient product development requires planning for the process to be followed. Planning for the design process is the topic of Chap. 5.

DEVELOP ENGINEERING REQUIREMENTS. The importance of developing a good set of requirements has become one of the key points in concurrent engineering. It has recently been realized that the time spent evolving complete requirements prior to developing concepts saves time and money and improves quality. A technique to help in developing requirements is covered in Chap. 6.

DEVELOP CONCEPTS. Chapters 7 and 8 focus on techniques for generating and evaluating new concepts. This is an important phase in the development of a product, as decisions made here affect all the downstream phases.

DEVELOP PRODUCT. Turning a concept into a manufacturable product is a major engineering challenge. Chapters 9–13 present techniques to make this a more reliable process. This phase ends with manufacturing specifications and release to production.

These first five phases all must take into account what will happen to the product in the remainder of its lifetime. When the design work is completed, the product is released for production and, except for engineering changes, the design engineers will have no further involvement with the product.

MANUFACTURE. Some products are just assemblies of existing components. For most products, unique components need to be formed from raw materials and thus require some manufacturing. In the over-the-wall design philosophy, design engineers sometimes consider manufacturing issues, but since they are not experts, they sometimes do not make good decisions. Concurrent engineering encourages having manufacturing experts on the design team to ensure that the product can be produced and can meet cost requirements. The specific consideration of *design for manufacturing* and product cost estimation is covered in Chap. 12.

ASSEMBLE. Considering how a product is to be assembled is a major consideration during the product design phase. Part of Chap. 12 is devoted to a technique called *design for assembly,* which focuses on making a product easy to assemble.

DISTRIBUTE. Although distribution may not seem like a concern for the design engineer, each product must be delivered to the customer in a safe and cost-effective manner. Design requirements may include the need for the product to be shipped in a prespecified container or on a standard pallet. Thus, the design engineers may need to alter their product just to satisfy distribution needs.

INSTALL. Some products require installation before the customer can use them. This is especially true for manufacturing equipment and building industry products. Additionally, concern for installation can also mean concern for how customers will react to the statement, "Some assembly required."

USE. Most design requirements are aimed at specifying the use of the product. Products may have many different operating sequences that describe their use. Consider as an example a common hammer that can be used to put in nails or take them out. Each use involves a different sequence of operations, and both must be considered during the design of a hammer.

Another aspect of a product's use is keeping it clean and maintaining it in usable condition. As shown in Fig. 1.7, to *maintain* a product requires that problems must be *diagnosed,* the diagnosis may require *tests,* and the product must be *repaired.* Every consumer has experienced the frustration of not being able to clean a product. This inability is seldom designed into the product on purpose; rather, it is usually simply the result of poor design.

RETIRE, DISASSEMBLE, REUSE, AND RECYCLE. The final phase in a product's life is its retirement. In past years designers did not worry about a product beyond its use. However, during the 1980s increased concern for the environment forced designers to begin considering the entire life of their products. In the 1990s some European countries have enacted legislation that makes the original manufacturer responsible for collecting and reusing or recycling its products when their usefulness is finished. This topic will be further discussed in Section 12.8.

This description of the life of a product gives a good basic understanding of the issues that will be addressed in this book. The rest of this chapter details the unique features of design problems and their solution processes.

1.5 THE MANY SOLUTIONS OF DESIGN PROBLEMS

Consider the following problem from a textbook on the design of machine components (see Fig. 1.8):

> What size SAE grade 5 bolt should be used to fasten together two pieces of 1045 sheet steel, each 4 mm thick and 6 cm wide, which are lapped over each other and loaded with 100 N?

In this problem the need is very clear, and if we know the methods for analyzing shear stress in bolts, the problem is easily understood. There is no necessity to design the joint because a design solution is already given, namely, a grade 5 bolt, with one parameter to be determined—its diameter. The product evaluation is straight from textbook formulas, and the only decision made is in determining whether we did the problem correctly.

In comparison, consider the following, only slightly different, problem:

> Design a joint to fasten together two pieces of 1045 sheet steel, each 4 mm thick and 6 cm wide, which are lapped over each other and loaded with 100 N.

The only difference between these problems is in their opening clauses. The second problem is even easier to understand than the first; we do not need to know how to design for shear failure in bolted joints. However, there is much more latitude in generating ideas for potential concepts here. It may be possible to use a bolted joint, a glued joint, a joint in which the two pieces are folded over each other, a welded joint, a joint held by magnets, a Velcro joint, or a bubble-gum joint. Which one is best depends on other, unstated factors. This problem is not as well defined as the first one. To evaluate proposed concepts, more information about the joint will be needed. In other words, the problem is not really understood at all. Some questions still need to be answered: Will the joint require disassembly? Will it be used at high temperatures? What tools are available to make the joint? What skill levels do the joint makers have?

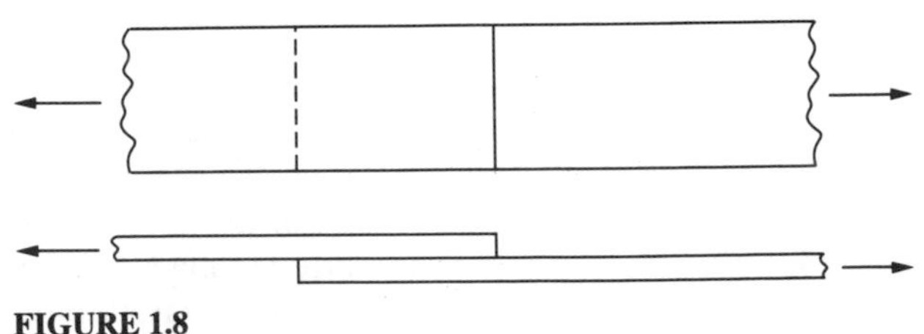

FIGURE 1.8
A simple lap joint.

The first problem statement describes an analysis problem. To solve it we need to find the correct formula and plug in the right values. The second statement describes a design problem, which is ill-defined in that the problem statement does not give all the information needed to find the solution. The potential solutions are not given and the constraints on the solution are incomplete. This problem requires us to fill in missing information in order to understand it fully. *All design problems are ill-defined.*

Another difference between the two problems is in the number of potential solutions. For the first problem there is only one correct answer. For the second there is no correct answer. In fact, there may be many good solutions to this problem, and it may be difficult if not impossible to define what is meant by the "best solution." Just consider all the different cars, televisions, and other products that compete in the same market. The goal in design is to find a good solution that leads to a quality product with the least commitment of time and other resources. *Most design problems have a multitude of satisfactory solutions and no clear best solution.* This is shown graphically in Fig. 1.9, where the factors that affect exactly what solution is developed are noted. Domain knowledge is developed through the study of engineering physics and other technical areas and through the observation of existing products. Design process knowledge is the subject of this book.

For mechanical design problems in particular, there is an additional characteristic: the solution must be a piece of working hardware—a product. Thus, mechanical

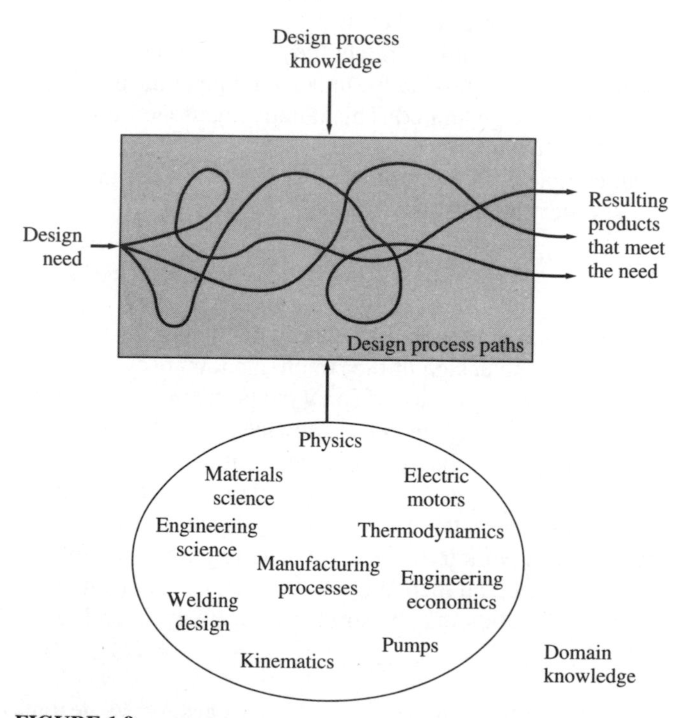

FIGURE 1.9
The many results of the design process.

design problems begin with an ill-defined need and result in a piece of machinery that behaves in a certain way, a way that the designers feel meets this need. This creates a paradox. *A designer must develop a machine that, by definition, has the capabilities to meet some need that is not fully defined.*

1.6 THE BASIC ACTIONS OF PROBLEM SOLVING

Regardless of what design problem we are solving, we always, consciously or unconsciously, take six basic actions:

1. *Establish* the need or realize that there is a problem to be solved.
2. *Plan* how to solve the problem.
3. *Understand* the problem by developing requirements and uncovering existing solutions for similar problems.
4. *Generate* alternative solutions.
5. *Evaluate* the alternatives by comparing them to the design requirements and to each other.
6. *Decide* on acceptable solutions.

This model fits design whether we are looking at the entire product (see the product life-cycle diagram, Fig. 1.7) or the smallest detail of it.

These actions are not taken in 1-2-3 order. In fact they are often intermingled with solution generation and evaluation improving the understanding of the problem, allowing new, improved solutions to be generated. This iterative nature of design is another feature that separates it from analysis.

The list of actions is not complete. If we want anyone else on the design team to make use of our results, a seventh action is also needed:

7. *Communicate* the results.

The need that initiates the process may be very clearly defined or ill-defined. Consider the problem statements for the design of the simple lap joint of two pieces of metal given earlier (Figure 1.8). The need was given by the problem statement in both cases. In the first statement, understanding is the knowledge of what parameters are needed to characterize a problem of this type and the equations that relate the parameters to each other (a model of the joint). There is no need to generate potential solutions, evaluate them, or make any decision, because this is an analysis problem. The second problem statement needs work to understand. The requirements for an acceptable solution must be developed, and then alternative solutions can be generated and evaluated. Some of the evaluation may be similar to the analysis problem.

Some important observations:

- New needs are established throughout the design effort because new design problems arise as the product evolves. Details not addressed early in the process must be dealt with as they arise; thus the design of these details poses new subproblems.

- Planning occurs mainly at the beginning of a project. Plans are always updated because understanding is improved as the process progresses.
- Formal efforts to understand new design problems continue throughout the process. Each new subproblem requires new understanding.
- There are two distinct modes of generation: concept generation and product generation. The techniques used in these two actions differ.
- Evaluation techniques also depend on the design phase; there are differences between the evaluation techniques used for concepts and those used for products.
- It is difficult to make decisions, as each decision requires a commitment based on incomplete evaluation. Additionally, since most design problems are solved by teams, a decision requires consensus, which is often difficult to obtain.
- Communication of the information developed to others on the design team and to management is an essential part of concurrent engineering.

We will return to these observations as the design process is developed through this text.

1.7 KNOWLEDGE AND LEARNING DURING DESIGN

When a new design problem is begun, very little is known about the solution, especially if the problem is a new one for the designer. As work on the project progresses, the designer's knowledge about the technologies involved and the alternative solutions increases, as shown in Fig. 1.10. Therefore, after completing a project, most designers want a chance to start all over in order to do the project properly now that

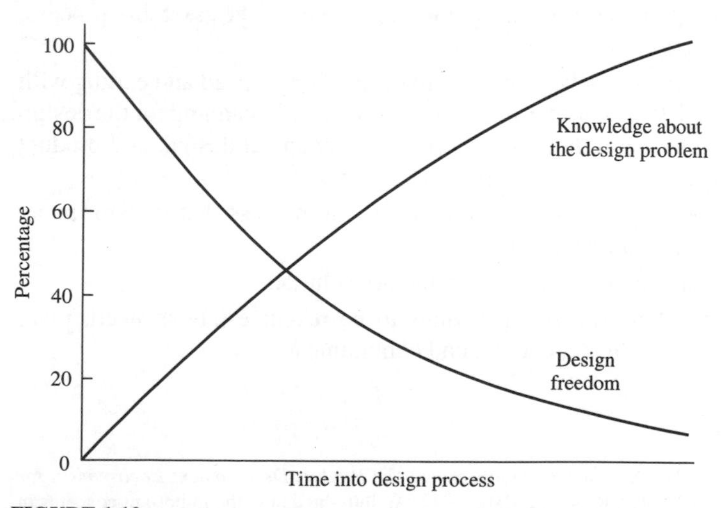

FIGURE 1.10
The design process paradox.

they fully understand it. Unfortunately, since time and cost—not the designer's sense of good-quality design—drive most design projects, few designers get the opportunity to redo their projects.

Throughout the solution process knowledge about the problem and its potential solutions is gained and, conversely, design freedom is lost. This can be seen in Fig. 1.10, where the time into the design process is equivalent to exposure to the problem. The curve representing knowledge about the problem is a learning curve; the steeper the slope, the more knowledge is gained per unit time. Throughout most of the design process the learning rate is high. The second curve in Fig. 1.10 illustrates the degree of design freedom. As design decisions are made, the ability to change the product becomes increasingly limited. At the beginning the designer has great freedom because few decisions have been made and little capital has been committed. But by the time the product is in production, any change requires great expense, which limits freedom to make changes. Thus, *the goal during the design process is to learn as much about the evolving product as early as possible in the design process because during the early phases changes are least expensive.*

1.8 SUMMARY

The design process is the organization and management of people and the information they develop in the evolution of a product.

* The success of the design process can be measured in the cost of the design effort, the cost of the final product, the quality of the final product, and the time needed to develop the product.
* Cost is committed early in the design process, so it is important to pay particular attention to early phases.
* Concurrent engineering integrates all the stakeholders from the beginning of the design process and emphasizes both the design of the product and concern for all processes—the design process, the manufacturing process, the assembly process, and the distribution process.
* All products have a life cycle beginning with establishing a need and ending with retirement. Although this book is primarily concerned with planning for the design process, engineering requirements development, conceptual design, and product design phases, attention to all the other phases is important.
* The mechanical design process is a problem-solving process that transforms an ill-defined problem into a final product.
* Design problems have more than one satisfactory solution.
* In problem solving there are seven actions to be taken: establish need, plan, understand, generate, evaluate, decide, and communicate.

1.9 SOURCES

Carter, D. E., and B. S. Baker: *Concurrent Engineering: The Product Development Environment for the 1990s,* Addison-Wesley, Reading, Mass. 1992. An introduction to the important concepts in concurrent engineering. It is easily readable.

Prasad, B.: *Concurrent Engineering Fundamentals,* Prentice Hall, Englewood Cliffs, N.J., 1996. A good overview of concurrent engineering issues.

Ulrich, K. T., and S. A. Pearson: "Does Product Design Really Determine 80% of Manufacturing Cost?" working paper 3601-93, Sloan School of Management, MIT, Cambridge, Mass., 1993. In the first edition of *The Mechanical Design Process* it was stated that design determined 80 percent of the cost of a product. To confirm or deny that statement, researchers at MIT performed a study of automatic coffeemakers and wrote this paper. The results show that the number is closer to 50 percent on the average (see Table 1.1) but can range as high as 75 percent.

1.10 EXERCISES

1.1. Change a problem from one of your engineering science classes into a design problem. Try changing as few words as possible.

1.2. Identify the basic problem solving actions for
(a) Selecting a new car
(b) Finding an item in a grocery store
(c) Installing a wall-mounted bookshelf
(d) Placing a piece in a puzzle

1.3. Find examples of products that are very different yet solve *exactly* the same design problem. Different brands of automobiles, cars, bikes, CD players, and personal computers are examples. For each, list its features, cost, and perceived quality. Compare the ease of maintenance and any obvious thoughts on the retirement of the products.

1.4. To experience the limitations of the over-the-wall design method, try the following. With a group of four to six people, have one person write down the description of some object that is not familiar to the others. This description should contain at least six different nouns that describe different features of the object. Without showing the description to the others, describe the object to one other person. This can be done by whispering or leaving the room. Limit the description to what was written down. The second person now conveys the information to the third person, and so on until the last person redescribes the object to the whole group and compares it to the original written description. The modification that occurs is magnified with more complex objects and poorer communication. (Professor Mark Costello of West Point originated this problem.)

CHAPTER 2

DESCRIBING MECHANICAL DESIGN PROBLEMS AND PROCESS

2.1 INTRODUCTION

One key feature of concurrent engineering is emphasis on communication. Communication depends on a shared understanding of terminology. All the parties involved must be using the same terms to describe the same objects, methods, and actions. In this chapter we discuss the terminology used to describe the mechanical design process. Although not complex, the terms defined here are the basis for the rest of the book.

2.2 DECOMPOSITION OF MECHANICAL SYSTEMS

For most of history, the discipline of mechanical design required knowledge only of mechanical parts and assemblies. But early in the twentieth century electrical components were introduced in mechanical devices. Since that time the discipline has undergone a steady transformation from purely mechanical to electromechanical products. However, no matter how "electronic" devices become, they still require mechanical machinery for manufacture and assembly and mechanical components for housing. Additionally, nearly all products require mechanical interface with humans.

In the 1960s and 1970s another discipline was added to electromechanical design, namely, software design. Many electromechanical products now have microprocessors as part of their control system. Consider, for example, cameras, office copiers, and our many "smart" toys. These products having mechanical, electronic, and software components are called *mechatronic* devices. What makes the design of these devices difficult is the necessity for domain and design process knowledge in three overlapping but clearly different disciplines.

As an example of a mechatronic system, consider the Kodak Cameo zoom camera (Fig. 2.1) and its shutter assembly (Fig. 2.2). The shutter assembly is part of the shutter system, one of many *systems* in a camera. A system is generally considered a conglomeration of objects that perform a specific *function.* The camera is a photographic system; its function is to record images. The shutter system is a subsystem whose purpose is to control the light coming through the lens. The shutter system is part of the exposure system, which also includes the light meter, the controller, and the auto focus system. Thus, we have *decomposed* the camera into three system levels, while still referring to the function of objects.

Another view of the camera is to look at it as an *assembly* of components in terms of the physical components or *form* of the camera. The camera assembly can be decomposed into subassemblies. As shown in Fig. 2.2, the shutter assembly can be further decomposed into individual *components,* or parts. In this book the term *component* will be used rather than *part,* which has too many other uses in English and is easily confused. Additionally, since there are subsystems of subsystems and subassemblies of subassemblies, to ease confusion, the terms *system* and *assembly* are used no matter where the object of interest falls in the decomposition. The prefix *sub* is only used to show one level of decomposition in a specific discussion.

FIGURE 2.1
A Kodak Cameo 300M zoom camera (by permission of Kodak Corp.).

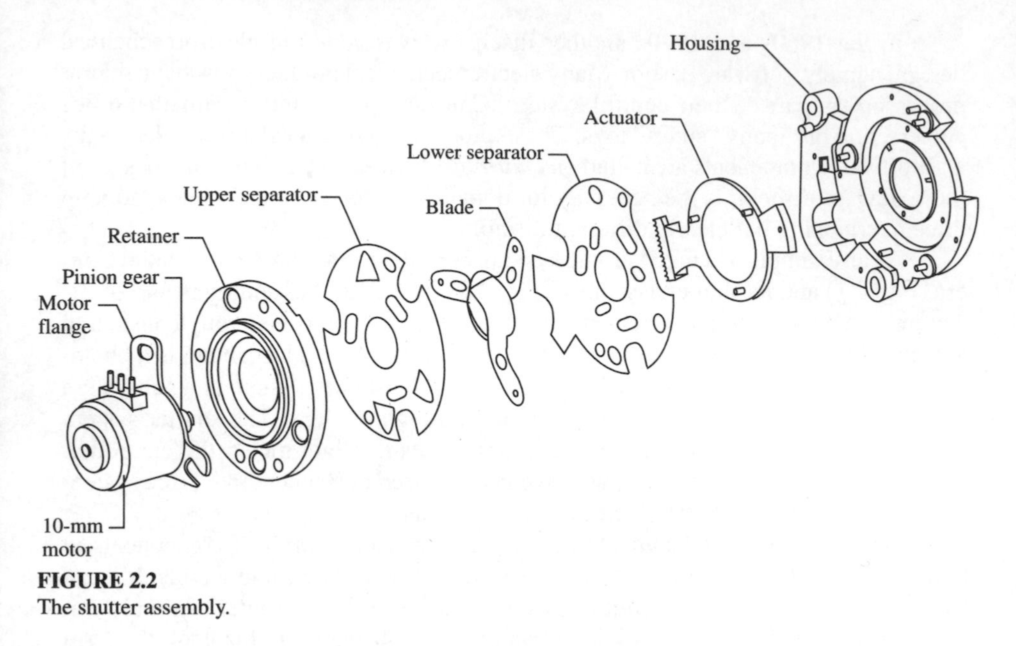

FIGURE 2.2
The shutter assembly.

In general, during the design process the function of the system and its decomposition are considered first. After the function has been decomposed into the finest subsystems possible, assemblies and components are developed to provide these functions. Thus there is a hierarchy of mechanical decomposition, as is shown in the top row of Fig. 2.3. Also shown in this figure is one further decomposition of mechanical objects. For systems, assemblies, or components we can use the term *feature* to refer to specific attributes that are important, such as dimensions, material properties, shapes, or functional details. For the shutter *system,* some of the features are the speed of opening and closing and the light pattern through the aperture.

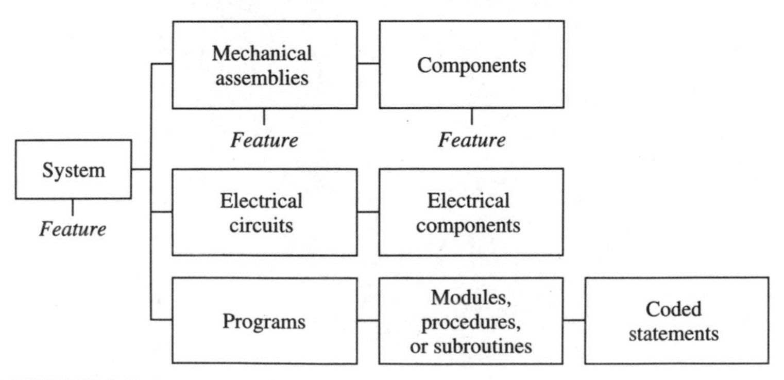

FIGURE 2.3
Decomposition of design disciplines.

For the *component* called the upper separator, the features are its dimensions and material properties. We must also note that on most cameras the shutter system has both electrical and mechanical components. The electrical systems provide energy transfer and control functions in the camera. The function of these electrical systems is fulfilled by circuits (electrical assemblies) that can be decomposed into electrical components, as shown in Fig. 2.3. Finally, some of the control functions are filled by microprocessors (the bottom row of Fig. 2.3). Physically these are electric circuits, but the actual control function is provided by a software program in the processor. This program is an assembly of coding modules composed of individual coding statements. It should be noted that the function of the microprocessor could be filled by an electrical or possibly even a purely mechanical system. During the early phases of the design process, when developing systems is the focus of the effort, it is often unclear whether the actual function should be met by mechanical assemblies, electrical circuits, software programs, or a mix of these elements.

2.3 IMPORTANCE OF PRODUCT FUNCTION, BEHAVIOR, AND PERFORMANCE

There are many synonyms for the word *function.* In mechanical engineering we commonly use the terms *function, operation,* and *purpose* to describe *what* a device does. A common way of classifying mechanical devices is by their *function.* In fact, some devices having only one main function are named for that function. For example, a screwdriver has the function of enabling a person to insert or remove a screw. The terms *drive, insert,* and *remove* are all verbs that tell what the screwdriver does. In telling what the screwdriver does, we have given no indication of how the screwdriver accomplishes its function. To answer how, we must have some information on the form of the device. The term *form* relates to any aspect of physical shape, geometry, construction, material, or size. As we shall see in Chap. 3, one of the main ways engineers mentally index their knowledge about the mechanical world is by function.

Earlier we physically decomposed mechanical systems into assemblies and components. Functional decomposition is often much more difficult than physical decomposition, as each function may use part of many components and each component may serve many functions. Consider the handlebars of a bicycle. The handlebars are a single component that serves many functions. It allows for steering (a verb that tells what the device does), and it supports upper-body weight (again, a function telling what the handlebars do). Further, it not only supports the brake levers but also transforms (another function) the gripping force to a pull on the brake cable. The shape of the handlebars and their relationship with other components determine how they provide all these different functions. The handlebars, however, are not the only component needed to steer the bike. Additional components necessary to perform this function are the front fork, the bearings between the fork and the frame, the front wheel, and miscellaneous fasteners. Actually, it can be argued that all the components on a bike contribute to steering, since a bike without a seat or rear wheel would be hard to steer. In any case, the handlebars perform many different functions,

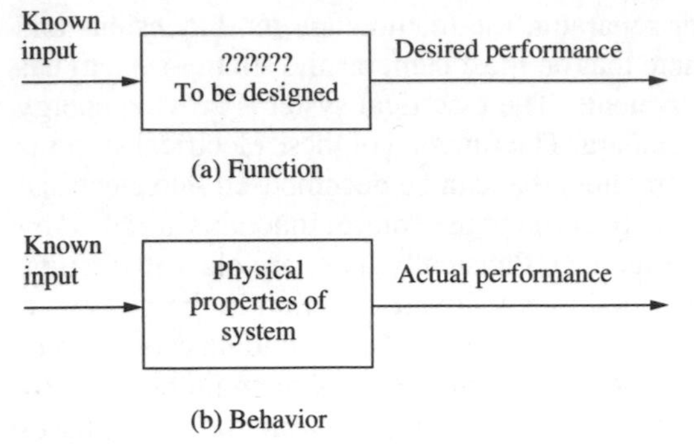

FIGURE 2.4
Function and behavior.

but in fulfilling these functions, the handlebars are only a part of various assemblies. This coupling between form and function makes mechanical devices hard to design.

Most common devices are cataloged by their function. If we want to specify a bearing, for example, we can search a bearing catalog and find many different styles of bearings (plain, ball, or tapered roller, for example). Each "style" has a different geometry—a different form—though all have the same primary function, namely, to reduce friction between a shaft and another object. Cataloging is possible in mechanical design as long as the primary function is clearly defined by a single piece of hardware, either a single component or an assembly. In other words, the form and function are decomposed along the same boundaries. This is true of many mechanical devices, such as pumps, valves, heat exchangers, gearboxes, and fan blades, and is especially true of many electrical circuits and components, such as resistors, capacitors, and amplifier circuits.

Two other terms often related to function are *behavior* and *performance. Function* and *behavior* are often used synonymously. However, there is a subtle difference, as shown in Fig. 2.4. In this figure there are two standard system blocks with an input represented by an arrow into the box, the system acted on by the input represented by the box, and the reaction of the system to the input represented by the arrow out of the box. The box in the upper part of the figure shows that *function is the desired output* from a system that is yet to be defined. When we begin to design a device, the device itself is unknown, but what we want it to do is known. If the system is known, as in the second part of the figure, then the behavior of the system can be found. *Behavior is the actual output,* the response of the system's physical properties to the input energy or control. Thus, the behavior can be simulated or measured, whereas function is only a desire.

Performance is the measure of function and behavior—how well the device does what it is designed to do. When we say that one function of the handlebars is to steer the bicycle, we say nothing about how well it serves this purpose. Before designing handlebars, we must develop a clear picture of their desired performance. For example, one design goal is that they must support 50 kg, a measurable desired performance for the handlebars. The development of clear performance measures is the focus of Chap. 6. Further, after designing the handlebars we can simulate their strength analytically or measure the strength of a prototype to find the actual performance for comparison to that desired. This comparison is a major focus of Chap. 11.

2.4 DIFFERENT TYPES OF MECHANICAL DESIGN PROBLEMS

Traditionally, we decompose mechanical engineering by discipline: fluids, thermodynamics, mechanics, and so on. In categorizing the types of mechanical design problems, this discipline-oriented approach is not appropriate. Consider, for example, the simplest kind of design problem, a selection design problem. Selection design means picking one (maybe more) item from a list such that the chosen item meets certain requirements. Common examples are selecting the correct bearing from a bearings catalog, selecting the correct lenses for an optical device, selecting the proper fan for cooling equipment, or selecting the proper heat exchanger for a heating or cooling process. The design process for each of these problems is essentially the same, even though the disciplines are very different. The goal of this section is to describe different types of design problems independently of the discipline.

Before beginning, we must realize that most design situations are a mix of various types of problems. For example, we might be designing a new type of consumer product that will accept a whole raw egg, break it, fry it, and deliver it on a plate. Since this is a new product, there will be a lot of *original design* work to be done. However, as the design process proceeds, we will find it necessary to *configure* the various parts; to analyze the heat conduction of the frying component, which will require *parametric design;* and to *select* a heating element and various fasteners to hold the components together. Further, if we are clever, we may be able to *redesign* an existing product to meet some or all of the requirements. Each of the italicized terms is a different type of design problem. It is rare to find a problem that is purely one type.

2.4.1 Selection Design

As was implied, selection design involves choosing one item (or maybe more) from a list of similar items. We do this type of design every time we choose an item from a catalog. It may sound simple, but if the catalog contains more than a few items and there are many different features to the items, the problem can be quite complex.

To solve a selection problem we must start with a clear need. The catalog or the list of choices then effectively generates potential solutions for the problem. We must evaluate the potential solutions with respect to our specific requirements to make the right choice. Consider the following example. During the process of designing a device, we must select a bearing to support a shaft. The known information is given in Fig. 2.5. The shaft has a diameter of 20 mm (0.787 in.). There is a radial force of 6675 N (1500 lb) on the shaft at the bearing, and the shaft rotates at a maximum of 2000 rpm. The housing to support the bearing is

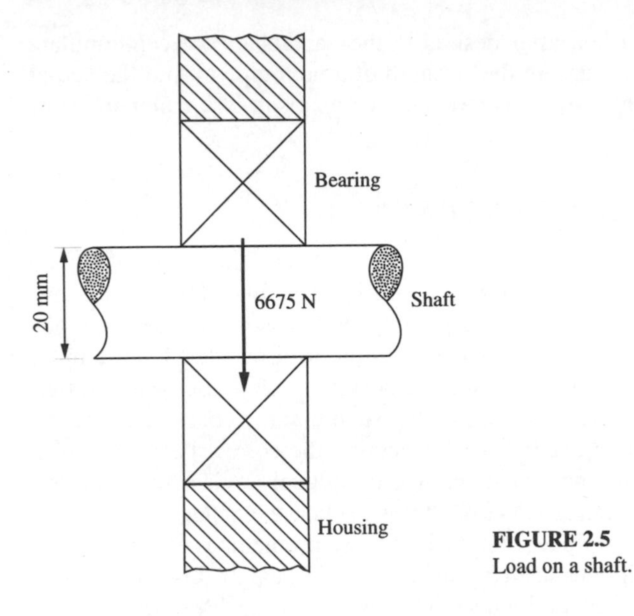

FIGURE 2.5
Load on a shaft.

still to be designed. All we need to do is select a bearing to meet the needs. The information on shaft size, maximum radial force, and maximum rpm given in bearing catalogs allows us to quickly develop a list of potential bearings (Table 2.1). This is the simplest type of design problem we could have, but it is still ill-defined. We do not have enough information to make a selection among the five possible choices. Even if a short list is developed—the most likely candidates being the 42-mm deep-groove ball bearing and the 24-mm needle bearing—there is no way to make a choice without more knowledge of the function of the bearing and of the engineering requirements on it.

2.4.2 Configuration Design

A slightly more complex type of design is called configuration design, or packaging. In this type of problem, all the components have been designed and the problem is how to assemble them into the completed product. Essentially, this type of design is similar to playing with an Erector set or other construction toy, or arranging living-room furniture.

Consider packaging electronic components in a laptop computer. A laptop computer has a keyboard built into the case, a power supply, a main circuit board, a hard-disk drive, a floppy-disk drive, and room for two extension boards. Each component is of known size and has certain constraints on its position. For example, the extension slots must be adjacent to the main circuit board and the keyboard must be in the front of the machine.

One methodology for solving design questions—How do we fit all the components in a case? Where do we put what?—is to randomly select a component from the list and position it in the case so that all the constraints on that component are

TABLE 2.1
Potential bearings for a shaft

Type		Outside diameter, mm	Width, mm	Load rating, lb	Speed limit, rpm	Catalog number
Deep-groove ball bearing		42	8	1560	18,000	6000
		47	14	2900	15,000	6204
		52	15	3900	9000	6304
Angular-contact ball bearing		47	14	3000	13,000	7204
		37	9	1960	34,000	71,904
Roller bearing		47	14	6200	13,000	204
		52	15	7350	13,000	220
Needle bearing		24	20	1930	13,000	206
		26	12	2800	13,000	208
Nylon bushing		23	Variable	290	10	4930
				⋮	⋮	
				8	500	

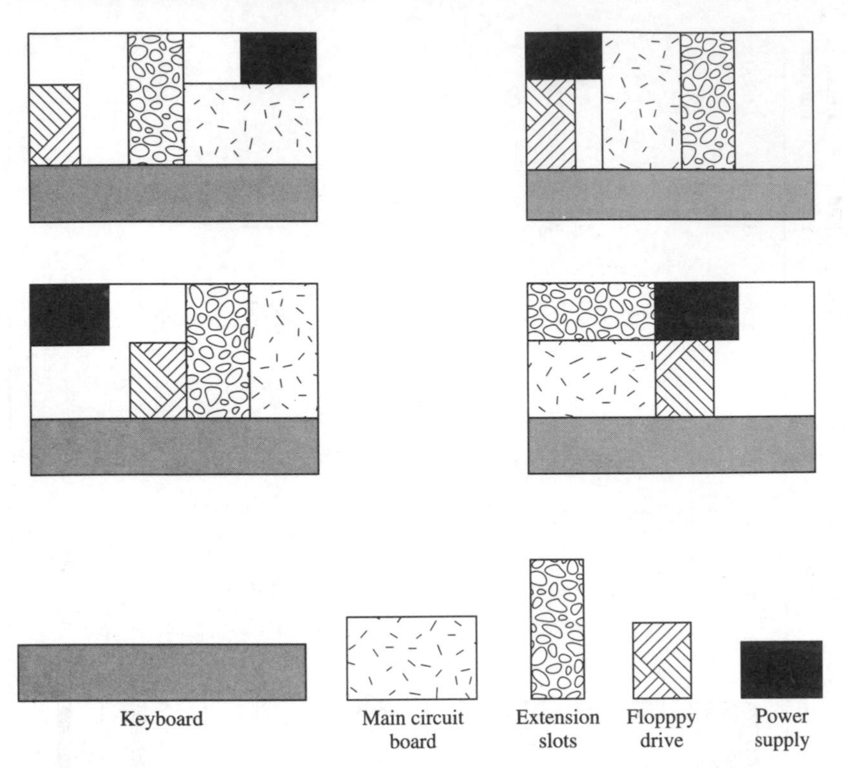

FIGURE 2.6
Possible configurations for a computer. (In each case the bottom of the sketch represents the front of the computer.)

met. The keyboard, if we start with that, has to be placed in the front. Then we select and place a second component. This procedure is continued until either we run into a conflict or all the components are in the case. If a conflict arises, we back up and try again. Using this logic and assuming a two-dimensional world, we eventually establish the potential configurations shown in Fig. 2.6.

Not all configuration problems are as well-defined as the computer example. For many problems some of the components to be fit into the assembly can be altered in size, shape, or function, giving the designer more latitude to determine potential configurations and making the problem solution more difficult.

2.4.3 Parametric Design

Parametric design involves finding values for the features that characterize the object being studied. This may seem easy enough—just find some values that meet the requirements. However, consider a very simple example. We want to design a cylindrical storage tank that must hold 4 m^3 of liquid. This tank is described by the parameters r, its radius, and l, its length. Thus the volume is determined by

$$V = \pi r^2 l$$

Given a volume equal to 4 m^3,

$$r^2 l = 1.273$$

We can see that an infinite number of values for the radius and length will satisfy this equation. To what values should the parameters be set? The answer is not obvious, nor even clearly defined with the information given. (This problem will be readdressed in Chap. 11, where the accuracy to which the radius and the length can be manufactured will be used to help find the best values for the parameters.)

Let us extend the concept further. It may be that instead of a simple equation, a whole set of equations and rules govern the design. Consider the instance in which a major manufacturer of copying machines had to design paper-feed mechanisms for each new copier. (A paper feed is a set of rollers, drive wheels, and baffles that move a piece of paper from one location to another in the machine.) Many parameters—the number of rollers, their positions, the shape of the baffles, etc.—characterize this particular design problem, but obviously there are certain similarities in paper feeders, regardless of the relative positions of the beginning and end points of the paper, the obstructions (other components in the machine) that must be cleared, and the size and weight of the paper. The company developed a set of equations and rules to aid designers in developing workable paper paths, and using this information, the designers could generate values for parameters in new products.

2.4.4 Original Design

Any time the design problem requires the development of a process, assembly, or component not previously in existence, it calls for original design. (It can be said that if we have never seen a wheel and we design one, then we have an original design.) Though most selection, configuration, and parametric problems are represented by equations, rules, or some other logical scheme, original design problems cannot be reduced to any algorithm. Each one represents something new and unique.

In many ways the other types of design problems—selection, configuration, and parametric—are simply constrained subsets of original design. The potential solutions are limited to a list, an arrangement of components, or a set of related characterizing values. Thus, if we have a clear methodology for performing original design, we should be able to solve any design problem with a more limited set of potential solutions.

2.4.5 Redesign

Most design problems solved in industry are for the redesign of an existing product. Suppose a manufacturer of hydraulic cylinders makes a product that is 0.25 m long. If the customer needs a cylinder 0.3 m long, the manufacturer might lengthen the outer cylinder and the piston rod to meet this special need. These changes may require only parameter changes, or they may require something more extensive. What if the materials are not available in the needed length, or cylinder fill time becomes too slow with the added length? Then the redesign effort may require much more than parameter changes. Regardless of the change, this is an example of *redesign,* the modification of an existing product to meet new requirements.

FIGURE 2.7
1890 Humber bicycle.

Many redesign problems are *routine;* the design domain is so well understood that the method used can be put in a handbook as a series of formulas or rules. The parameter changes in the example of the hydraulic cylinder are probably routine for the manufacturer.

The hydraulic cylinder can also be used as an example of a *mature design,* in that it has remained virtually unchanged over many years. There are many examples of mature designs in our everyday lives: pencil sharpeners, hole punches, and staplers are a few found on the average desk. For these products, knowledge about the design problem is complete (Fig. 1.10). There is nothing more to learn.

However, consider the bicycle. The basic configuration of the bicycle—the two tensioned, spoked wheels of equal diameter, the diamond-shaped frame, and the chain drive—was fairly refined late in the last century. While the 1890 Humber shown in Fig. 2.7 looks much like a modern bicycle, not all bicycles of this era were of this configuration. The Otto dicycle (*di* and *bi* both mean *two*), shown in Fig. 2.8, had two spoked wheels and a chain; stopping and steering this machine must have been a challenge. In fact, the technology of bicycle design was so well developed by the end of the nineteenth century that a major book on the subject, *Bicycles and Tricycles: An Elementary Treatise on Their Design and Construction,* was published in 1896.[1] The only major change in bicycle design since the publication of that book was the introduction of the derailleur in the 1930s.

However, in the 1980s the traditional bicycle design began to change again. For example, the concept bike shown in Fig. 2.9 no longer has a diamond frame, the

[1] The book, written by Archibald Sharp, has recently (1977) been reprinted by the MIT Press, Cambridge, Mass.

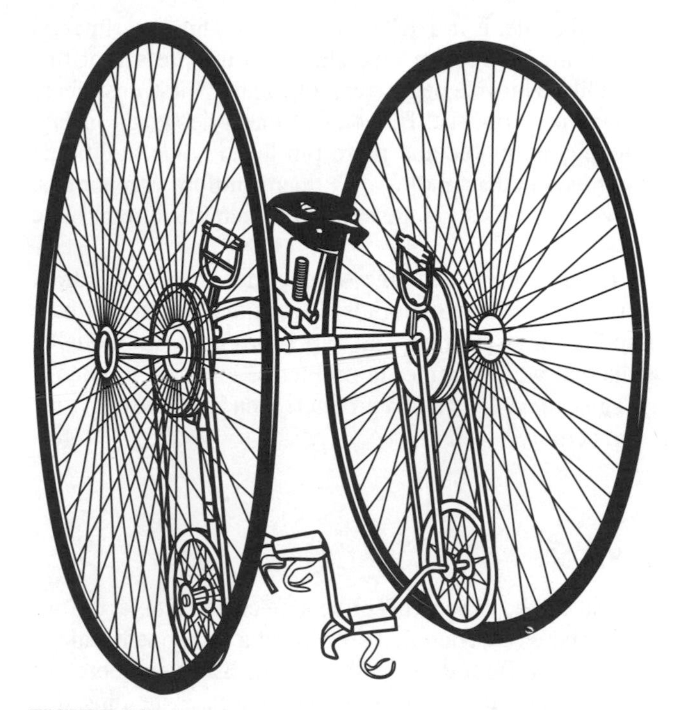

FIGURE 2.8
The Otto dicycle.

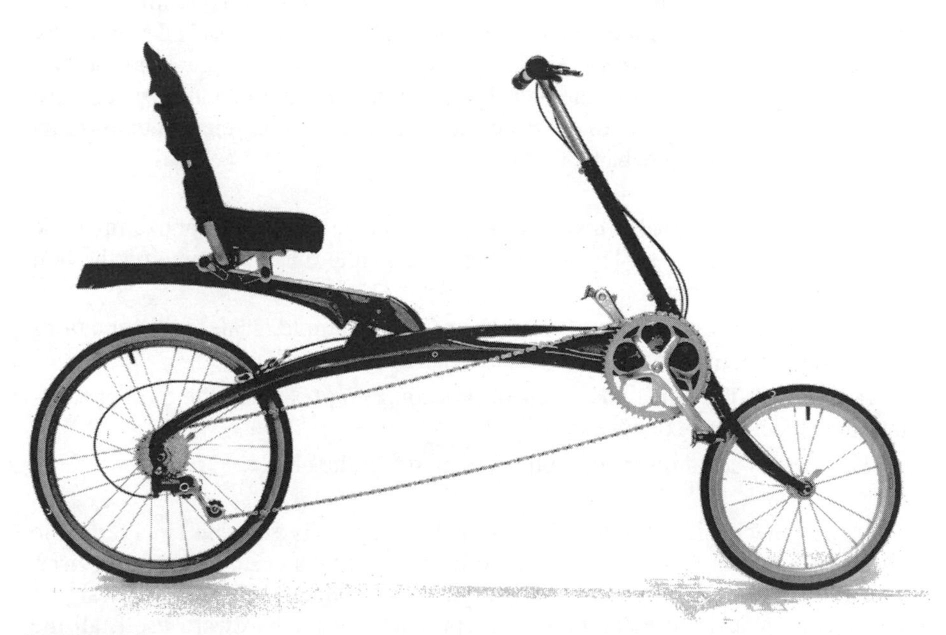

FIGURE 2.9
The FantasE from BikeE Corporation (designed by the author and his colleagues).

frame has no steel in it, and even the rider is in a different position. This experimental bike does not differ much from the production bike shown in the preface of this book. Why did a mature design like a bicycle begin evolving again? First, customers are always looking for improved performance. Bicycles of the style shown in Fig. 2.9 and in the preface are more comfortable and more fun to ride than traditional bicycles. Second, there is improved understanding of human comfort, ergonomics, and aerodynamics. Third, bicycles of this style are simpler to manufacture in large numbers.

The point is that even mature designs change to meet new needs, to attract new customers, or to take advantage of new materials. Part of the design of a new bicycle like the one shown in Fig. 2.9 is routine, and part is original. Additionally, there were many subproblems that were parametric problems, selection problems, and configuration problems. Thus, the redesign of a product, even a mature one, may require a wide range of design activity.

2.5 LANGUAGES OF MECHANICAL DESIGN

There are many "languages" that can be used to describe a mechanical object. Consider for a moment the difference between a detailed drawing of a component and the actual hardware that *is* the component. Both the drawing and the hardware represent the same object; however, they represent it in different languages. Extending this example further, if the component we are discussing is a bolt, then the word *bolt* is a textual (semantic or word) description of the component. Additionally, the bolt can be represented through equations that describe its functionality and possibly its form. For example, the ability of the bolt to carry shear stress (a function) is described by the equation $\tau = F/A$; the shear stress τ is equal to the shear force F on the bolt divided by the stress area A of the bolt. Based on the paragraph above we can use four different representations to describe the bolt. These same representations can be used to describe any mechanical object:

Semantic. The verbal or textual representation of the object—for example, the word *bolt,* or the sentence, "The shear stress is equal to the shear force on the bolt divided by the stress area."

Graphical. The drawing of the object—for example, scale representations such as orthogonal drawings, sketches, or artistic renderings.

Analytical. The equations, rules, or procedures representing the form or function of the object—for example, $\tau = F/A$.

Physical. The hardware or a physical model of the object.

In most mechanical design problems the initial need is expressed in a semantic language as a written specification or a verbal request by a customer or supervisor. The final result of the design process is a physical product. Although the designer produces a graphical representation of the product, not the hardware itself, all the languages will be used as the product is refined from its initial, abstract semantic representation to its final physical form.

2.6 CONSTRAINTS, GOALS, AND DESIGN DECISIONS

The progression from the initial need (the design problem) to the final product is made in increments punctuated by *design decisions.* Each design decision changes the *design state,* which is like a snapshot of all the information known about the product being designed at any given time during the process. In the beginning the design state is just the problem statement. During the process the design state is a collection of all the knowledge, drawings, models, analyses, and notes thus far generated.

Two different views can be taken of how the design process progresses from one design state to the next. One view is that products evolve by a continuous comparison between the design state and the *goal,* that is, the requirements for the product given in the problem statement. This philosophy implies that all the requirements are known at the beginning of the design problem and that the difference between them and the current design state can be easily found. It is this difference that controls the process. This is the case in some simple selection and configurational design problems. However, for most design problems this view of design toward a known goal is too simplistic; because the problem is ill defined, the goal cannot be completely known.

Another view of the design process is that when a new problem is begun, the design requirements effectively constrain the possible solutions to a subset of all possible product designs. As the design process continues, other *constraints* are added to further reduce the potential solutions to the problem, and potential solutions are continually eliminated until there is only one final design. In other words design is the successive development and application of constraints until only one unique product remains. This model best represents most design problems.

Beyond the constraints in the original problem specifications, constraints added during the design process come from two sources. The first is from the designer's knowledge of mechanical devices and the specific problem being solved. If a designer says, "I know bolted joints are good for fastening together sheet metal," this piece of knowledge constrains the solution to bolted joints only. Since every designer has different knowledge, the constraints introduced into the design process make each designer's solution to a given problem unique.

The second type of constraint added during the design process is the result of design decisions. If a designer says, "I will use 1-cm-diameter bolts to fasten these two pieces of sheet metal together," the solution is constrained to 1-cm-diameter bolts, a constraint that may affect many other decisions—clearance for tools to tighten the bolt, thickness of materials used, etc. During the design process a majority of the constraints are based on the results of design decisions. Thus the individual designer's ability to make well informed decisions throughout the design process is essential.

Regardless of the source of the constraints, each decision requires two pieces of information—constraints and alternative solutions—and two activities—comparison between the constraints and the alternatives and, based on the results of the comparison, the selection of the best alternative. The comparison activity is often called *evaluation* and is the focus of Chaps. 8, 11, and 12.

2.7 THE VALUE OF INFORMATION

We live in the age of information. For this statement to have any use for us, it is important for us to understand the value of information, especially engineering information. Consider Fig. 2.10. In this figure four classes of information are shown. The simplest form of information is raw data—parameters or values for variables. If we can develop relationships between these data, then we have models. These relationships may be derived equations, empirically developed relationships, or just qualitative models. These models are only static relationships among the data. To gain knowledge, we must understand or interpret the behavior of the models. Finally, when the knowledge is sufficiently understood, we can make decisions using judgment based on the knowledge.

During the course of our education, we solve many problems that are models of parameters. By solving these problems we gain knowledge used to make decisions. Good decisions require good knowledge, which requires models of the variables involved.

Look at the drawing of the gear set in Fig. 2.11. The lines and dimensions on the drawing are the data. The relationships between the lines make shapes that, if you understand the behavior of the shapes, you will recognize as a gear set. However, this knowledge about what gears look like is useful only if you can use them in a power transmission or other system. Judgments are necessary for making decisions on the proper use of the gears. This example is included to both convince you that the model in Fig. 2.10 is a usable way to show the relationships between the classes of information and to demonstrate that understanding a drawing requires a model of

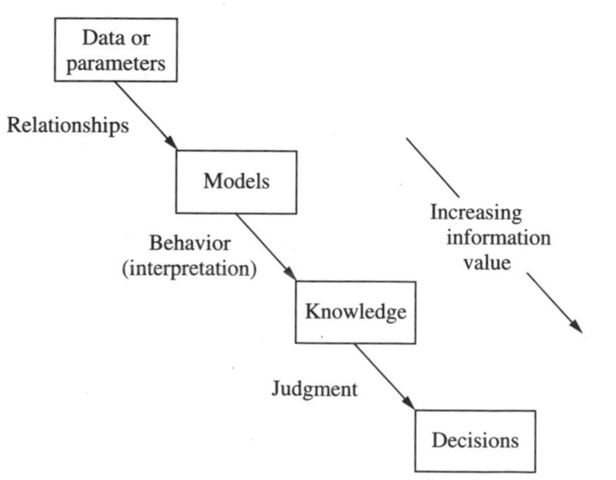

FIGURE 2.10
The value of information.

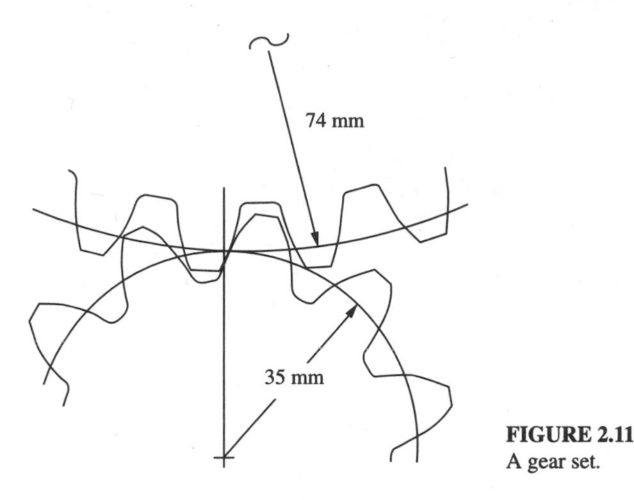

FIGURE 2.11
A gear set.

what the data represents and interpretation of that model. This model is also consistent with that for modeling how people process information in their minds (Section 3.2).

2.8 DESIGN AS REFINEMENT OF ABSTRACT REPRESENTATIONS

Consider the two drawings for a single product, shown in Fig. 2.12. The upper figure is a rough sketch, which gives only abstract information about the component. The bottom figure is a detailed drawing of the final component. In progressing from the sketch to the final drawing, the *level of abstraction* of the device is *refined.*

Some design process techniques are better used on abstract levels and others on more concrete levels, though in actuality there are no true levels of abstractions but rather a continuum on which the form or function can be represented. Descriptions of three levels of abstraction in each of the four languages are given in Table 2.2. The object we call a bolt is used as an example in Table 2.3.

The process of making an object less abstract (or more concrete) is called *refinement.* Mechanical design is a continuous process of refining the given needs to the final hardware. The refinement of the bolt in Table 2.3 is illustrated on a left-to-right continuum. In most design situations the beginning of the problem appears in the upper left corner of such a chart and the final product in the lower right. The path connecting these is a mix of the other representations and levels of abstractions.

2.9 SUMMARY

* A product can be divided into functionally oriented operating systems. These are made up of mechanical assemblies, electronic circuits, and computer programs. Mechanical assemblies are built of various components.
* The important form and function aspects of mechanical devices are called *features.*

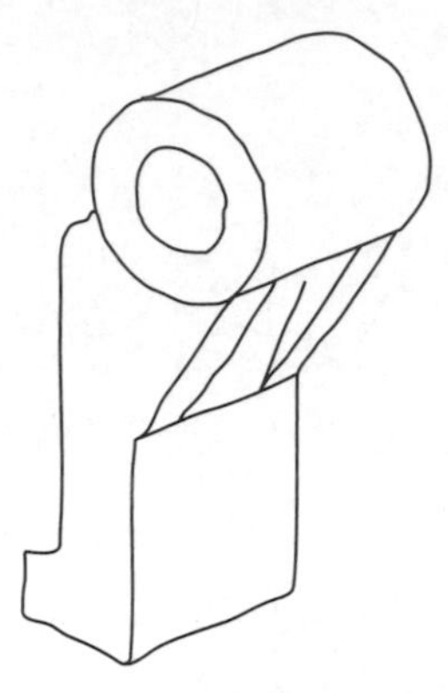

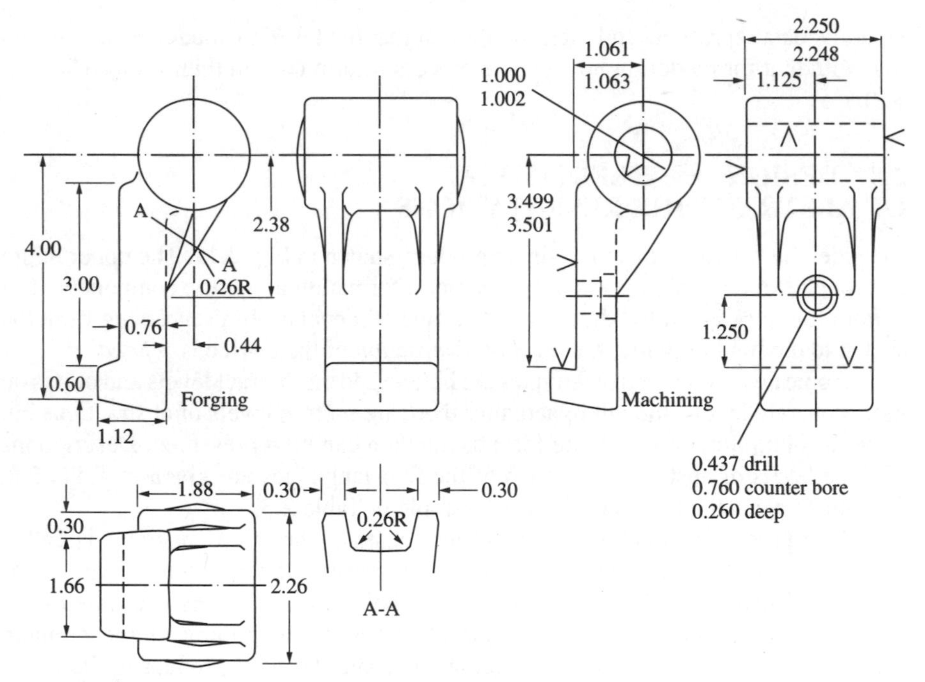

Draft angle 7°
Unspecified radii 0.10

FIGURE 2.12
Abstract sketch and final drawing of a component.

* Function and behavior tell *what* a device does; form describes *how* it is accomplished.
* Function relates desired behavior.
* One component may play a role in many functions, and a single function may require many different components.

TABLE 2.2
Levels of abstraction in different languages

	Level of abstraction		
Language	**Abstract**	⟶	**Concrete**
Semantic	Qualitative words (e.g., *long, fast, lightest*)	Reference to specific parameters or components	Reference to the values of the specific parameters or components
Graphical	Rough sketches	Scale drawings	Detailed drawings with tolerances
Analytical	Qualitative relations (e.g., *left of*)	Back-of-the-envelope calcuations	Detailed analysis
Physical	None	Models of the product	Final hardware

* There are many different types of mechanical design problems: selection, configuration, parametric, original, redesign, routine, and mature.
* Mechanical objects can be described semantically, graphically, analytically, or physically.
* The design process is a continuous constraining of the potential product designs until one final product evolves. This constraining of the design space is made through repeated comparison with the design requirements.

TABLE 2.3
Levels of abstraction in describing a bolt

	Level of abstraction		
Language	**Abstract**	⟶	**Concrete**
Semantic	A bolt	A short bolt	A 1" 1/4-20 UNC Grade 5 bolt
Graphical		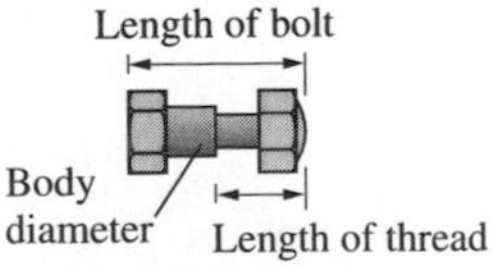 Length of bolt; Body diameter; Length of thread	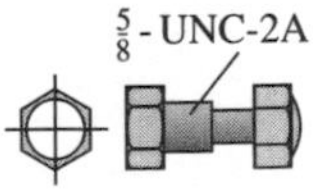 $\frac{5}{8}$-UNC-2A
Analytical	Right-hand rule	$\tau = F/A$	$\tau = F/A$
Physical	—	—	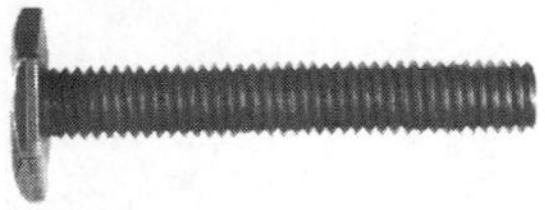

* Mechanical design is the refinement from abstract representations to a final physical artifact.
* The most valuable information is the decisions that are communicated to others.

2.10 SOURCES

Love, S.: *Planning and Creating Successful Engineered Designs: Managing the Design Process,* Advanced Professional Development, Los Angeles, 1980. One of the first books to specifically address itself to the mechanical design process.

Pahl, G., and W. Beitz: *Engineering Design,* 2nd edition, Springer-Verlag, 1996 (original German text, Springer-Verlag, 1977). One of the first and most complete efforts at structuring the mechanical design process; many good domain-specific designs are included in the text.

Tjavle, E.: *A Short Course in Industrial Design,* Newnes-Butterworths, London, England, 1979. Relates industrial design to the functional basis of mechanical engineering.

2.11 EXERCISES

2.1. Decompose a simple system such as a home appliance, bicycle, or toy into its assemblies, components, electrical circuits, etc. Figures 2.3 and 4.4 will help. List all the important features of one component.

2.2. Select a fastener from a catalog that meets the following requirements:

Can attach two pieces of 14-gauge sheet steel (0.075 in., 1.9 mm) together
Is easy to fasten with a standard tool
Can only be removed with special tools
Can be removed without destroying either base materials or fastener

2.3. Sketch at least five ways to configure two passengers in a new three-wheeled commuter vehicle that you are designing.

2.4. You are a designer of diving boards. A simple model of your product is a cantilever beam. You want to design a new board so that a 150-lb (67-kg) woman deflects the board 3 in. (7.6 cm) when standing on the end. Parametrically vary the length, material, and thickness of the board to find five configurations that will meet the deflection criterion.

2.5. Find five examples of mature designs. Also, find one mature design that has been recently redesigned. What pressures or new developments led to the change?

2.6. Describe your pencil or pen in each of the four languages at the three levels of abstraction, as was done with the bolt in Table 2.3.

CHAPTER 3

DESIGNERS AND DESIGN TEAMS

3.1 INTRODUCTION

Since the time of the early potter's wheel, mechanical devices have become increasingly complex and sophisticated. This sophistication has evolved without much concern for how humans solve design problems. Throughout history people who were just naturally good at design were trained, through an apprentice program, to be masters in their art. The design methods they used and the knowledge of the domain in which they worked was refined through their personal experiences and passed, in turn, to their apprentices. Much of this experience was gained through experiments, through building prototypes and then going "back to the drawing board" to iterate toward the next design. The results of these experiments taught the designers what worked and what did not and pointed the way to the next refinement. With this methodology, products took many generations to be refined to the point of mature design.

However, as systems grew more complex and the world community grew more competitive, this mode of design became too time-consuming and too expensive. Designers recognized the need to find ways to deal with larger, more complex systems, to speed the design process, and to ensure that the final design be reached with a minimum use of resources and time. In this book we discuss design techniques that meet these goals. To understand how these techniques help streamline the design process, it is important to understand how designers progress from abstract needs to final, detailed products.

To put this chapter in context, it is important to realize that design is the confluence of technical processes, cognitive processes, and social processes. In this chapter we begin our discussion of how humans design mechanical objects by describing a cognitive model of how memory is structured in the individual designer. The types of information that are processed in this structure are explored, and the term *knowledge* is defined. Once we understand the information flow in human memory, we develop the different types of operations that a designer must perform in memory during the design process, and we explore creativity.

Based on this model of the individual's cognitive process, the chapter moves to the social aspect of design—working in teams. First, team goals are discussed with emphasis on how a team is different from an individual. This is followed by itemizing team roles and how they relate to individual problem-solving style. Finally, guidelines for keeping teams functioning smoothly are presented.

3.2 A MODEL OF HUMAN INFORMATION PROCESSING

Solving design problems is a decision-making process. There is approximately one decision made each minute during design activity. These decisions are based on processing information in the designer's mind. The study of human problem-solving abilities is called *cognitive psychology.* Although this science has not yet fully explained the problem-solving process, psychologists have developed models that give us a pretty good idea of what happens inside our heads during design activities. A simplification of a generally accepted model is shown in Fig. 3.1. This model, called

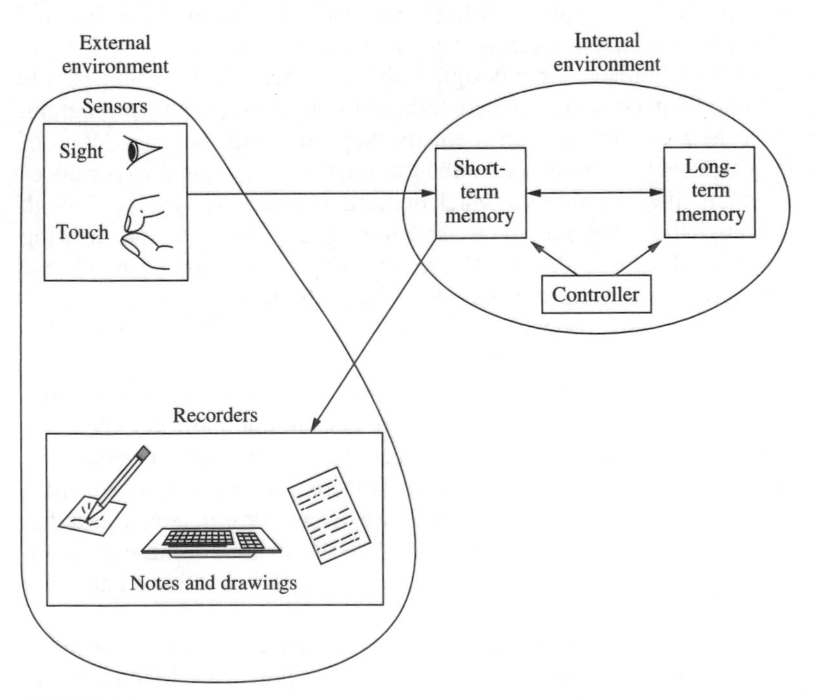

FIGURE 3.1
The human problem solver.

the *information-processing system* and developed in the late 1950s, describes the mental system used in the solution of any type of problem. In discussing that system here, we give special emphasis to the solution of mechanical design problems.

Information processing takes place through the interaction of two environments: the *internal environment* (information storage and processing inside the human brain) and the *external environment.* The external environment comprises paper and pencil, catalogs, computer output, and whatever else is used outside the human body to aid in the solution of the problem.

In the internal environment, that is, within the human mind, there are two different types of memory: *short-term memory,* which is similar to a computer's operating memory (its random access memory or RAM) and *long-term memory,* which is like a computer's disk storage. Bringing information into this system from the external environment are *sensors,* such as the eyes, ears, and hands. Taste and smell are less often used in design. Information is output from the body with the use of the hands and the voice. There are other means of output, such as body position, that are less often used in design. Additionally, as part of the internal processing capability, there is a *controller* that manages the information flow from the sensors to the short-term memory, between the short-term and the long-term memory, and between the short-term memory and the means of output.

Before describing short-term and long-term memory and the control of information flow, we need to describe the *information* that is processed in this system. In a computer the information is in terms of bytes, or binary digits (0s and 1s), but in the human brain information is much more complex.

In recent experiments, an orthographic drawing of a power transmission system consisting of shafts, gears, and bearings was shown to mechanical engineering students and professional engineers. The students were lower-level undergraduates who had not studied power transmission systems. The drawing was shown briefly and then removed, and the subjects were asked to sketch what they had seen. The students tended to reconstruct the drawings from the line segments and simple shapes they had seen in the original drawing. Not understanding the complexities of geared transmissions, they could not remember anything more complicated. They remembered and drew only the basic form of the system. On the other hand, the professional engineers were able to remember components grouped together by their function. In recalling a gear set, for example, the experts knew that two meshed gears and their associated shafts and bearings provide the function of changing the rpm and torque in the system. They also knew what geometry or line segments were needed to represent the form of a gear set. Thus, the experienced engineers were able to include substantially more information than the students in their sketches.

The line segments remembered by the students and the functional groupings remembered by the experienced designers are called *chunks of information* by cognitive psychologists. The greater the expertise of the designer, the more content there is in the chunks of information processed. Exactly what types of information are in these chunks, however, is not always clear. Types of knowledge that might be in a chunk include the following:

- *General knowledge,* information that most people know and apply without regard to a specific domain. For example, red is a color, the number 4 is bigger than the number 3, an applied force causes a mass to accelerate—all exemplify general knowledge. This knowledge is gained through everyday experiences and basic schooling.
- *Domain-specific knowledge,* information on the form or function of an individual object or a class of objects. For example, all bolts have a head, a threaded body, and a tip; bolts are used to carry shear or axial stresses; the proof stress of a grade 5 bolt is 85 kpsi. This knowledge comes from study and experience in the specific domain. It is estimated that it takes about 10 years to gain enough specific knowledge to be considered an expert in a domain. Formal education sets the foundation for gaining this knowledge.
- *Procedural knowledge,* the knowledge of what to do next. For example, if there is no answer to problem X, then decomposing X into two independent subproblems, X_1 and X_2, would illustrate procedural knowledge. This knowledge comes from experience, but some procedural knowledge is also based on general knowledge and some on domain-specific knowledge. For solving mechanical design problems, we must often make use of procedural knowledge.

In mechanical engineering the term *feature* is synonymous with *chunks of information.* Since a design feature is some important aspect of a component, assembly, or function, the gear set discussed in the example above is both a chunk and a feature.

The exact language in which chunks of information are encoded in the brain is unknown. They might be dealt with as semantic information (text), graphical information (visual images), or analytical information (equations or relationships). Psychologists believe that most mechanical designers process information in terms of visual images and that these images are three-dimensional and are readily manipulated in the short-term memory.

3.2.1 Short-Term Memory

The short-term memory is the main information processor in the human brain. It has no known specific anatomic location, yet it is known to have very specific attributes.

One important attribute of the short-term memory is its quickness. Information chunks can be processed in the short-term memory in about 0.1 second. The term *processed* implies such actions as comparing one chunk of information to another, modifying a chunk by decomposing it into smaller parts, combining two or more chunks into one new one, changing a chunk's size or distorting its shape, and making a decision about the chunk. It is unknown how much of the short-term memory is actually used to process the information. We do know that the harder it is to solve the problem, the more short-term memory is used for processing.

The capacity of the short-term memory was first described in a paper titled "The Magical Number Seven, Plus or Minus Two" (see Section 3.7), which reported that the short-term memory is effectively limited to seven chunks of information (plus or minus two). This is like having a computer RAM with only seven memory locations. These approximately seven chunks—these seven unique things—are all that a person can deal with at one time. For example, let us say we are working on a design problem and have an idea (a chunk of information, maybe just a word or maybe a visual image) that we want to compare to some constraints on the design (other chunks of information). How many constraints can we compare to the idea in our head? Only two or three at a time, since the idea itself takes one slot in the short-term memory and the constraints take two or three more. That does not leave much memory to do the processing necessary for comparison. Add any more constraints and the processing stops; the short-term memory is simply too full to make any progress on solving the problem.

A couple of quick experiments are convincing about the limits of the short-term memory. Open a phone book and randomly choose a phone number in which the seven digits are unrelated to each other. (A number such as 754-2000 is not acceptable because the last four digits can be lumped together as a single chunk—two thousand.) After looking at the number briefly, close the phone book, walk across the room, and dial the number. Most people can manage to do this task if they are not interrupted or do not think about anything else. The same experiment can be tried with two unrelated phone numbers. Few people would be able to remember them long enough to dial them both since they require dialing 14 pieces of information, which is beyond the capacity of the short-term memory. Granted, these 14 numbers can be memorized, or stored in long-term memory, but that would take some study time.

Another example of the size limitations of short-term memory is more mechanical in nature. Consider the four-bar linkage of Fig. 3.2. It is made up of four

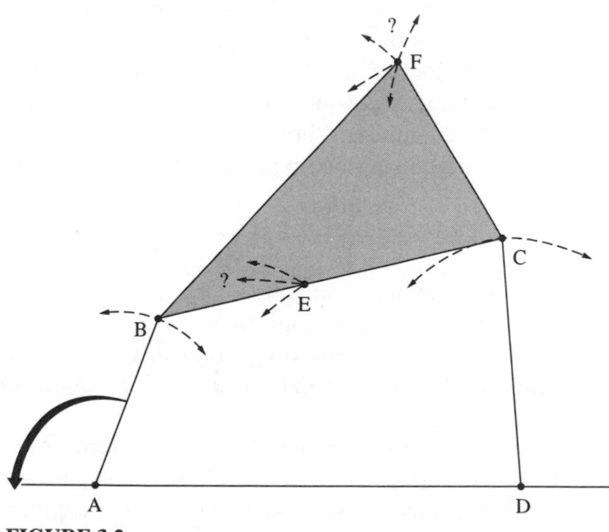

FIGURE 3.2
A four-bar linkage.

elements: the driver A-B, the link B-C, the follower C-D, and the base D-A. It is not difficult for most engineers to visualize the follower rocking back and forth as the driver is rotated. Point B makes a circle, and point C moves in an arc about point D. An expert on linkages would only use a single chunk to encode this mechanism. But a novice in the domain of four-bar linkages would need to visualize four line segments, using four chunks plus others for processing the motion. To make the task more difficult, trace the path of point E on the link. This requires more short-term memory. Harder still is tracing the path of point F. In fact, this requires so many different parameters to track that only a few linkage experts can visualize the path of point F.

Another feature of the short-term memory is the fading of information stored there. The phone number remembered above is probably forgotten within a few minutes. To keep from forgetting short-term information, like the phone number, many people keep repeating the information over and over. With such continuous refreshing, it is possible to retain certain objects or parts of objects within the short-term memory and to let only the unimportant information fade to make room for the processing of new chunks of information.

Last, it is impossible for us to be aware of what is happening in our short-term memory while we are solving problems. To follow our own thoughts, we need to use some of that memory to monitor and understand the problem-solving process, making that space no longer available for problem solving.

3.2.2 Long-Term Memory

The long-term memory was earlier compared with the disk storage in a computer; like disk storage, it is for permanent retention of information. Let us look at the four major characteristics of long-term memory. First, long-term memory has seemingly unlimited capacity. Despite the cartoon in Fig. 3.3, there is no documented case of anybody's brain becoming "full," regardless of head size. It is hypothesized that as we learn more we unconsciously find more efficient ways to organize the information by reorganizing the chunks in storage. Reconsider the difference between the student's and the expert's ways of remembering information about the power transmission system. The expert's information storage was more efficient than the student's.

The second characteristic of the long-term memory is that it is fairly slow in recording information. It takes two to five minutes to memorize a single chunk of information. This explains why studying new material takes so long.

The third characteristic is the speedy recovery of information from long-term memory. Retrieval is much quicker than storage, the time depending on the complexity of the information and the recentness of its use. It can be as fast as 0.1 second per chunk of information.

The fourth characteristic is that the information stored in the long-term memory can be retrieved at different levels of abstraction, in different languages, and with different features. For example, consider the knowledge an average engineer can retrieve about a car (Fig. 3.4). The sample data ranges from images of entire

"Mr. Osborne, may I be excused? My brain is full."

FIGURE 3.3
Long-term memory problems. (From Gary Larson, *The Prehistory of the Far Side,* Andrews and McMeel, New York, 1988.)

vehicles to semantic rules and equations for diagnosing problems. Human memory is very powerful in matching the form of the data retrieved to that which is needed for processing in the short-term memory.

3.2.3 Control of the Information Processing System

During problem solving, the controller (Fig. 3.1) allows us to encode outside information obtained through our senses or retrieve information from long-term memory for processing in the short-term memory. Some of the information in the short-term memory is allowed to fade, and new information is input as it is needed and becomes available. Additionally, the controller can help extend the short-term memory by making notes and sketches; these need to be done quickly so that they do not bog down the problem-solving process. When we have completed manipulating the

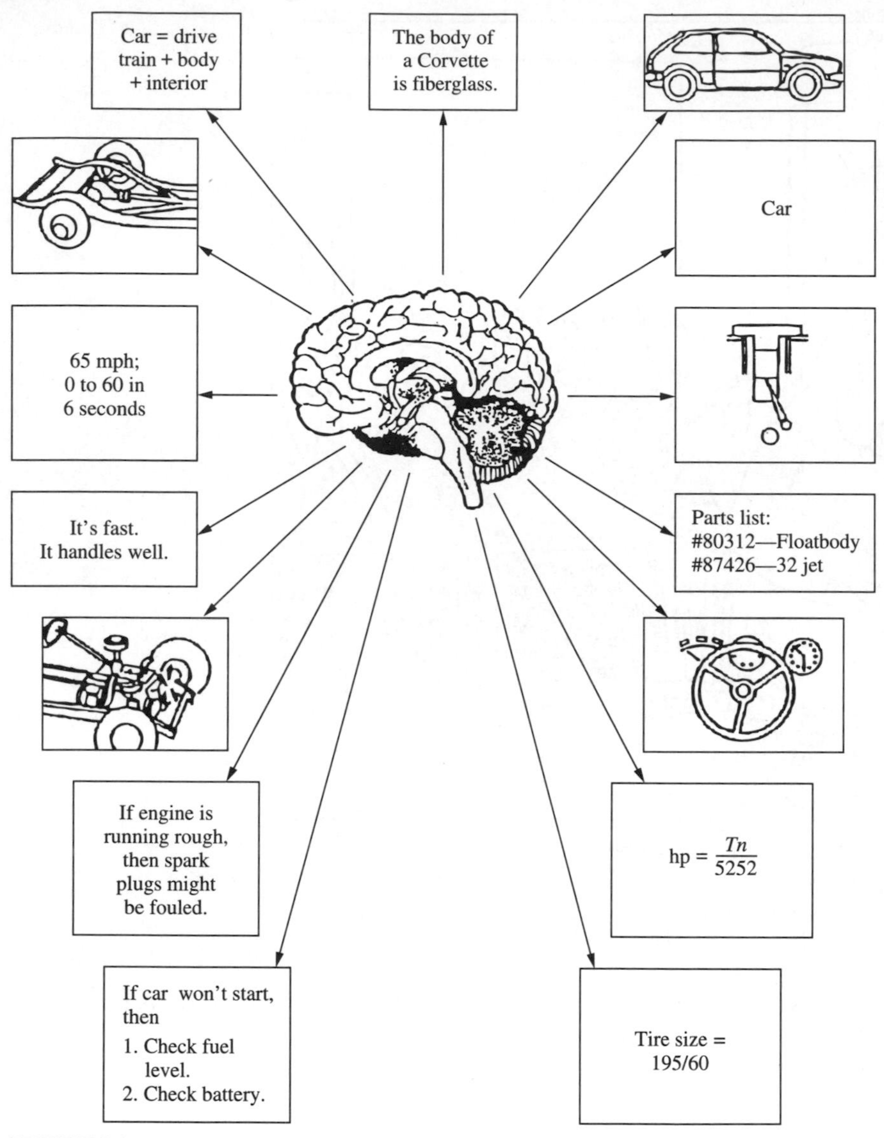

FIGURE 3.4
Knowledge stored in memory about cars.

information, the controller can store the results in long-term memory, or in the external environment by describing it in text, verbally, or in graphic images.

3.2.4 External Environment

The external environment—paper and pencil, computers, books—plays a number of roles in the design process: it is a source of information; it is an analytical capability;

it is a documentation/communication facility; and, most importantly for designers, it is an extension for the short-term memory. The first three of these roles seems evident; however, the last role, as an extension for the short-term memory, needs some discussion.

Because the short-term memory is a space-limited central processor, human problem solvers utilize the external environment as a short-term memory extension. This is accomplished by making notes and sketches of ideas and other information needed in problem solving. In order to be useful to the short-term memory, any extension must share the characteristics of being very fast and having a high information content. Watch any design engineer trying to solve a problem. He or she will make sketches even when not trying to communicate. These sketches serve as aids in generating and evaluating the ideas by serving as additional chunks of information to be processed. Sketches are fast to make and are information-rich.

3.2.5 Implications of the Model

One of the implications of the information processing model of human problem solving is that the size of the short-term memory is a major limiting factor in the ability to solve problems. Our thinking process has evolved so that, as we solve problems, our expertise about the constraints and potential solutions increases and our configuration of chunks becomes more efficient. This helps offset the "magic number" seven, but human designers are still quite limited. It would almost seem that these limitations would preclude our ability to solve complex problems. As will be discussed in the sections below, processing speed and flexibility of information storage and recovery enable designers to develop very complex products.

3.3 MENTAL PROCESSES THAT OCCUR DURING DESIGN

We can now describe what happens when a designer faces a new design problem. The problem may be the design of a large, complex system or of some small feature on a component. We will focus on how a designer understands new information such as the problem statement, how ideas are generated, and how they are evaluated.

3.3.1 Understanding the Problem

Consider what happens when a new problem is broached. If we think of its design state as a blackboard on which is written or drawn everything known about the device being designed, then the blackboard is initially blank, i.e., the design state is empty. Let us return to the fastening problem presented in Chap. 1 (see Fig. 1.8):

> Design a joint to fasten together two pieces of 1045 sheet steel, each 4 mm thick and 6 cm wide, that are lapped over each other and loaded with 100 N.

Before any information about the problem is put on the design-state blackboard, the problem statement must be understood. If the problem is outside the realm of experience (the designer does not know what the term *lapped* means, for example), then the problem cannot be understood.

But how do we "understand" a problem? Most likely in this way: As the problem is read, it is "chunked" into significant packets of information. This happens in the short-term memory, where we naturally parse the sentence into phrases like "design a joint," "to fasten together," and so on. These chunks are compared with long-term memory information to see if they make sense, and then most are allowed to fade. The goal of this first pass through the problem is to try and retain only the major functions of the needed device. Usually a problem will be read or sensed a number of times until the major function(s) is identified. Unfortunately there is no guarantee that, from the usually incomplete data that exist at the beginning of a design problem, the most important functions will be identified. In our example there is no ambiguity. The prime function is to transfer a load from one sheet of steel to another through a lapped joint.

What is important to realize is that a problem is "understood" by comparing the requirements on the desired function to information in the long-term memory. Thus, every designer's understanding of the problem is different, because each designer has different information stored in the long-term memory. (In Chap. 6 we develop a method to ensure that the problem is fully understood with minimal bias from the designer's own knowledge.)

3.3.2 Generating a Solution

We have seen that in trying to understand a design problem, we compare the problem to information from the long-term memory. In order to retrieve information from the long-term memory, we need a way to index the knowledge stored there. We can index that information in many ways (Fig. 3.4). As in the gearbox example at the beginning of this chapter, the most efficient indexing method is by function. What are recalled and downloaded to the short-term memory are specific (usually abstract) visual images from past experience. Thus, we search by function and recall form or graphical representations. This is not always true: we can also index our memory by shape, size, or some other form feature. However, in solving design problems, function is usually the primary index. For some problems the information recalled meets all the design requirements and the problem is solved.

If, in understanding a problem, we must recall images of previous designs, we have a predisposition to use these designs. Some designers get stuck on these initially recalled images and have difficulty evaluating them objectively and generating other, potentially better ideas. Many of the techniques discussed in Chaps. 7 and 10 are specifically designed to overcome this tendency.

On the other hand, what happens if the problem being solved is new and we find no solution to it in the long-term memory? We then use a three-step approach: decompose the problem into subproblems, try to find partial solutions to the subproblems, and finally recombine the subsolutions to fashion a total solution. The subproblems are generally functional decompositions of the total problem. The creative part of this activity is in knowing how to decompose and recombine cognitive chunks.

3.3.3 Evaluating the Solution

Often people generate ideas but have no ability to evaluate them. Evaluation requires two actions: comparison and decision making. Before we make a decision on the ability of a concept to solve a design problem, we must compare the performance of the concept with the laws of nature, the capability of technology, and the requirements of the design problem itself. Comparison, then, necessitates modeling the concept to see how it performs with respect to these measures. The ability to model is usually a function of knowledge in the domain. We will address evaluation techniques in Chaps. 8, 11, and 12.

3.3.4 Controlling the Design Process

To understand how designers progress through a design problem, subjects were videotaped as they worked. In the study of these videotapes, it became evident that the path from initial problem presentation to solution was not very straightforward. It seemed like an almost random process—efforts on a subproblem made the designer aware of another subproblem, and the designer then focused attention on this second problem without having solved the first. No model for the control of focus was found. However, it was clear that the process for some designers is so chaotic that they never find solutions to their problems, while other designers rapidly proceed through the design effort. The techniques discussed in this book are intended to give structure to the design process so that the path from problem statement to solution is as controlled and direct as possible.

3.3.5 Problem-Solving Behavior

Everybody has a unique manner of problem solving. A person's problem-solving behavior affects how problems are solved individually and has a significant effect on team effectiveness. The discussion below is centered around four personal problem-solving dimensions.[1] These four are useful for describing how an individual solves a design problem because they describe an individual's information management and decision-making preferences. Since all the team members bring their individual problem-solving processes to team activities, it is the interaction of all the individuals' solution processes that determines the team's health. For each of the four dimensions, suggestions for how to counteract extreme behavior are given. Some of these are useful to the individual working alone, and all are important in team situations.

The first dimension describes an individual's problem-solving style. If problem solving is done internally, i.e., the team member is reflective, is a good listener, thinks and then speaks, and enjoys having time alone to solve problems, then this person is an *introvert*. If the person's energy comes from outside through interactions with others, i.e., the member is sociable and tends to speak and then think, then this person

---

[1]These dimensions are based on the Myers-Briggs Type Indicator method. Three books listed in Section 3.7 are focused on this measure.

is an *extrovert.* About 75 percent of all Americans and 48 percent of engineering students and top executives are extroverts.

An individual may behave in some situations as an introvert and in others as an extrovert. In team settings both introverts and extroverts have characteristics that are essential to the team and may cause difficulty—the extroverts tend to overwhelm the introverts, who are reluctant to share their ideas. Here are some suggestions to keep the extroverts productive but not domineering:

- Extroverts need to allow others time to think. Point out to them that it is not necessary to fill in all the pauses with words.
- Extroverts need to practice listening to the ideas and suggestions of others and pausing before they react. Brainstorming or other creativity-support activity can help here (see Section 7.4.4).
- Encourage extroverts to recap what has been said to make sure they have heard the contributions of others.
- Extroverts need to realize that silence does not always mean consent. Sometimes an extrovert will overwhelm the introverts, who will become quiet rather than argue the point.

Here are some suggestions to assist introverts in getting their ideas out for consideration:

- Encourage introverts to share more than their final response. There is value in thinking out loud, as even the most trivial idea may be part of a good solution. The process will judge the value of the ideas.
- Try suggesting techniques that allow introverts to have an equal say in selecting ideas and plans, such as the techniques in Chaps. 5–13.
- Encourage introverts to develop some nonverbal, body-language signals that indicate assent or dissent. Make sure that these signals are understood by other team members.
- Encourage introverts to restate their ideas. This restating signifies to the introvert that his or her ideas count and forces the extroverts to listen.
- Get introverts to push extroverts for more clarity and meaning.

The second dimension reflects whether the individual prefers to work with *facts* or *possibilities.* People who deal with facts and details are literal, practical, and realistic and appreciate the here and now. Those who think in terms of possibilities like concepts and theories. They are looking for relationships between pieces of information and the meaning of the information. About 75 percent of Americans are fact-oriented, as are 66 percent of top executives; yet only 34 percent of all engineering students are fact-oriented. This is interesting in light of the heavy emphasis on math and science that is the focus of an engineering education.

This problem-solving dimension is the cause of most miscommunication, misunderstanding, and team problems. Design requires working with both facts and possibilities. Thus, both types of thinking are essential on a design team. However, individuals with a strong tendency toward either extreme may need help in the team setting. Some suggestions for fact-oriented team members are as follows:

- Encourage fact-oriented team members to fantasize, think wildly, and allow others to think wildly. Wild ideas can lead to good ideas. Brainstorming (Section 7.4.4) and thinking out loud (rambling) bring out such ideas.
- Encourage fact-oriented team members to allow the team to set goals rather than dive right into the problem.

Here are some suggestions for team members who think in terms of possibilities:

- Encourage possibility-oriented team members to deal with details. The best idea will never reach maturity if the details are not attended to. It is frustrating to them but possibly worthwhile to have them take on the responsibility of a detail task.
- Force possibility-oriented team members to be specific and avoid generalities. They should be encouraged to try to enumerate the exact items they want to address instead of making sweeping general statements.
- Remind possibility-oriented team members to stick to the issues. Other team members can control the flow of the problem solving by clearly stating the issues being addressed. Other issues that arise during discussion should be recorded and then shelved for later consideration.

The third dimension reflects the objectivity with which decisions are made. Some people take an *objective* approach, others a *subjective* approach. People who are logical, detached, and analytical take an objective approach to making decisions. Conversely, people who make decisions based on an interpersonal involvement, circumstances, and the "right thing to do" take a subjective approach to decision making. About 51 percent of Americans are objective decision makers, as are 68 percent of engineering students and 95 percent of top executives.

As it is important to have a variety of information-collection approaches on a design team, it is equally important to have a range of decision-making styles. Although engineers are trained to make decisions based on objective measures, the greatest number of decisions faced in every design problem have incomplete, inconsistent, qualitative information requiring subjective evaluation. For objective decision makers the following may help in working with the team:

- Encourage objective decision makers to pay attention to the feelings of others. Gut feelings are often right, and sometimes a lack of information forces one to rely on these feelings.
- Help objective decision makers understand that how the team functions is as important as what is accomplished. If there is acrimony, no decisions will be made.
- Remind objective decision makers that not everyone likes to discuss a topic merely for the sake of argument. Others may drop out from exhaustion and be taken to be conceding the point.

- Encourage objective decision makers to express how they feel about the outcome once in a while. Objective decision makers may have trouble expressing feelings.

Subjective decision makers are in a minority on most design teams. Thus, they must develop techniques to get their opinions heard and not get their sensitivities hurt. Here are some ideas:

- Help subjective decision makers to realize that it is all right to disagree and argue.
- Reassure subjective decision makers that while harmony is important, not every resolved issue will satisfy everyone even if consensus is reached.
- Reinforce to subjective decision makers that discussions about ideas are not personal attacks.

The fourth, and final, personality dimension relates to the need to make decisions. Some team members are *decisive,* and others are *flexible.* If a team member makes decisions with a minimum of stress and likes an environment that is ordered, scheduled, controlled, and deliberate, then he or she is decisive. If, on the other hand, the person goes with the flow, is flexible, adaptive, and spontaneous, and finds making and sticking with decisions difficult, then the team member is considered flexible. About half of all Americans are decisive, as are 64 percent of engineering students and 88 percent of top executives.

One characteristic of flexible decision makers is that they have a tendency to procrastinate because they want to remain adaptive. This makes working in a team with them difficult because decisions must be reached. The following are some suggestions for flexible decision makers on the team:

- Give flexible decision makers plans in advance so that they can think about them in their own time.
- Acknowledge the flexible decision maker's contribution as a step toward moving to closure. Remind them that problems are solved one step at a time.
- Set clear decision deadlines in advance.
- Encourage feedback from flexible decision makers so that they can think about the direction of their thoughts.
- Encourage flexible decision makers to settle on something and live with it a while before redesigning. Encourage them to take a clear position and stick to it. This may be difficult for them to do.

A contrary characteristic of decisive people is that they tend to jump to conclusions. This too can adversely affect teamwork, as many ideas may be generated and consensus may be needed to reach a decision. Here are some suggestions for slowing down decisive people:

- Ask decisive people questions about their decision process. Remind them that most problems need to be subdivided into smaller problems to be solved.

- Let decisive people organize the data collection and review process.
- Utilize techniques, such as brainstorming, that suppress judgment. Do not let them settle on the first good idea they hear.
- Remind decisive people that they are not always right.

3.4 CHARACTERISTICS OF A CREATIVE DESIGNER

Some people seem naturally more creative than others. In exploring this statement, we address two questions: What are the characteristics of a creative design engineer, and can creativity be enhanced?

First, let us clarify what we mean by "creative." A creative solution to a problem must meet two criteria: it must solve the problem in question, and it must be original. Solving a problem involves understanding it, generating solutions for it, evaluating the solutions, deciding on the best one, and determining what to do next. Thus, creativity is more than just coming up with good ideas. The second criterion, originality, depends on the knowledge of the designer and of society as a whole. What is new and original to one person may be old hat to another. If someone who has never experienced a wheel before designs one, then it is original for that person. But it is society that assesses "originality" and labels a solution or a person "creative."

As discussed earlier, all humans have the same cognitive, or problem-solving, structure. Why is it, then, that some engineers can generate ingenious ideas while others, who may be brilliant at complex analysis, cannot come up with new concepts, no matter how hard they try? There has been a lot of research on creativity, yet this trait is still not very well understood. The best way to understand the results of the research to date is in terms of creativity's relationship to other attributes.

Creativity and intelligence. There appears to be little correlation between creativity and intelligence.

Creativity and visualization ability. Creative engineers have good ability to visualize, to generate and manipulate visual images in their heads. We have seen before that people represent information in their minds in three ways: as semantic information (words), as graphical information (visual images), and as analytical information (equations or relationships). Words and equations convey serial information. They are generally understood on the basis of word order or the order of variables and constants. Pictures, or visual images, on the other hand, contain parallel information. One can decompose them in many different ways and focus on only the important features. For example, consider Fig. 3.5, a drawing of a small aircraft's fuselage and tail. Information on the structure (a truss), the configuration (cockpit in front, tail in the rear), the landing gear (only wheels and struts), and a specific joint are easily recovered by looking at the drawing. To describe all these features in text would require, at a minimum, many sentences. This is why designers often make sketches during problem-solving (using the sketches as an external extension of their short-term memory) and make textual notes less frequently.

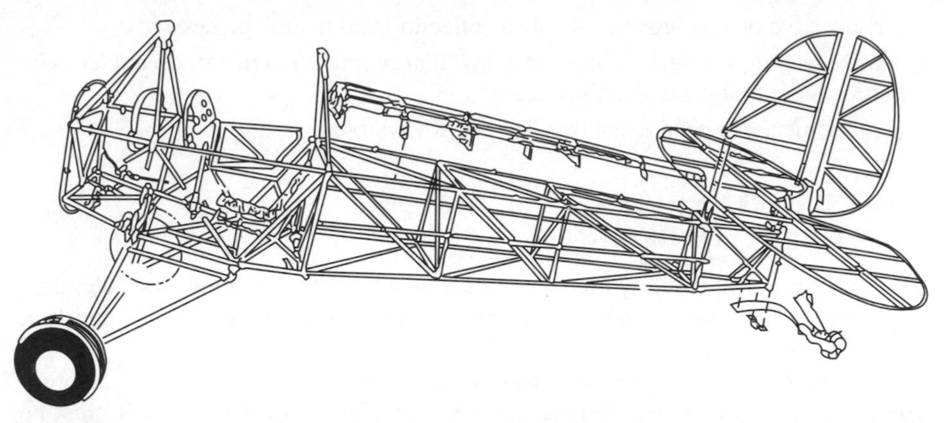

FIGURE 3.5
Airframe of a small aircraft.

There is little difference between individuals in the ability to visualize very simple images. However, it appears that the ability to manipulate complex images of mechanical devices can be improved with practice. This may be related to the formation of more information-rich chunks having functional information or to some other mechanism.

Creativity and knowledge. The model of the information-processing system implies that all designers start with what they know and modify this to meet the specific problem at hand. At every step of the way, the process involves small movements away from the known, and even these small movements are anchored in past experience. Since creative people form their new ideas out of bits of old designs, they must retain a storehouse of images of existing mechanical devices in their long-term memory. Thus, in order to be a creative mechanical designer, a person must have knowledge of existing mechanical products.

Additionally, part of being creative is being able to evaluate the viability of ideas. Without knowledge about the domain, the designer cannot evaluate the design. Knowledge about a domain is only gained through hard work in that domain. Thus, a firm foundation in engineering science is essential to being a creative designer of mechanical devices. During World War II many people sent ideas for weapons to the Department of War. Some were very far-fetched ideas for death rays or for building five-mile-high walls or domes over Europe to stop the bombers. These were very original but unworkable and were therefore not creative. The "inventors" had good intentions but lacked the knowledge to develop creative solutions to the war problems.

Creativity and partial solution manipulation. Since new ideas are born from the combination of parts of existing knowledge, the ability to decompose and manipulate this knowledge seems to be an important attribute of a creative designer. This attribute, more than any other so far discussed, appears to become stronger with exercise. Although there is no scientific evidence to support this contention, anecdotal evidence does support it.

Creativity and risk taking. Another attribute of creative engineers is the willingness to take an intellectual chance. Fear of making a mistake or of spending time on a design that in the end does not work is characteristic of a noncreative individual. Edison tried hundreds of different lightbulb designs before he found the carbon filament.

Creativity and conformity. Creative people also tend to be nonconformists. There are two types of nonconformists: constructive nonconformists and obstructive nonconformists. Constructive nonconformists take a stand because they think they are right. Obstructive nonconformists take a stand just to have an opposing view. The constructive nonconformist might generate a good idea; the obstructive nonconformist will only slow down the design progress. Creative engineers are constructive nonconformists who may be hard to manage since they want to do things their own way.

Creativity and technique. Creative designers have more than one approach to problem-solving. If the process they initially follow is not yielding solutions, they turn to alternative techniques. A number of books listed in Section 3.7 give methods to enhance creativity. Many of the techniques covered in these books are woven into the mechanical design techniques presented in the remainder of this book. This is especially true in the chapters on concept and product generation (Chaps. 7 and 10).

Creativity and environment. If the work environment allows risk taking and nonconformity and encourages new ideas, creativity will be higher. Further, if teammates and other colleagues are creative, the environment for creativity is greatly enhanced. In the discussion of teams in the next section, it is stated that, on a team, the sum is greater than the parts. This is especially true for creativity.

Creativity and practice. Creativity comes with practice. Most designers find that they have creative phases in their careers—periods when they have many good ideas. During these times the environment is supportive and one good idea builds on another. However, even with a supportive environment, practice enhances the number and quality of ideas.

To summarize, the creative designer is generally a person of average intelligence, a visualizer, a hard worker, and a constructive nonconformist with knowledge about the domain and the ability to dissect things in his or her head. However, even those designers who do not have a strong natural ability can develop creative methods by using good problem-solving techniques to help decompose the problem in ways that maximize the potential for understanding it, for generating good solutions, for evaluating the solutions, for deciding which solution is best, and for deciding what to do next.

One final comment: There are many design tasks that require talents very different from those used to describe a creative person. Design requires much attention to detail and convention and demands strong analytic skills. Therefore there are many good designers who are not particularly creative individuals; a design project requires people with a variety of skills and talents.

3.5 ENGINEERING DESIGN TEAMS

Most of the above material describes an individual designer. However, because of the complexity of most problems, design work is generally done by teams or groups. As shown in Fig. 3.6, the complexity of mechanical devices has grown rapidly over the last two hundred years. Gone are the days when a single individual could design an entire product. For example, the Boeing 747 aircraft, which has over five million components, required over 10 thousand person-years of design time. Thousands of designers worked over a three-year period on the project. Obviously, a single designer could not approach this effort.

Teams solve design problems in the same way an individual does—understanding, generating, evaluating, and decision making. However, there are some important differences. First, there are the social aspects of team work, including problems that arise in any social activity. Second, each team member has a different understanding of the problem, different alternatives for solving it, and different knowledge for evaluating it. This is both good and bad as it gives the potential for more solutions and for more confusion.

To address what is special about teams, in this chapter we first itemize team goals. Second, the unique roles people play in teams will be discussed. Finally, some guidelines for maintaining team performance will be presented. Additionally, many of the best practices given in future chapters can help support team activities.

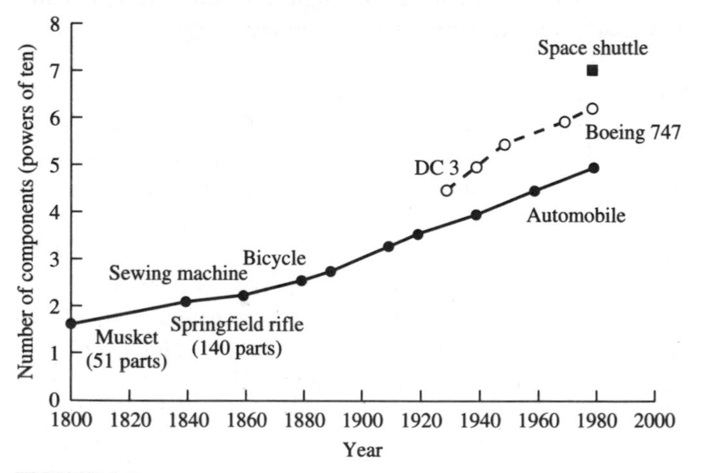

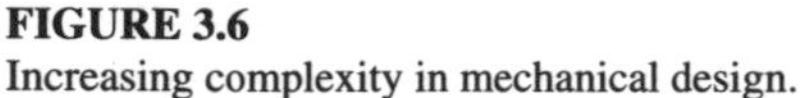
FIGURE 3.6
Increasing complexity in mechanical design.

3.5.1 Team Goals

Modern design problems require a design *team*—a small number of people with complementary skills who are committed to a common purpose, common performance goals, and a common approach for which they hold themselves mutually accountable. An effective team however, is more than the sum of it parts. Beyond the goal of solving a design problem, teams have another set of goals that need attention:

- Team members must learn how to *collaborate* with each other. Collaboration means more than just working together—it means getting the most out of other team members. The suggestions in Sec. 3.3.5 help develop a collaborative team.
- Teams are generally empowered to make decisions. Since these are team decisions, members must *compromise* to reach them. Empowering teams to make these decisions means that management takes a risk in giving up responsibility for them. Further, developing decisions by *consensus* rather than by authority leads to more robust decisions.
- Team members must establish *communication* to support real-time problem-solving. Further, members need to ensure that the others have the same understanding of design ideas and evaluations that they have. It is very difficult for people with different areas of expertise to develop a shared vision of the problem and its potential solutions. Developing this shared vision requires the development of a rich understanding of the problem.
- It is important that team members and management be *committed* to the good of the team. If they are not, it will be difficult reaching the other team goals.

3.5.2 Team Roles

The primary reason a person is made a member of a team is the person's technical expertise. However, a member also has a secondary role that is dependent on his or her problem-solving style and on that of the other members of the team.[2] An ideal team will have individuals filling all of the eight roles listed below. Depending on the makeup of the team, individuals will fill roles most closely associated with their problem-solving style. The eight secondary team roles are discussed in the following.

Coordinator. The coordinator is a team member who is mature, confident, and trusting. The coordinator is good at clarifying goals and promoting decision making but is not necessarily creative or clever. This person makes a good chairperson for the team.

Creator. This person is imaginative and can solve difficult problems. Such people are sometimes called the "plant" because they spread seeds of new ideas for others to develop. Creators often are impractical, have a disregard for protocol, and would rather work with possibilities than facts.

---

[2]This model is based on the work of R. M. Belbin. See Section 3.7.

Resource-investigator. This team member is good at exploring opportunities and developing contacts. Such people are extroverts and are very enthusiastic but may lose interest when the details are reached.

Shaper. This person is dynamic, outgoing, and assertive. This team member makes things happen by finding ways around obstacles, tends to be impatient with vagueness, and makes logical, objective decisions.

Monitor-evaluator. The monitor-evaluator is good at seeing the big picture and judging outcomes accurately. Such people are not inspirational leaders, but they are intelligent and shrewd.

Team worker. This person is a consensus builder who is concerned about avoiding friction on the team. Team workers are subjective decision makers and thus do not like disagreement on the team.

Implementer. This person turns ideas into practical action. Such people are disciplined, reliable, and efficient. They are often seen as decisive and resistant to change and are consider "company workers."

Completer-finisher. This team member is conscientious and detail-oriented and delivers results on time. Such people are reluctant to delegate authority and are inclined to worry about progress. They would rather work with facts than with ideas.

No person plays exactly the same role on each team. However, in all situations some of the roles are more consistent with each individual's problem-solving style and are therefore easier for him or her to fill.

3.5.3 Building Team Performance

It can be very exciting being part of a team that is productive and is making good use of all the members. Conversely, it can be very frustrating working on a team that is not functioning very well. Listed below are some guidelines for developing productive teams.

Guideline 1: Keep the team productive. For a team to be productive, there must be both urgency and direction. These can be established by establishing that

- All members understand the purpose.
- The members feel it is exciting.
- The goals are clear, simple, and measurable.
- The goals are realistic.
- The approach is clear.

Guideline 2: Select team members on the basis of skills in both primary and secondary roles. This means that the team members should have not only technical expertise but also a personality balance that allows all the secondary team roles to be easily filled.

Guideline 3: Establish clear rules of behavior. The behaviors of individuals on the team can be interpreted by understanding their problem-solving styles. Suggestions in Sec. 3.3.5 can help when behavior problems occur.

Guideline 4: Set and seize upon a few immediate performance-oriented goals and tasks. It is very difficult to get everyone working on the same problem if it has not been clearly articulated. Techniques in Chaps. 5 and 6 will help with this guideline.

Guideline 5: Spend time together. Infrequent team meetings will not lead to a successful outcome.

3.6 SUMMARY

* The human mind uses the long-term memory, the short-term memory, and a controller in the internal environment in problem solving.
* Knowledge can be considered composed of chunks of information that are general, domain-specific, or procedural in content.
* The short-term memory is a small (seven chunks, features, or parameters) and fast (0.1 second) processor. Its properties determine how we solve problems. We use the external environment to augment the size of the short-term memory.
* The long-term memory is the permanent storage facility in the brain. It is slow to remember, it is fast to recall (sometimes), and it never gets full.
* Creative designers are people of average intelligence; they are visualizers, hard workers, constructive nonconformists with knowledge about the problem domain. Creativity takes hard work and can be aided by a good environment, practice, and design procedures.
* Because of the size and complexity of most products, design work is usually accomplished by teams rather than by individuals.
* Working in teams requires attention to every team member's problem-solving style (including yours)—introverted or extroverted, fact- or possibility-oriented, objective or subjective, decisive or flexible.

3.7 SOURCES

Adams, J. L: *Conceptual Blockbusting,* Norton, New York, 1976. A basic book for general problem solving that develops the idea of blocks that interfere with problem solving and explains methods to overcome these blocks; methods given are similar to some of the techniques in this book.

Glegg, G. L.: *The Development of Design,* Cambridge University Press, New York, 1981. A very entertaining book of anecdotes about one designer's experiences and approach to design; insightful and educational.

———: *The Design of Design,* Cambridge University Press, New York, 1969. Similar to the preceding title.

Koberg, D., and J. Bagnall: *The Universal Traveler: A Systems Guide to Creativity, Problem Solving and the Process of Reaching Goals,* Kaufman, Los Altos, Calif., 1976. A general book on problem solving that is easy reading.

Mayer, R: *Thinking, Problem Solving and Cognition,* Freeman, San Francisco, 1983. An easy-to-read text on cognitive psychology.

Miller, G. A.: "The Magical Number Seven, Plus or Minus Two: Some Limits on Our Capacity for Processing Information," *Psychological Review,* vol. 63, 1956, pp. 81–97. The classic study of short-term memory size, and the paper with the best title ever.

Newell, A., and H. Simon: *Human Problem Solving,* Prentice Hall, Englewood Cliffs, N.J., 1972. This is the major reference on the information processing system. A classic psychology book.

Weisberg, R. W: *Creativity: Genius and Other Myths,* Freeman, San Francisco, 1986. Demystifies creativity; the view taken is similar to the one in this book.

The next five titles are all good books on developing and maintaining teams.

Belbin, R. M.: *Management Teams,* Heinemann, New York, 1981.

Cleland, D. I., and H. Kerzner: *Engineering Team Management,* Van Nostrand Reinhold, New York, 1986.

Johansen, R., et al.: *Leading Business Teams,* Addison Wesley, New York, 1991.

Katzenbach, J. R., and D. Smith: *The Wisdom of Teams,* Harvard Business School Press, 1993.

Scholtes, P. R.: *The Team Handbook,* Joiner, 1988.

The problem-solving dimensions in Section 3.3.5 are based on the Myers-Briggs Type Indicator. The following titles give more details on this method.

Keirsey, D., and M. Bates: *Please Understand Me,* 5th ed., Prometheus Nemesis, 1978.

Kroeger, O., and J. M. Thuesen: *Type Talk at Work,* Delta, 1992.

———: *Type Talk,* Delta, 1989.

3.8 EXERCISES

3.1. Develop a simple experiment to convince a colleague that the short-term memory has a capacity of about seven chunks.

3.2. Think of a simple object and write about it and sketch it in as many ways as possible. Refer to Table 2.2 and Fig. 3.4 to encourage a range of language and abstraction.

3.3. Describe a mechanical design problem to a colleague. Be sure to describe only its function. Have the colleague describe it back to you in different terms. Did your colleague understand the problem the same way as you? Was the response in terms of previous partial solutions?

3.4. During work on a team, identify the secondary roles each person is playing. Can you identify who fills each role?

3.5. For a new team begin with the following team-building activities.

(a) **Paired introductions.** Get to know each other by asking questions such as

What is you name?
What is your job (class)?
Where did you grow up (go to school)?
What do you like best about your job (school)?
What do you like least about your job (school)?
What are your hobbies?
What is your family like?

(b) **Third party introductions.** Have one member of the team tell another the information in (a). Then the second member introduces the first member to the rest of the team using all the information that he or she can remember. It makes no difference if the team heard the initial introduction.

(c) **Talk about first job.** Have each member of the team tell the others about his or her first job or other professional experience. Information such as the following can be included.

What did you do?
How effective was your manager?
What did you learn about the real world?

(d) **"What I want for myself out of this."** Have each member of the team tell the others for three to five minutes what his or her goals are for participation in the project. What do they want to learn or do, and why? Consider personal goals such as getting to know other people, feeling good about oneself, learning new skills, and other nontask goals.

(e) **Team name.** Have each person write down as many potential team names as possible (at least five). Discuss the names in the team, and choose one. Try to observe who plays which secondary role.

CHAPTER 4

THE DESIGN PROCESS

4.1 INTRODUCTION

We now set the stage for the rest of the book, where we will look at techniques, or best practices, that help achieve quality products. To understand how to make the best use of the techniques, it is important to look at them in the context of the overall design process. Thus, the progress of a product from need to production is explored by means of examples that demonstrate the flexibility of this process. The emphasis is on the early phases of the process, since they are the most critical to design success, and on the importance of design documentation. A design problem is introduced and is used as a case study throughout the remainder of the text.

4.2 OVERVIEW OF THE DESIGN PROCESS

Developing a manufacturable product from an initial need is not an easy job. Even though the process varies from product to product and industry to industry, we can construct a generic diagram of the activities that must be accomplished, as shown in Fig. 4.1. The first five phases in a product's life cycle (Fig 1.7), those of concern to the designer, are expanded in this figure.

Before the design of a product can begin, the need for that product must be established. As shown in Fig. 4.1, there are two sources for design projects: the market or the development of a new technology. About 80 percent of new product development is market-driven. Without a customer for the product, there is no way to recover the costs of design and manufacture. Thus, the most important part in understanding the design problem lies in assessing the market, i.e., establishing what the customer wants in the product. Even if market-driven, new products must contain the latest technology if they are to be perceived as being high quality (see Table 1.2 for what consumers mean by "high quality").

Often a company wants to develop a product utilizing a new technology. Developing new technologies usually requires an extensive amount of capital investment

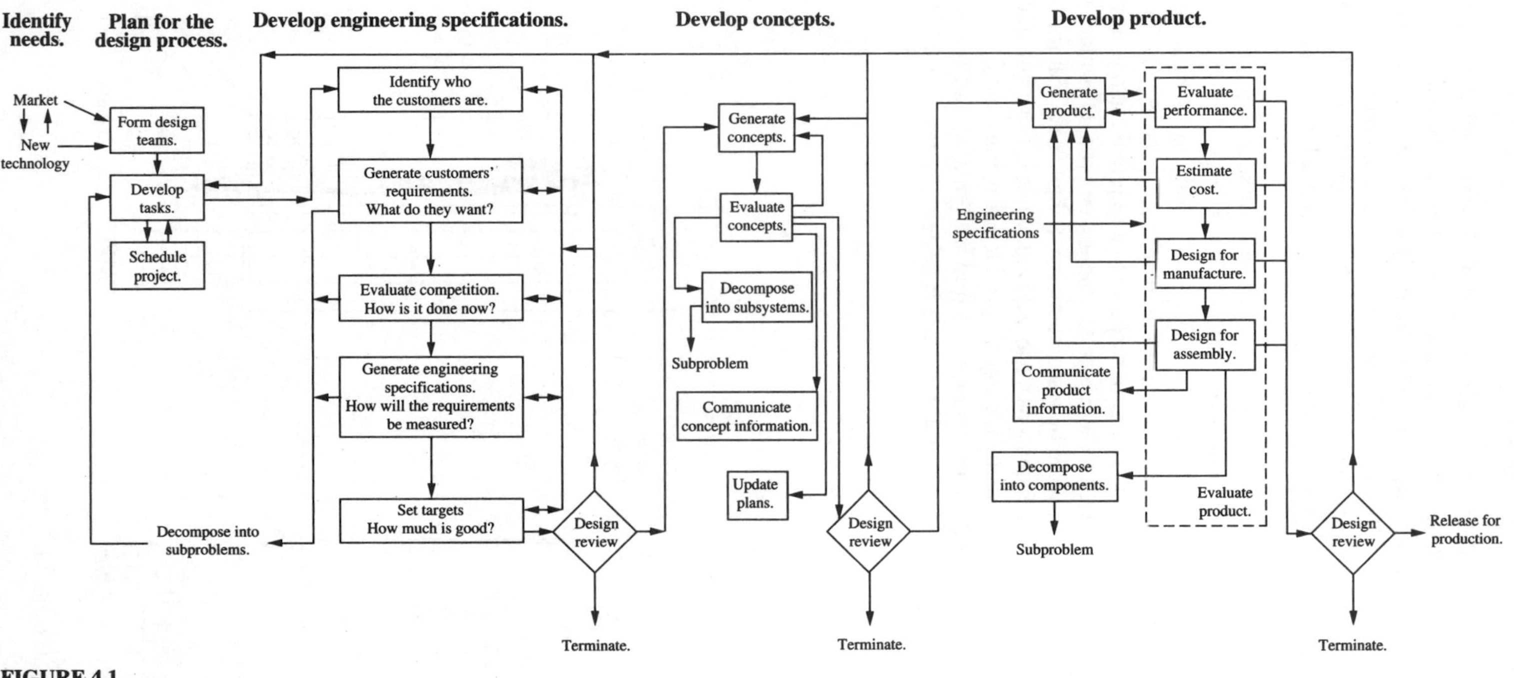

FIGURE 4.1
The mechanical design process.

and possibly years of scientific and engineering time. Even though the resulting ideas may be innovative and clever, they are useless unless they can be matched to a market need or a new market can be developed for them. Of course, electronic devices such as the electric carving knife and the Walkman serve as examples of products that have been successfully introduced without an obvious market need. While these types of products have high financial risk, they can reap a large profit because of their uniqueness.

Additionally, a design project for a new product or some feature of a product can be initiated by the desire to *redesign* it. Redesign is fostered by market demand for a new model or the desire to include a new technology in an existing product. Redesign can also be initiated to fix a problem with an existing product, reduce product cost, simplify manufacturing, or respond to a required change of materials, or for many other reasons. Often the desire to change the product is market-driven: the customers want the product to be less expensive, to have new features, or to last longer (see Table 1.2 for other reasons). Most design work is redesign, and thus the design process often begins with an existing product. It is important to realize that the process in Fig. 4.1 and the techniques given in the remainder of the book apply whether the project is for a new product or the redesign of even a simple feature of an existing product.

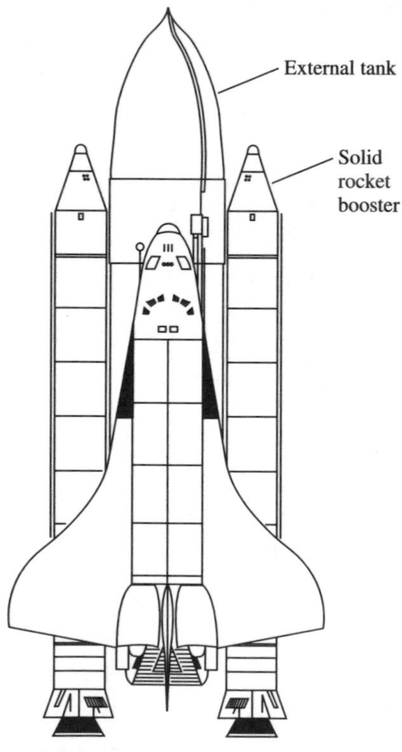

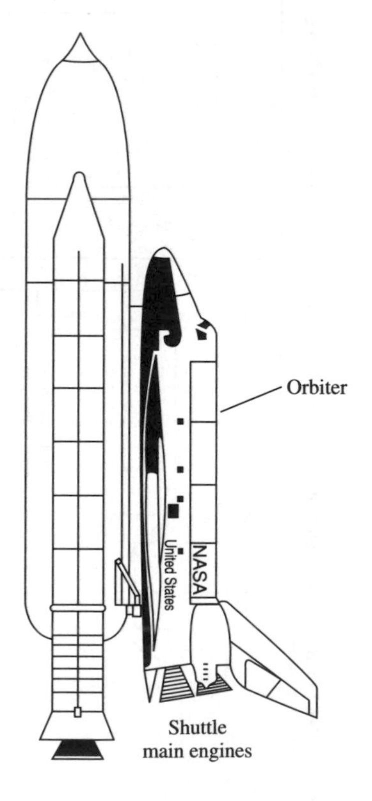

FIGURE 4.2
Space shuttle ready for launch.

The design process shown can also be applied to the *subproblems* that result from the decomposition of a higher-level system. Consider, for example, the design of a system such as the space shuttle (Fig. 4.2). The design of a system this large is a tremendous undertaking requiring thousands of design and manufacturing engineers, materials scientists, technicians, purchasing agents, drafters, and quality-control specialists, all working over many years. In a project such as this, the overall system is broken down into the design of many smaller subsystems.

One of the subsystems in the space shuttle project is the rocket booster, shown in Fig. 4.3, where some of the major subsystems of the booster itself are identified. The relationships of these systems, subsystems, and sub-subsystems are shown in the tree diagram of Fig. 4.4. In this figure the solid rocket booster is further decomposed to the level of the aft field joint used to attach two of the propellant segments together. The importance of this joint in the history of the U.S. space program will be seen later in this chapter. The design of the aft field joint, as was true with the design of every other component or assembly of the space shuttle, was a separate design problem that had to be solved during the product design process. This specific design problem arose during the conceptual design of the total space shuttle system, when the decision to utilize reusable solid boosters was made. The requirement of reusability meant that the booster had to be disassembled, repacked with fuel, shipped to the launch site in segments, and assembled in the field. Out of this decision came the design problem of developing field joints to fasten the segments of the booster together.

Further, as the booster subsystem was being solved, many individual components were identified. Each of these was developed through the flow of activities shown in Fig. 4.1.

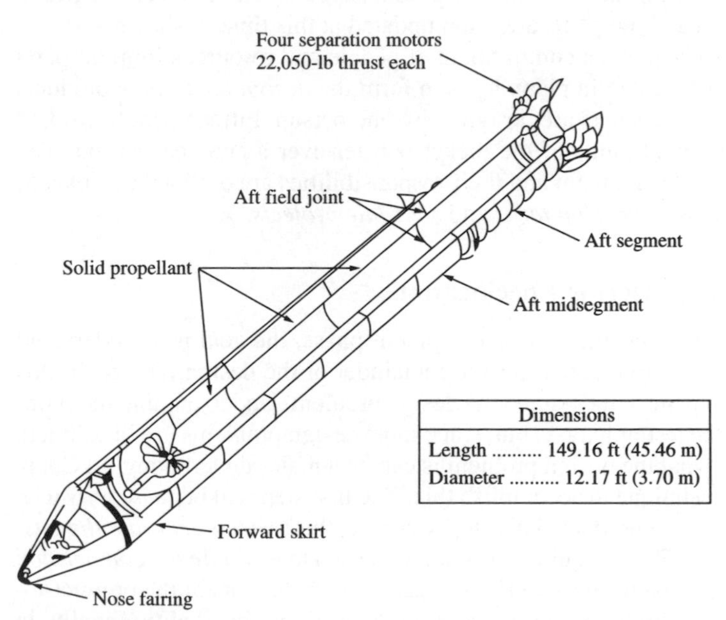

Dimensions
Length .... 149.16 ft (45.46 m)
Diameter .... 12.17 ft (3.70 m)

FIGURE 4.3
Major parts of the solid rocket booster.

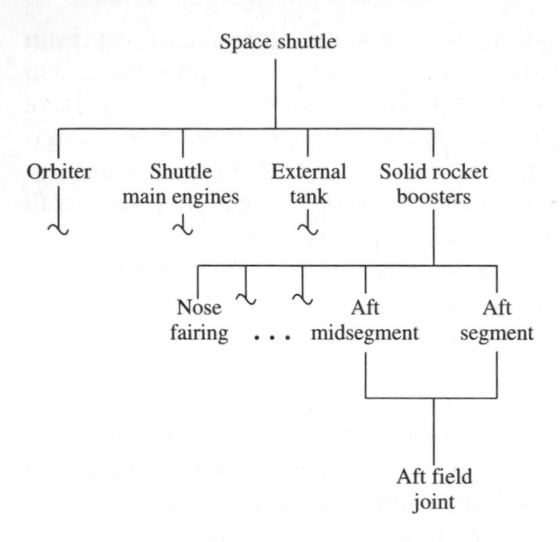

FIGURE 4.4
Space shuttle partial design decomposition.

4.2.1 Plan for the Design Process

The design process shown in Fig. 4.1 is an ideal flow chart of activities. The second phase of activities is to plan for the design process in order to allocate the company's resources of money, people, and equipment to accomplish the design activities. Planning needs to precede any commitment of resources; however, as with much design activity, this requires speculating about the unknown. This makes the planning for a project that is similar to an earlier project easier than that for a totally new one. The demands on resources, especially with new products, are much more certain after a concept is developed and the plans are often updated at this time as shown.

Since planning requires a commitment of people and resources from all parts of the company, the first step in planning is to form the *design team.* Few products or even subsystems of products are designed by one person. Fifteen people worked on the design of the field joints of the rocket booster over a one-year period. The members of a typical design team and their responsibilities are outlined in Chap. 5, which also covers how to *develop tasks* and *schedule projects.*

4.2.2 Develop Engineering Specifications

During the engineering specification development phase, the goal is to understand the problem and lay the foundation for the remainder of the design project. In this phase the design team must understand the design problem. Understanding the problem may appear to be a simple task, but, since most design problems are ill-defined, finding out exactly what the design problem is can be a major undertaking. In Chap. 6 we will look at a technique to accomplish this. The first step will be to *identify who the customers are* for the product. This step serves as the basis to *generate the customers' requirements.* These requirements are then used to *evaluate the competition* and to *generate engineering specifications,* measurable behaviors of the product-to-be that will help, later in the design process, in determining its quality. Finally, in order to measure the "quality" of the product, we *set targets for its performance.*

The engineering specification development phase, like all others, may be iterative with previous phases. As each phase is accomplished, new knowledge is gained that may refine an earlier decision. In Fig. 4.1 this is represented by an arrow pointing back to planning.

At the end of this phase of the design process, as in all the others, there is a *design review,* a formal meeting during which the members of the design team report their progress to management. Depending on the results of the design review, management then decides either to continue the development of the product or to *terminate* the project before any more resources are expended.

Often, the results of the activities in this phase determine how the design problem is decomposed into smaller, more manageable design subproblems (Fig. 4.4). Sometimes not enough information is known yet about the problem, and decomposition occurs later in the design process.

4.2.3 Develop Concepts

Designers use the results of the planning and specification definition phases to generate and evaluate concepts for the product. When we *generate concepts,* the customer's requirements serve as a basis for developing a functional model of the product. The understanding gained through this functional approach is essential for developing conceptual designs that lead to a quality product. Techniques for concept generation are given in Chap. 7.

When we *evaluate concepts,* the goals are to compare the concepts generated to the requirements developed during specification development and to select the best concepts for refinement into products. Techniques helpful in this evaluation are given in Chap. 8.

As shown in Fig. 4.1, techniques for generating and evaluating concepts are used iteratively. As design concepts are evaluated, more ideas are generated and need to be evaluated. Because iteration is less expensive during this phase than in the product design phase, it should be encouraged here, before the product is too well developed.

As a result of knowledge gained during the conceptual design phase, the problem is often *decomposed into subsystems* for individual design efforts. Thus, what began as a single design problem may now be many subsystems. The concepts generated for each subsystem can now be developed into a manufacturable product. It is very important that each new subsystem be considered from the beginning as a new project requiring planning and problem understanding.

Additionally, as shown in Fig. 4.1, one of the challenges the design team faces is to *communicate the concept information* to the right people at the right time. This is a key feature of concurrent engineering.

4.2.4 Develop Product

After concepts have been generated and evaluated, it is time to refine the best of them into actual products. (The product design phase is discussed in detail in Chaps. 9–13). Unfortunately, many design projects are begun here, without benefit of prior specification or concept development. This design approach often leads to poor-quality products and in many cases causes costly changes late in the design process. It cannot be overemphasized: *Starting a project with a single conceptual design in mind, without concern for the earlier phases, is poor design practice.*

The conceptual design phase leads quite naturally to *product generation.* Techniques for generating products, discussed in Chap. 10, emphasize the importance of the concurrent design of the product and the manufacturing process. As products are generated, they are evaluated for *manufacture, assembly, behavior,* and *cost.* As the product is increasingly refined, more evaluation methods become available. (In Chaps. 11 and 12 we look at techniques for product evaluation.) In Fig. 4.1 product generation and evaluation are shown as synergistic; they form an iterative loop. The evaluation of proposed components and assemblies leads naturally to their generation and improvement. As products are refined, new features are developed that did not exist in the original concept. These require careful attention ensuring that they are understood and that good concepts are developed to fill the needs. Product evaluation may require that the design process return to developing engineering requirements or new concepts especially as components are developed.

As with the earlier phases, there is critical need to communicate product information to others. Also, there is the potential for decomposition and the need for design reviews. Even during this phase of the design process, the design review can result in the termination of the project or return to an earlier phase.

4.3 THE DESIGN PROCESS: AN ORGANIZATION OF TECHNIQUES

Table 4.1 presents an itemization of techniques generally considered as best practice and discussed in this text. They appear in the order in which they are generally

TABLE 4.1
Best practices presented in this text

Plans for the design process (Chap. 5)	**Develop product**
Forming design teams	Generating the product (Chap. 10)
Generating a product development plan	Form generation from function
Develop engineering specification (Chap. 6)	Material and process selection
Understanding the design problem	Evaluating the product (Chaps. 11 and 12)
Developing customer requirements	Evaluating functional changes
Assessing the competition	Evaluating performance
Generating engineering requirements	Sensitivity analysis
Establishing engineering targets	Robust design
Develop concepts	Design of experiments
Generating concepts (Chap. 7)	Design for cost
Functional decomposition	Value engineering
Generating concepts from functions	Design for manufacture
Evaluating concepts (Chap. 8)	Design for assembly
Judging feasibility	Design for reliability
Assessing technology readiness	Design for the environment
Go/no-go screening	**Launching the product** (Chap. 13)
Using the decision matrix	Applying for a patent
	Working with vendors

applied to a typical design problem. However, each design problem is different, and some techniques may not be applicable to some problems. Additionally, even though the techniques are described in an order that reflects sequential and specific design phases, they are often used in different order and in different phases. Understanding the techniques and how they add to the design process aids in selecting the best technique for each situation.

The techniques described in this text compose a design strategy that will help in the development of a quality product that meets the needs of the customer. Though the techniques will consume time early in the design process, they may eliminate expensive changes later. The importance of this design strategy is clearly shown in Fig. 4.5, a reprint of Fig. 1.4. Here you can see that Company A structures its design process so that changes are made early, while Company B is still refining the product after it has been released to production. At this point changes are expensive, and early users are subjected to a low-quality product. The goal of the design process is not to eliminate changes but to manage the evolution of the design so that most changes come through iterations early in the process. The techniques listed in Table 4.1 also help in developing creative solutions to design problems. This may sound paradoxical, as lists imply rigidity and creativity implies freedom. However, creativity does not spring from randomness. Thomas Edison, certainly one of the most creative designers in history, expressed it well: "Genius," he said, "is 1 percent inspiration and 99 percent perspiration." The inspiration for creativity can only occur if the perspiration is properly directed and focused. The techniques presented here help the perspiration occur early in the design process so that the inspiration does not occur when it is too late to have any influence on the product. Inspiration is still vital to good design. The techniques that make up the design process are only an attempt to organize the perspiration.

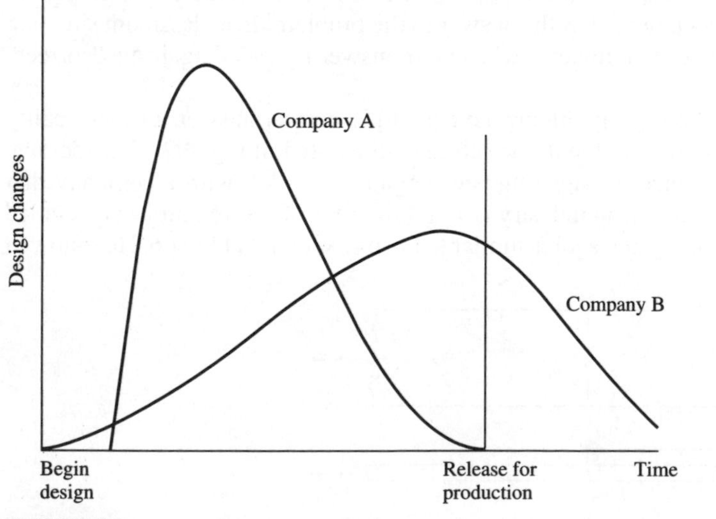

FIGURE 4.5
Engineering changes during automobile development.

The techniques also force documentation of the progress of the design, requiring the development of notes, sketches, informational tables and matrices, prototypes, and analysis—records of the design's evolution that will be useful later in the design process.

4.4 SIMPLE DESIGN PROCESS EXAMPLES

We will now look at two simple problems to see how different problems require different design processes. Recall the design problem statements from Chap. 1 (see Fig. 4.6):

> What size SAE grade 5 bolt should be used to fasten together two pieces of 1045 sheet steel, each 4 mm thick and 6 cm wide, which are lapped over each other and loaded with 100 N?

and

> Design a joint to fasten together two pieces of 1045 sheet steel, each 4 mm thick and 6 cm wide, which are lapped over each other and loaded with 100 N.

The solution of the first joint design problem is fairly straightforward (Fig. 4.7). It is fully defined, and understanding the problem is not hard. Since the problem statement actually defines the product, there is no need to generate and evaluate concepts or to generate a product design. The only real effort involved in this design problem is to evaluate the product. This is done using standard equations from a text on machine component design or using company or industrial standards. In a component-design text we find analysis methods for several different failure modes: the bolt can shear, the sheet steel can crush, and so on. After completing the analysis, you will make a decision as to which of the failure modes is most critical and then specify the smallest size of bolt that will not permit failure. This decision, part of the evaluation, is documented as the answer to the problem. In a classroom situation you will undergo a "design review" when your answer is graded against a "correct" answer.

Very few real design problems have a single correct answer. In fact, reality can cause quite a shift from the design process illustrated in Fig. 4.7. Consider one example: An experienced design engineer began a new job with a company that manufactured machines in an industry new to him. One of his first projects included the subproblem of designing a joint similar to the one shown in Fig. 4.6. He followed

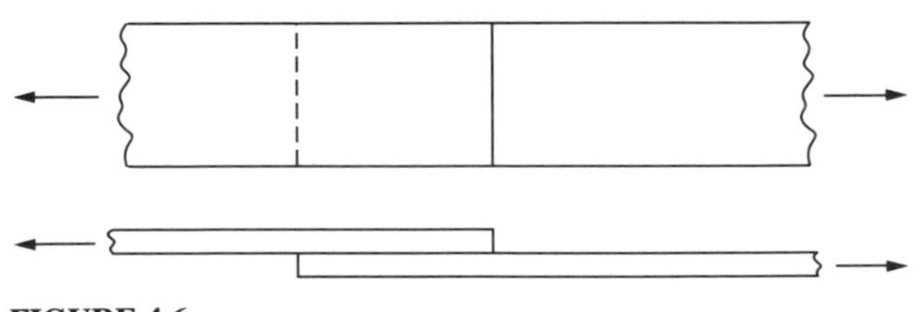

FIGURE 4.6
Design of a simple lap joint.

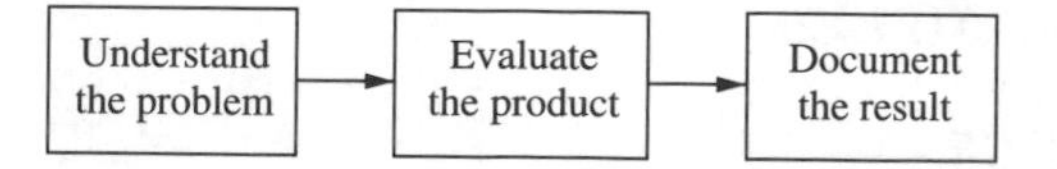

FIGURE 4.7
Design process for a simple lap joint.

the process in Fig. 4.7 and documented his results on an assembly drawing of the entire product. His analysis told him that a $\frac{1}{4}$-in.-diameter bolt would carry the load with a generous factor of safety. However, his manager, an experienced designer in the industry, on reviewing the drawing, crossed off the $\frac{1}{4}$-in. bolt and replaced it with a $\frac{1}{2}$-in. bolt, explaining to the new designer that it was an unwritten company standard never to use bolts of less than $\frac{1}{2}$-in. diameter. The standard was dictated by the fact that service personnel could not see anything smaller than that in the dirty environment in which the company's equipment operated. On all subsequent products the designer specified $\frac{1}{2}$-in. bolts without performing any analysis.

For the second joint design problem, the process is more complex. A reading of the problem statement can generate a number of concepts that might serve the fastening function. Typical options include using a bolt, welding the pieces together, using an adhesive, or folding the metal to make a seam. You might perform an analysis on each of these options, but that would be a waste of time because the results would still provide no clear way of knowing which joint design might be best. What is immediately evident is that the requirements on this joint are not well understood. In fact, if they were, perhaps none of the above concepts would be acceptable.

So the first step in solving this problem should be specification development for the joint. Various questions should be addressed: Does the joint need to be easily disassembled or leak-resistant? Does it need to be less than a certain thickness? Can it be heated? After all the specifications are understood, it will be possible to generate concepts (maybe ones previously thought of, maybe not), evaluate these concepts, and limit the potential designs for the joint to one or two concepts. Thus, before performing analysis on all the joint designs (evaluating the product), it may be possible to limit the number of potential concepts to one or two. With this logic, the design process would follow the flow of Fig. 4.8, a process similar to that in Fig. 4.1, except there is no need to generate product. The problem solved here is so mature that the concepts developed are fully embodied products. The concept, a "welded lap joint," is fairly refined. The only missing details are the materials, the weld depth, the length of the weld leg, and other details requiring expertise in welding design. However, if the requirements on the joint were out of the ordinary, then the concepts generated might be more abstract and have many possible product embodiments.

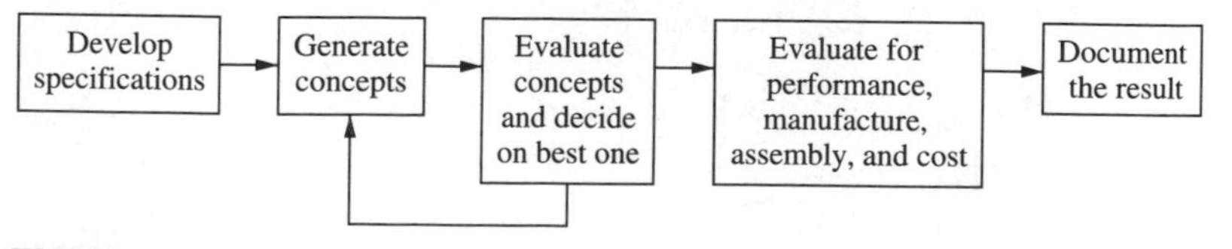

FIGURE 4.8
Design process for a more complex lap joint.

4.5 A MORE COMPLEX EXAMPLE: DESIGN FAILURE IN THE SPACE SHUTTLE *CHALLENGER*

On January 28, 1986, the space shuttle *Challenger* exploded during launch, killing the crew and virtually stopping U.S. exploration of space for two years. The presidential commission formed to study the disaster concluded in its report that the principal cause of the explosion was the failure of an O-ring seal in the aft field joint of the right-hand solid booster. In fact, the commission blamed the failure on the design of the joint:

> The Space Shuttle's Solid Rocket Booster problem *began with the faulty design* of its joint and increased as both NASA and contractor management first failed to recognize it as a problem, then failed to fix it and finally treated it as an acceptable flight risk. [Italics added.]

To demonstrate how a good design method may have averted the problem, the failure of the aft field joint and the original design process that developed it will be further explored.

The booster that failed is one of two solid-fuel boosters designed to help the shuttle reach the velocity needed for orbit. Their positions on the shuttle are shown in Fig. 4.2. The booster itself is shown in a cutaway view in Fig. 4.3. The booster is shipped to the launch site in sections and is assembled on site. The aft field joint is one of the joints made during the final field assembly. The solid propellant, cast inside the booster, burns from the center outward. The burning fuel expands, causing great pressure inside the booster and giving the booster its thrust. Essentially the booster is a thin-walled, large-diameter pressure vessel. As the fuel is burned, the temperatures of the outer wall and the joint also increase.

The solid rocket booster segments are joined by means of a tang-and-clevis arrangement, with two O-rings to seal the joint and 177 steel pins around the circumference to hold the joint together (Fig. 4.9). The *clevis* is the U-shaped section of the right segment into which the *tang* from the left segment fits. The zinc chromate putty acts as an insulation that under pressure, according to the intended design, would behave plastically and move toward the O-rings. This would pressurize the air in the gap between the two segments, which would in turn cause the first O-ring

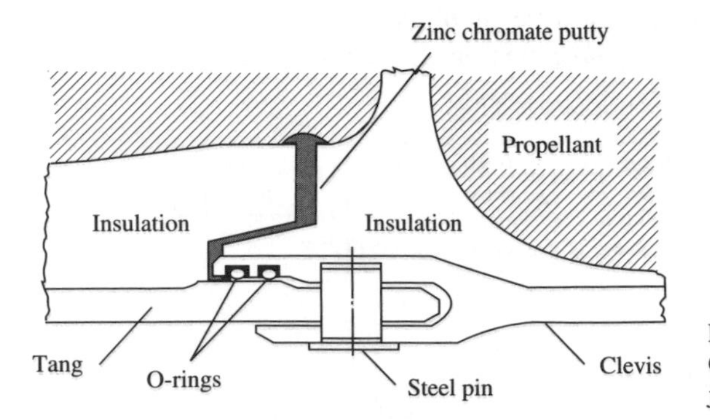

FIGURE 4.9
Cross section of the booster field joint.

to extrude into the gap between the clevis and the tang, sealing the joint. If the first O-ring failed, the second one could take the pressure.

This design was based on similar joints designed for the very reliable Titan III rocket. But there were some critical differences between the two rockets:

1. The *Challenger* booster was larger in diameter than the Titan.
2. The *Challenger* booster was reusable–the joint would be assembled and disassembled repeatedly–while the Titan was a single-use rocket.
3. The *Challenger* booster's O-rings took the pressure of combustion, whereas the single O-ring in the Titan did not. In the Titan the insulation was tight-fitting and the O-ring had only to take the pressure of any leakage through the insulation.
4. The tang on the *Challenger*'s booster joint was longer and flexed under pressure more than that on the Titan.
5. The O-ring on the *Challenger* was made from sections glued together, whereas that on the Titan was molded as one piece.

Early experimental tests of the shuttle's joint showed that the flexibility of the joint (difference 4 above) caused the tang to rotate in the clevis, opening the joint at the O-rings, as shown in Fig. 4.10. This condition was amplified when the O-rings were cold, as they lost resiliency and could not change from their compressed shape (Fig. 4.9) to the shape needed to fill the opening (Fig. 4.10).

With this background, let us review the design process that resulted in this joint. We can find the origin of the trouble in the engineering specification development phase, where the goal is to understand the customer's requirements and translate these requirements into engineering specifications. The presidential commission's report makes evident that this phase of the design process left much to be desired. In fact, the very first recommendation from the commission specified that "the joints should be fully understood, tested and verified."

In all fairness to the engineers on this project, we must make two points. First, as presented in Chap. 1, three factors affect the design process: cost, quality, and time. It appears from the commission's report that low cost won the original shuttle contract. (The proposal said that the designers need only modify an existing Titan-quality design.) Time pressures, as well as cost, kept the joint from being totally redesigned. Second, the engineers probably understood the operation of the Titan joint, but, as changes were made during the development of the shuttle joint, that level of understanding eroded.

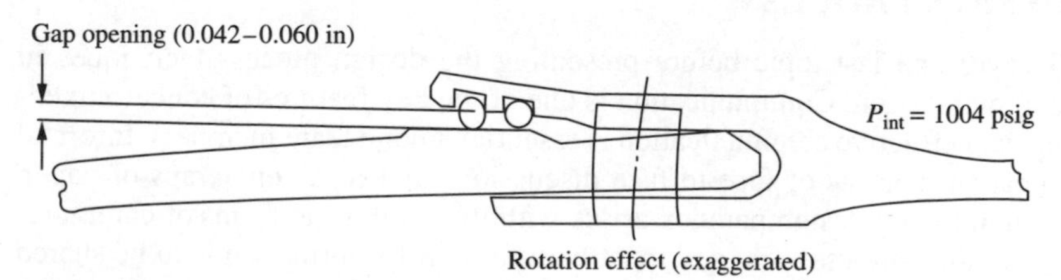

FIGURE 4.10
Pressurized field joint.

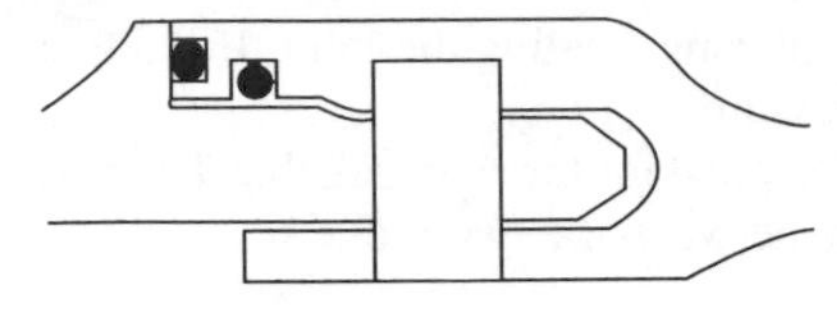

FIGURE 4.11
Alternative design for O-ring seals.

In the second phase of the design process, the conceptual development phase, the shuttle booster concept was taken from the Titan rocket with relatively little modification. Though it is estimated that a team of eight engineers and analysts spent four months developing the concept from the Titan design, which seems like a lot of effort, it is not clear that it was effort well spent. It is good design practice to utilize past concepts—as discussed in Sec. 3.3, all design ideas are pieces of prior experience. However, great care must be taken to ensure that the past concept is really applicable to the new problem. The differences between the shuttle and the Titan made use of the existing joint design questionable. In the shuttle design, the critical functions of load carrying and sealing were combined in the joint. Although having a feature perform more than one function is often good design, great care must be taken to realize and control the complexity created.

As the shuttle design was refined from the concept to an actual product, there was continued evidence of potential problems. First, the concept taken from the Titan design was used to generate two configurations for the joint. The configuration used is shown in Fig. 4.9; the second had one O-ring on the face of the clevis and another designed to compress axially, as shown in Fig. 4.11. In this design the axial seal still bore the major part of the combustion pressure, and it was found during evaluation that the tolerances required to make this seal could not be held.

To evaluate a product design, one must fully understand the stress and strain (deflection) in a load-carrying member. Analysis during design evaluation of the shuttle showed that there were high stress concentrations around the joint pins. Additionally, although calculations pointed to a satisfactory design, experiments showed that the joint rotated considerably when loaded, affecting the seal performance. High stress concentration and large deflection do not lead to quality design; these potential problems should have been corrected during the design process.

In conclusion, the design process being flawed in nearly all phases—planning, requirements development, conceptual design, and product design—resulted in a poor product and the loss of life, property, and prestige.

4.6 COMMUNICATION DURING THE DESIGN PROCESS

We will cover one last topic before presenting the design process techniques in detail: communication. Communication is one of the key features of concurrent engineering, and effective communication is essential among team members. Informal communication consists of face-to-face discussions and notes on scraps of paper. Another engineering design paradox arises with these informal forms of communication. First, they are essential and must be informal if information is to be shared and progress to be made. Second, for the most part the information is not in a form that is documented for future use. In other words, the information and arguments used to reach decisions are not recorded as part of any permanent design record and can be lost or easily misinterpreted.

Formal communication generally is in the form of *design records, communications to management,* and *communication of the final design to downstream phases*—production, use, and product retirement. Each of these three types is discussed in the sections below.

4.6.1 Design Records

Each technique discussed in this book produces documents that will become part of a design file for the product. The company keeps this file as a record of the product's development for future reference, perhaps to prove originality in case of patent application or to demonstrate professional design procedures in case of a lawsuit. However, a complete record of the design must go beyond these formal documents.

In solving any design problem, it is essential to keep track of the ideas developed and the decisions made in a *design notebook.* Some companies require these, with every entry signed and dated for legal purposes. In cases where a patent may be applied for or defended against infringement, it is necessary to have complete documentation of the birth and development of an idea. A design notebook with sequentially numbered, signed, and dated pages is considered good documentation; random bits of information scrawled on bits of paper are not. Additionally, a lawsuit against a designer or a company for injury caused by a product can be won or lost on the basis of records that show that state-of-the-art design practices were used in the development of the product. Design notebooks also serve as reference to the history of the designer's own work. Even in the case of a simple design, it is common for designers to be unable to recall later why they made a specific decision. Also, it is not uncommon for an engineer to come up with a great idea only to discover it in earlier notes.

The design notebook is a diary of the design. It does not have to be neat, but it should contain all sketches, notes, and calculations that concern the design. Before starting a design problem, be sure you have a bound notebook—one with lined paper on one page and graph paper on the other is preferable. The first entry in this notebook should be your name, the company's name, and the title of the problem. Follow this with the problem statement, as well as it is known. Number, date, and sign each page. If test records, computer readouts, and other information are too bulky to be cut and pasted into the design notebook, enter a note stating what the document is and where it is filed.

4.6.2 Documents Communicating with Management

During the design process periodic presentations to managers, customers, and other team members will be made. These presentations are usually called *design reviews* and are shown as decision points in Fig. 4.1. Although there is no set form for design reviews, they usually require both written and oral communication. Whatever the form, the following guidelines are useful in preparing material for a design review.

Make it understandable to the recipient. Clear communication is the responsibility of the sender of the information. It is essential in explaining a concept to others that you have a clear grasp of what they already know and do not know about the concept and the technologies being used.

Carefully consider the order of presentation. How should a bicycle be described to someone who has never seen one? Would you describe the wheels first, then the frame, the handle bars, the gears, and finally the whole assembly? Probably not, as the audience would understand very little about how all these bits fit together. A three-step approach is best: (1) Present the whole concept or assembly and explain its overall function. (2) Describe the major parts and how they relate to the whole and its function. (3) Tie the parts together into the whole. This same approach works in trying to describe the progress in a project: *Give the whole picture; detail the important tasks accomplished; then give the whole picture again.* There is a corollary to this guideline: *New ideas must be phased in gradually.* Always start with what the audience knows and work toward the unknown.

Be prepared with quality material. The best way to make a point, and to have any meeting end well, is to be prepared. This implies (1) having good visual aids and written documentation, (2) following an agenda, and (3) being ready for questions beyond the material presented.

Good visual aids include diagrams and sketches specifically prepared to communicate a well-defined point. In cases in which the audience in the design review is familiar with the design, mechanical drawings might do, but if the audience is composed of nonengineers who are unfamiliar with the product, such drawings communicate very little. It is always best to have a written agenda for a meeting. Without an agenda, a meeting tends to lose focus. If there are specific points to be made or questions to be answered, an agenda ensures that these items are addressed.

4.6.3 Documents Communicating the Final Design

The most obvious form of documentation to result from a design effort is the material that describes the final design. Such materials include drawings (or computer data files) of individual components (detail drawings) and of assemblies to convey the product to manufacturing. They also include written documentation to guide manufacture, assembly, inspection, installation, maintenance, retirement, and quality control. These topics will be covered in Chaps. 9 and 13.

4.7 INTRODUCTION OF A SAMPLE DESIGN PROBLEM

We now introduce a design problem that will be used as example throughout the remainder of the book. A major bicycle manufacturer makes mountain bikes, as shown in Fig. 4.12. These heavy-duty bicycles are designed to be ridden on rough, often muddy trails. They have no fenders, since mud and debris would be easily trapped

FIGURE 4.12
Mountain bicycle. (Courtesy Schwinn Bicycle Company.)

between tire and fender. When riding on trails, cyclists generally do not care if they get muddy. However, if the mountain bike is also to be used for street transportation, for instance, in commuting to work or school, there will be a problem on a rainy day. On the street the rider would naturally prefer to stay dry, but because there are no fenders, both rider and baggage will be splattered with water and mud.

The bicycle manufacturer has found from market surveys that there is need for an easily removable device to protect bike rider and baggage from road water and has initiated a project that will concentrate on controlling the spray generated by the rear wheel. They have named the project the Splashgard. A future project will address the front wheel.

4.8 SUMMARY

* There are specific design process techniques to support the planning, specification development, conceptual design, and product design phases of the design cycle. They are listed in Table 4.1 and will be developed throughout the rest of this book.
* The techniques help the design effort in its earliest stages, when the major decisions are made. Additionally, the techniques encourage communication, force documentation, and encourage data gathering to support creativity.
* Communication is an integral part of the design process . The techniques described in this book produce design records; less formal work on the product needs to be kept in a design notebook.
* The design of a simple product that will keep water off the rider of a mountain bike will be used as an example in the following chapters.

4.9 SOURCES

Andreasen, M. M.: *Integrated Product Development,* Springer-Verlag, New York. A text on the theory of design; some of the techniques presented here are taken from this book.

Presidential Commission on the Space Shuttle *Challenger* Accident: *Report of the Presidential Commission on the Space Shuttle* Challenger *Accident,* GPO, Washington D.C., 1986.

4.10 EXERCISES

4.1. Select an original design problem to use throughout the remainder of the book as an exercise. This problem should meet the following criteria to be solvable in 8–12 weeks:

(i) The problem concerns an area in which some of the design team members have some knowledge or training.

(ii) The final product will have fewer than three original parts per team member if hardware is to be developed or six original parts per team member if only drawings are to be produced. This estimate can be based on examining existing products that provide similar function.

(iii) If hardware is to be built, either team members or support staff have expertise to make components.

4.2. Select an existing product to use for redesign throughout the remainder of the text. Items such as mechanical toys, home appliances, automotive subsystems, or other devices with three to six components per team member are of suitable size. An example of the product is essential. Parts lists, mechanical drawings, user manuals, assembly drawings, etc., are all useful.

CHAPTER 5

PLANNING FOR THE DESIGN PROCESS

5.1 INTRODUCTION

Every company has a design process; for many it is unstated, part of the way things are done, part of the company's culture. In fact, companies that have not analyzed themselves may not know exactly how new products are developed within their own walls. Some company executives see engineering activity as described in Fig. 1.5—requirements are thrown over-the-wall to the engineering designers, who then throw the product specifications over-the-wall to manufacturing. In this view the design process is reduced to a simple linear series of activities. All of the engineering activities occur without any knowledge of other people and organizations outside the engineering function that may have a stake in the product.

Traditionally, *planning is the process used to develop a scheme for scheduling and committing the resources of time, money, and people.* Planning is one of the many processes that are important in concurrent engineering. Also, *planning results in a map showing how product design process activities are scheduled.* The activities shown in Fig. 4.1—*develop engineering specifications, develop concept,* and *develop product*—must be scheduled and have resources committed to them. The flow shown in the figure is only intellectual; it is not sufficient for allocating resources or for developing a schedule.

Finally, *planning generates a procedure for developing needed information and distributing it to the correct people at the correct time.* Important information includes product requirements, concept sketches, system functional diagrams, component drawings, assembly drawings, material selections, and any other representation of decisions made during the development of the product. This is a concurrent engineering point of view on planning.

The activity of planning results in a blueprint for a process. The terms *plan* and *process* are often used interchangeably in industry. Most companies have a generic process (i.e., a master plan) that they customize for specific products. This master plan is called the *product development process, product delivery process, new product development plan,* or *product realization plan.* In this book we will refer to this generic process as the *product development process* and use the acronym PDP.

Changing the design process in a company requires breaking down the way things have always been done before. Although it can be quite difficult, many companies have done so over the last decade. Generally, companies that have enjoyed good markets for their products and have begun to see these markets erode begin to look at their product development process as part of their effort to reengineer themselves to meet the competition. A key feature of concurrent engineering is the continuous improvement of both the product and the process for developing the product. The material in this chapter not only supports planning for a single project but is the basis for structuring and improving the PDP. It is divided into three major topics: background material, the steps taken in planning a design project, and examples of project plans.

5.2 BACKGROUND FOR DEVELOPING A DESIGN PROJECT PLAN

A plan shows how a project will be initiated, organized, coordinated, and monitored. These activities are traditionally associated with management. However, in concurrent engineering, all team members are a part of these activities. How a plan accounts for these management activities is developed in this and the following section. Here we look at how design projects are initiated, organized, and coordinated. In the following section we develop how drawings, prototypes, and other models are both design tools and planned deliverables that are used to monitor progress.

5.2.1 Types of Design Projects

A design project and thus the need to plan for design activity is initiated by the desire for a new or improved product. Plans range from simple, one-page descriptions for simple product changes to hundreds of pages of detailed requirements for the development of new products. To understand what initiates a design project and to better organize the types of plans, consider the four different types of design projects:

Variation of an existing product. This could be a change in a single or a few parameters such as the length of a cylinder or the power of a motor. This type of project takes little engineering. In fact, some products are designed with parameters that can be easily changed by the salesperson to meet the customer's needs.

Improvement of existing product. The need to redesign some of the features of an existing product comes from the following sources.

- The customers want a new feature or improved performance from existing features.
- A vendor can no longer supply materials or components used in the product or has recommended improved ones.
- Manufacturing, assembly, or another downstream phase in the product's life cycle has identified a quality, time, or cost improvement.
- New technology or new understanding of an existing technology allows for improvement in performance or reduction in manufacturing cost. This type of project may take only minor redesign of some feature or may require developing a virtually new product. This type of design project is hard to plan for.

Development of a new product for a single or small run. Many products are only made once or at most a few times. Planning for manufacture and assembly is different for these products than for those that are mass produced. Specifically, for a small run there is less latitude for choosing manufacturing methods. Methods such as forging metal, injection-molding plastics, and manufacturing custom control circuits require mass production to amortize the tooling costs required. This restricts the types of components that can be designed. Often the first item built is both a prototype and the final product delivered to the customer. There is more tendency to buy off-the-shelf components for short-run products. There is also less concern for assembly time than for mass-produced products.

Development of a new product for mass production. Designing products for mass production requires careful planning for manufacture and assembly. These projects give the design engineer more flexibility in selecting materials and manufacturing processes and increase the project's dependence on manufacturing engineers.

5.2.2 Members of Design Teams

A plan must organize the people and resources needed to fulfill the goal of the project. Below we provide a list of individuals who might fill a role on a product design team. Their role on the design team will vary with product development phase and from product to product, and their titles will vary from company to company. Each position on the team is described as if filled by one person. In a large design project there may be many persons filling that role, whereas in a small project one individual may fill many roles.

PRODUCT DESIGN ENGINEER. The major design responsibility is carried by the product design engineer (hereafter referred to as the *design engineer*). This individual must be sure that the needs for the product are clearly understood and that engineering requirements are developed and met by the product. This usually requires both creative and analytical skills. The design engineer must bring knowledge about the design process and knowledge about specific technologies to the project. The person who fills this position usually has a four-year engineering degree. In smaller companies he or she may be a nondegreed designer who has extensive experience in the product area. For most product design projects more than one design engineer will be involved.

PRODUCT MANAGER. In many companies this individual has the ultimate responsibility for the development of the product and represents the major link between the product and the customer. Because the product manager is accountable for the success of the product in the marketplace, he or she is also often referred to as the *marketing manager* or the *product marketing manager.* The product manager also represents the interests of sales and service.

In order to initiate a design project, management must appoint the nucleus of a design team—at a minimum, a design engineer and a product manager.

MANUFACTURING ENGINEER. Design engineers generally do not have the necessary breadth or depth of knowledge about various manufacturing processes to fully support the design of most products. This knowledge is provided by the manufacturing or industrial engineer, who must have a grasp not only of in-house manufacturing capabilities but also of what the industry as a whole has to offer.

DETAILER. In many companies the design engineer is responsible for specification development, planning, conceptual design, and the early stages of product design. The project is then turned over to detailers (often called *designers*), who finish detailing the product and developing the manufacturing and assembly documentation. Detailers and drafters (see below) usually have two-year technology degrees.

DRAFTER. A drafter aids the design engineer and detailer by making drawings of the product. In many companies the detailer and the drafter are the same individual.

TECHNICIAN. The technician aids the design engineer in developing test apparatus, performing experiments, and reducing data in the development of the product. The insights gained from the technician's hands-on experience are usually invaluable.

MATERIALS SPECIALIST. In some products the choice of materials is forced by availability. In others, materials may be designed to fit the needs of the product. The more a product moves away from the use of known, available materials, the more a materials specialist is needed as a member of the design team. This individual is usually a degreed materials engineer or a materials scientist. Often the materials specialist will be a vendor's representative who has extensive knowledge about the design potential and limitations of the vendor's materials. Many vendors actually provide design assistance as part of their service.

QUALITY CONTROL/QUALITY ASSURANCE SPECIALIST. A quality control (QC) specialist has training in techniques for measuring a statistically significant sample to determine how well it meets specifications. This inspection is done on incoming raw materials, incoming products from vendors, and products produced in-house.

A quality assurance (QA) specialist makes sure that the product meets any pertinent codes or standards. For example, for medical products, there are many FDA (Food and Drug Administration) regulations that must be met. Often QC and QA are covered by one person.

INDUSTRIAL DESIGNER. Industrial designers are responsible for how a product looks and how well it interacts with consumers; they are the stylists who have a background in fine arts and in human factors analysis. They often design the envelope within which the engineer has to work.

ASSEMBLY MANAGER. Where the manufacturing engineer is concerned with making the components from raw materials, the assembly manager is responsible for putting the product together. As you will see in Chap. 12, concern for the assembly process is an important aspect of product design.

VENDOR'S OR SUPPLIER'S REPRESENTATIVE. Very few products are made entirely in one factory. In fact, many manufacturers outsource (i.e., have suppliers provide) 70 percent or more of their product. Usually there will be many suppliers of both raw and finished goods. There are three types of relationships with suppliers: (1) partnership—the supplier takes part beginning with requirements and concept development; (2) mature—the supplier relies on the parent company's requirements and concepts to develop needed items; and (3) parental—the supplier only builds what the parent company specifies. Often it is important to have critical suppliers on the design team, as the success of the product may be highly dependent on them.

As Fig. 5.1 illustrates, having a design team made up of people with varying views may create difficulties, but teams are essential to the success of a product. It is the breadth of these views that helps in developing a quality design.

5.2.3 Structure of Design Teams

Since projects require team members with different domains of expertise, it is valuable to look at the different structures of teams in an organization. This is important because concurrent engineering requires coordination across the functions of the product and across the phases in the product's development process. Listed below are the five types of project structures. The number in parentheses is the percentage of development projects that use that type. These results are from a study of 540 projects in a wide variety of industries (see the Larson and Gobeli article in Section 5.9).

Functional organization (13 percent). Each project is assigned to a relevant functional area or group within a functional area. A functional area focuses on a single discipline. For Boeing and other aircraft manufacturers, for example, the main

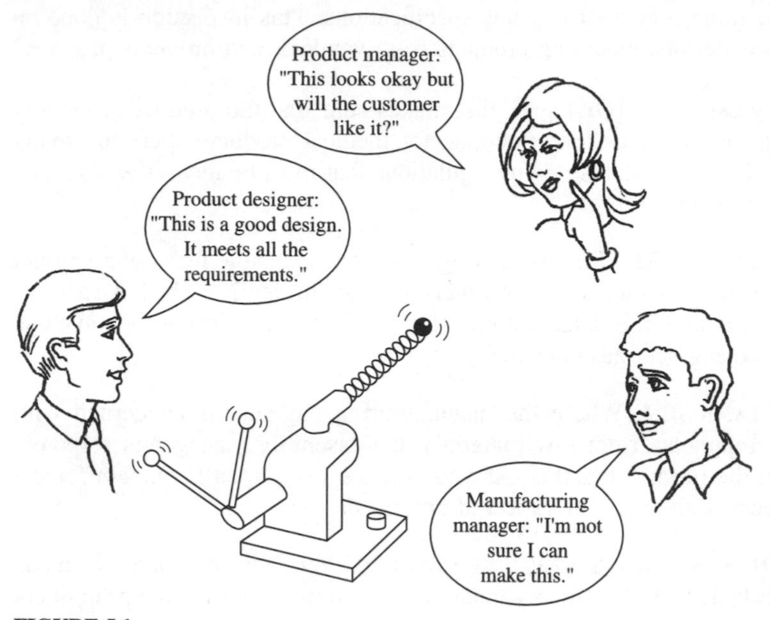

FIGURE 5.1
The design team at work.

functions are aerodynamics, structures, payload, propulsion, etc. The project is coordinated by functional and upper levels of management.

Functional matrix (26 percent). A project manager with limited authority is designated to coordinate the project across different functional areas or groups. The functional managers retain responsibility and authority for their specific segments of the project.

Balanced matrix (16 percent). A project manager is assigned to oversee the project and shares with the functional managers the responsibility and authority for completing the project. Project and functional managers jointly direct many workflow segments and jointly approve many decisions.

Project matrix (28 percent). A project manager is assigned to oversee the project and has primary responsibility and authority for completing the project. Functional managers assign personnel as needed and provide technical expertise.

Project team (16 percent). A project manager is put in charge of a project team composed of a core group of personnel from several functional areas or groups, assigned on a full-time basis. The functional managers have no formal involvement. Project teams are sometimes called "Tiger teams," "SWAT teams," or some other aggressive name, because this structure is high-energy and is disbanded after the project is completed.

What is important about these structures is that some of them are more successful than others. Structures with mixed functional areas are more successful than those built around the functional areas in the company (Fig. 5.2). Thus, when planning for a design project, organize the talent around the project whenever possible.

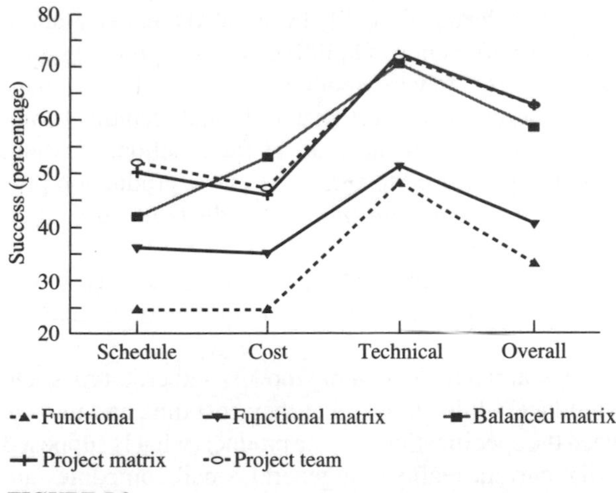

FIGURE 5.2
Project success versus team structure.

5.3 PLANNING FOR DELIVERABLES

Progress in a design project is measured by deliverables such as drawings, prototypes, bills of materials (e.g., parts lists), results of analysis, test results, and other representations of the information generated in the project. This information not only serves as part of the engineering effort but is also part of coordinating and monitoring the project's progress. Information included in drawings, physical prototypes, and analytical models is especially critical to project success and is a major factor in determining project cost.

During product development many models (i.e., design information representations) will be made of the evolving product. Some of these models will be analytical—quick calculations on a bit of paper or complex computer simulations; some will be graphical representations—simple sketches or orthographic mechanical drawings; and some will be physical models. Physical models of products are often called *prototypes.* Characteristics of prototypes that must be taken into account when planning when to use them and what types to use are their *purpose,* the *phase* in the design process when they are used, and the *media* used to build them.

There are four *purposes* for prototypes: proof-of-concept, proof-of-product, proof-of-process, and proof-of-production. These terms are traditionally only applied to physical models; however, solid models in CAD systems can often replace these prototypes with less cost and time. A proof-of-concept prototype focuses on developing the function of the product for comparison with the customers' requirements or engineering specifications. This kind of prototype is intended as a learning tool, and exact geometry, materials, and manufacturing process are usually not important. Thus, it can be built of paper, wood, parts from children's toys, parts from a junkyard, or whatever is handy. A proof-of-product prototype is developed to help refine the components and assemblies. Function is less important than geometry, materials, and manufacturing process. The recent development of *rapid prototyping,* using stereolithography or other methods to form a part rapidly from a CAD representation, has greatly improved the time and cost efficiency of building proof-of-product prototypes. A proof-of-process prototype is used to verify both the geometry and the manufacturing process. For these prototypes the exact materials and manufacturing processes are used to manufacture samples of the product for functional testing. Finally, a proof-of-production prototype is used to verify the entire production process. This prototype is the result of *preproduction run,* the products manufactured just prior to production for sale.

Table 5.1 lists many types of models and prototypes that can be used in developing a product. These are listed by the medium used to build the model and the phase in the design process.

There are trade-offs to be considered in developing models and prototypes: On one hand they help verify the product, while on the other they cost time and money. Further, there is a tension between the specifications for the product (what is supposed to happen) and the prototype (the current reality). In general, small companies are prototype-driven; they develop many prototypes and work from one to the next, refining the product. Large companies, ones that coordinate large volumes of information, tend to try to meet the specification through analytical modeling, building only a few prototypes.

An important decision made during planning is how many models and prototypes to schedule in the design process. There is currently a strong move toward replacing prototypes with computer models because simulation is cheaper and faster. This move will become stronger as virtual reality and rapid prototyping are further developed. Toyota has resisted these technologies in favor of developing physical prototypes, especially in the design of components that are primarily visual (e.g., car bodies). In fact, Toyota claims that through the use of many simple prototypes, it can develop cars with fewer people and less time than companies that rely heavily on computers. The number of prototypes to schedule is dependent on the company culture and the ability to produce usable prototypes rapidly.

Finally, in planning for prototypes be sure to set realistic goals for the time required and the information learned. One company had a series of four prototypes

TABLE 5.1
Types of models

	Medium		
Phase	**Physical (form and function)**	**Analytical (mainly function)**	**Graphical (mainly form)**
Concept	Proof-of-concept prototype	Back-of-the-envelope analysis	Sketches
↓	Proof-of-product prototype	Engineering science analysis	Layout drawings
Final product	Proof-of-process and proof-of-production prototypes	Finite element analysis; detailed simultation	Detail and assembly drawings; solid models

in its product development plan. But it turned out that the engineers were designing the second prototype (P2) while P1 was still being tested. Further, they developed P3 while P2 was being tested, and they developed P4 while P3 was being tested. Thus, what was learned from P1 influenced P3 and not P2, and what was learned from P2 influenced only P4. This waste of time and money was caused by a tight time schedule developed in the planning stage. The engineers were developing the prototypes on schedule, but since the tasks were not planned around the information to be developed, they were not learning from them as much as they should have. They were meeting the schedule for deliverable prototypes, not for the information that should have been gained.

5.4 THE FIVE STEPS IN PLANNING

A project plan is a document that defines the tasks necessary to be completed during the design process. For each task, the plan states the objectives; the personnel requirements; the time requirements; the schedule relative to other tasks, projects, and programs; and, sometimes, cost estimates. In essence, a project plan is a document used to keep that project under control. It helps the design team and management to know how the project is actually progressing relative to the progress anticipated when the plan was first established or last updated. There are five steps to establishing a plan. They are discussed below and further refined in the rest of the chapter.

STEP 1: IDENTIFY THE TASKS. As the design team gains an understanding of the design problem, the tasks needed to bring the problem from its current state to a final product become more clear. Tasks are often initially thought of in terms of the activities that need to be performed (e.g., "develop concepts" or other terms used in Fig. 4.1). The tasks should be made as specific as possible. In some industries the exact tasks to be accomplished are clearly known from the beginning of the project. For example, the tasks needed to design a new car are similar to those that were required to design the last model; the auto industry has the advantage of beginning with a clear picture of the tasks needed to complete a new design.

STEP 2: STATE THE OBJECTIVE FOR EACH TASK. Each task must be characterized by a clearly stated objective. This objective is to take some existing information about the product—the input—and, through some activity, refine it for output to other tasks. Even though tasks are often initially conceived as activities to be performed, they need to be refined so that the results of the activities are the stated objectives. Although the output information can only be as detailed and refined as the present understanding of the design problem, each task objective must be:

- Defined as information to be refined or developed and communicated to others, not as activities to be performed. This information is contained in deliverables, such as completed drawings, results of calculations, information gathered, or tests performed.
- Easily understood by all on the design team.

- Specific in terms of exactly what information is to be developed. If concepts are needed, then tell how many are sufficient.
- Feasible, given the personnel, equipment, and time available. See the next step.

STEP 3: ESTIMATE THE PERSONNEL, TIME, AND OTHER RESOURCES NEEDED TO MEET THE OBJECTIVES. For each task it is necessary to identify who on the design team will be responsible for meeting the objectives, what percentage of their time will be required, and over what period of time they will be needed. In large companies it may only be necessary to specify the job title of the workers on a project, as there will be a pool of workers, any of whom could perform the given task. In smaller companies or groups within companies, specific individuals might be identified.

Many of the tasks require virtually a full-time commitment; others require only a few hours per week over an extended period of time. For each person on each task, it is necessary to estimate not only the total time requirement but the distribution of this time. Finally, the total time to complete the task must be estimated. Some guidance on how much effort and how long a design task might take is given in Table 5.2. (The values given are only for guidance and can vary greatly.)

Similar comments apply to other resources needed to complete the task, especially those used for simulation, testing, and prototype manufacture. These resources and personnel are the means to complete the task.

Notice in Table 5.2 that no entry estimates the required time to be less than a week. *Design takes time.* Usually it takes twice as long as the original estimate, especially if the design project is not routine or new technologies are used. Some pessimists claim that after making the best estimate of time required, the number should be doubled and the units increased one step. For example, an estimate of one

TABLE 5.2
The time it takes to design

Task	Personnel/time
Design of elemental components and assemblies. All design work is routine or requires only simple modifications of an existing product.	One designer for one week
Design of elemental devices such as mechanical toys, locks, and scales, or complex single components. Most design work is routine or calls for limited original design.	One designer for one month
Design of complete machines and machine tools. Work involved is mainly routine, with some original design.	Two designers for four months
Design of high-performance products that may utilize new (proven) technologies. Work involves some original design and may require extensive analysis and testing.	Five designers for eight months

day should really be two weeks. Confidence in the time estimate is further addressed in the next step.

STEP 4: DEVELOP A SEQUENCE FOR THE TASKS. The next step in working out the plan is to develop a task sequence. Scheduling tasks can be complex. The goal is to have each task accomplished before its result is needed and, at the same time, to make use of all of the personnel, all of the time. Additionally, it is necessary to schedule design reviews. As shown in Fig. 4.1, design reviews need to come at the end of all the phases and may be scheduled more frequently if necessary.

For each task it is essential to identify its *precessors,* which are the tasks that must be done before it, and the *successors,* the tasks that can only be done after it. By clearly identifying this information, the sequence of the tasks can be determined. A method called the *CPM (critical path method)* helps determine the most efficient sequence of tasks. The CPM is not covered in this book. See the source list in Section 5.9.

Often tasks are interdependent—two tasks need decisions from each other in order to be completed. Thus it is important to explore how tasks can be started with incomplete information from precessors and how they can supply incomplete information to successors.

The best way to develop a schedule for a fairly simple project is to use a bar chart, shown in Fig. 5.3 for the Splashgard. (This type of chart is often called a *milestone* or *Gantt chart.*) On the bar chart, (1) each task is plotted against a time scale (time units are usually weeks, months, or quarters of a year); (2) the total personnel requirement for each time unit is plotted; and (3) the schedule of design reviews is shown.

PERT (program evaluation and review technique) aids in finding the best estimate of the total time the project will take. This method is based on three estimates: an optimistic estimate o, a most likely estimate m, and a pessimistic estimate p. From these three the best estimate of the task time is

$$\text{time estimate} = \frac{o + 4m + p}{6}$$

See the sources in Section 5.9 for more details on PERT.

STEP 5: ESTIMATE THE PRODUCT DEVELOPMENT COSTS. The planning document generated here can also serve as a basis for estimating the cost of designing the new product. Even though design costs are only about 5 percent of the manufacturing costs of the product (Fig. 1.1), they are not trivial.

One way to help organize these steps is through a modeling technique generated during the 1980s called IDEF.[1] As shown in Fig. 5.4, each task is represented by a box. The task activity is written in the box; the information needed by the activity comes in from the left; and the information output, the objective of the task, comes out the right side. The means of accomplishing the task, the needed people and other resources, are represented by arrows coming into the bottom of the box. The tasks

[1]This name has no important meaning. It is a military acronym.

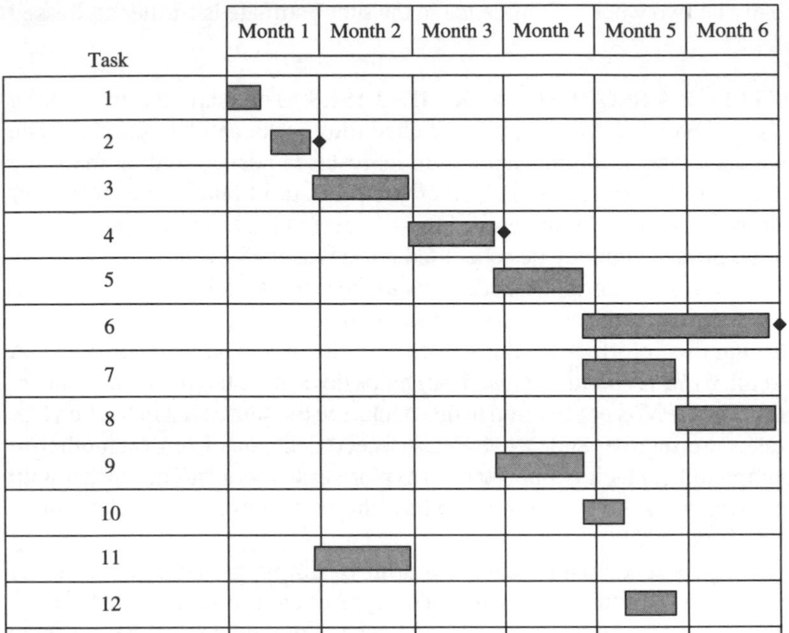

Task	Month 1	Month 2	Month 3	Month 4	Month 5	Month 6
1						
2						
3						
4						
5						
6						
7						
8						
9						
10						
11						
12						

	Month 1	Month 2	Month 3	Month 4	Month 5	Month 6	Person-months
Design engineer	100	110	50	70	70	20	4.20
Product manager	25	10	10		20	70	1.35
Technician	50	50	50	50	10	10	2.20
Manufacturing manager	5			50			0.55
Drafter		30			100		1.30
Machinist			100	200			3.00
Manufacturing				50			0.50
Materials specialist		20		20			0.40
Industrial designer		50			50		1.00
Total person months							14.50

Note: Time estimates are percentage of full time.

◆Design review

FIGURE 5.3
The Splashgard design schedule.

can be written on separate pieces of self-stick note paper and easily rearranged to find the best order.

Although the planning activities just described occur early in the design process and at the system level, the plan developed may be refined as the project progresses. This may be necessary because design is learning, and the further the progress into the project, the clearer the picture of what lies ahead. Additionally, the project may be decomposed after more is learned. This may require a new plan for each subsystem.

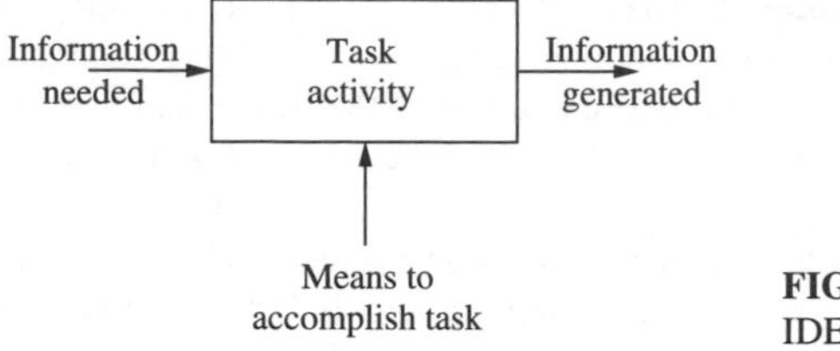

FIGURE 5.4
IDEF task representation.

There are a number of computer software packages available to support the steps outlined here. One of these is given in Section 5.9.

The remainder of this chapter contains examples of plans for different types of projects.

5.5 A DESIGN PLAN FOR LOW-VOLUME PRODUCTS

MechTool, a company that makes custom machine tools, is used as an example of planning for low-volume products. This company has 65 employees, half of whom are in manufacturing. The product development process is typical of that used by small companies.

All new projects within this company are based on what the company calls the "New Product Development Process" (NPDP). This process is contained in a seven-page document (shortened here), developed and managed by the Steering Team. This guidance team represents the enterprise and has eight members representing engineering, manufacturing, marketing, and service. It oversees teams assigned to specific projects. This company uses a project matrix type of structure for the project teams.

The tasks in the company's NPDP are itemized below along with the objectives and personnel on the project team. Since each project requires different elapsed time and schedule depending on other work in progress, these are project-specific. When a new project is begun, exact times are added, a schedule is developed, and costs are determined.

Task 1. Evaluate possible new products. Select a specific new product from the list of possible new products. Define preliminary product parameters, including expected price range, on the basis of available information. Using available data, determine if the new product has a reasonable chance of meeting customer's expectations. A quick, preliminary market research study may be needed, in some cases, to make this determination. Reject any new product proposals that cannot be supported by preliminary marketing data. Establish Product Development Team (PDT) and schedule for project.

Task 2. Develop conceptual design proposal. PDT will determine the design feasibility of the proposed PNP. Engineers will develop conceptual designs proposals for the new product. Manufacturing will evaluate concepts for manufacturability within the company's capability. They will present three concepts to Steering Team with recommended first choice.

Product design review with Steering Team approval or rejection of concept.

Task 3. Produce a layout design of concept. The Product Development Team will prepare standard overall layout drawings, a cost estimate range, and a delivery schedule for the prototype(s). These proposals will include an outline of the theoretical specifications and performance parameters. Manufacturability will be assessed in detail.

Task 4. Evaluate layout for new product design. The layout will be evaluated by the Steering Team for how well it fulfills the customer's expectations and the corporation's objectives. If needed, further, more specific marketing research will be done in order to predict market acceptance and to quantify projected sales volume. This is effectively a product design review with Steering Team approval or rejection of the layout for the product.

Task 5. Develop prototype. Full detail and assembly drawings will be developed by the PDT. A working prototype will be developed for testing.

Task 6. Test prototype. The Product Development Team *will* test the prototype machine in-house to determine how well it meets the customer's expectations, marketing's expectations (including costs), and manufacturability considerations. If deemed necessary, the prototype will be tested by outside evaluators to gather unbiased data on performance and market acceptance.

Task 7. Develop market launch plan. A market launch plan will be prepared by the Steering Team. It will include types of media and schedule for promotion, pricing, collateral production schedule, and expected market launch date. Using the prototype machine, preliminary collateral materials and news releases will be prepared according to the requirements of the market launch plan. In-depth training in setup and operation of the new product will be done to ensure that the sales engineers have a good grasp of the major features and benefits of the machine.

Product design review with Steering Team approval or rejection of new product. If accepted, the new product is released for manufacture.

This plan is fairly linear. It is sufficient for a small company where communication involves only a few people who all work in the same building. In a large company the product development plan is much more complex, as detailed in the following section.

5.6 A DESIGN PLAN FOR MASS-PRODUCED PRODUCTS: XEROX CORPORATION

In this and the next section, the design process for mass-produced products is examined. Two examples are given, the process used by Xerox and a plan for the Splashgard example. The Xerox example is of their product development plan, the overall company process.

The Xerox Corporation has been one of the leaders in refining its product delivery process. As pointed out in Chapter 1 (see Section 1.2 and Fig. 1.3), this leadership was born of necessity. In response to the problems in the early 1980s, Xerox instituted its PDP, Fig. 5.5. The process is broken into seven phases. Design reviews or

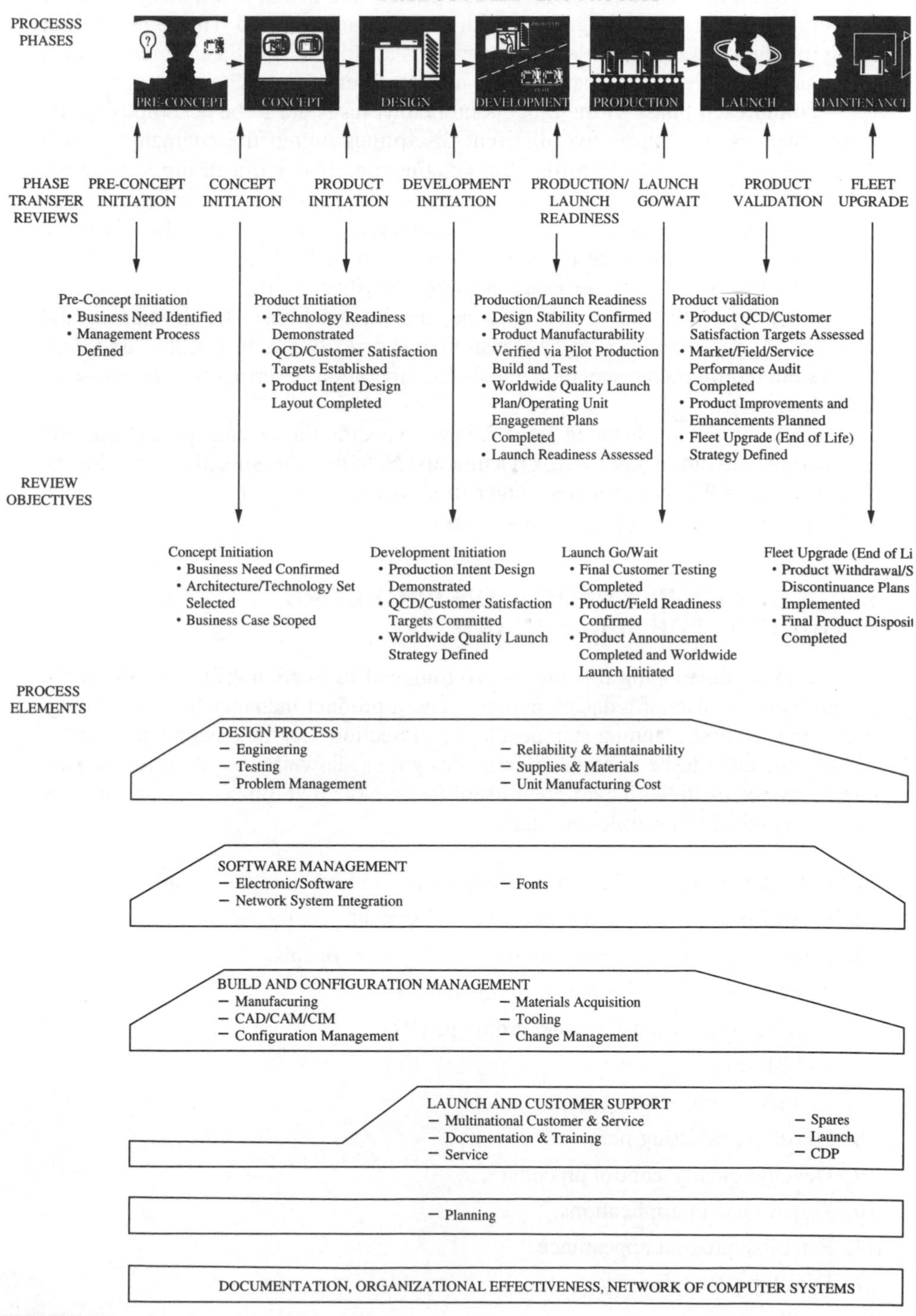

FIGURE 5.5
The Xerox product delivery process.

"Phase Transfers," the transitions between phases, are listed, and the objectives or deliverables approved for progress to the next phase are itemized. The bottom half of the figure shows the process elements—projects at work during the different phases. Note that the height of each bar signifies amount of effort in each phase.

Within each phase of the process numerous tasks are to be accomplished by representatives of each of five different disciplines within the company: launch (those concerned with the transition from engineering to manufacturing, marketing, sales, and support), marketing, planning, engineering, and manufacturing.

In Table 5.3 the Xerox phases are compared with those used in this book. The number of tasks in each phase is shown in brackets. There are a total of 231 different tasks. Each of these tasks is fully described by the information that is needed for input, the activity to be done, the information to be developed, and to whom the information is to be transmitted. For example, task C25, "Issue initial system requirements specification/system operation description," is detailed in Table 5.4.

Each of the items listed in Table 5.4 has a specific format and contains specific information laid out in other Xerox documents. Note that this specification is for the program that will generate a new copier (the system). Each subsystem is a project, and within each project there are many tasks.

5.7 A DESIGN PLAN FOR MASS-PRODUCED PRODUCTS: THE SPLASHGARD

The company developing this product (introduced in Section 4.7) is small, so the design team consists of a design engineer and a product manager from marketing. Following the first planning step developed in Section 5.4, the design team drafted a list of the tasks to be undertaken in its design, as shown below. A different team might have come up with a different list of tasks. Note that this was the preliminary list; it was refined in subsequent steps.

1. Collect and evaluate customer requirements and competition samples.
2. Establish two concepts for product development.
3. Develop first prototype (P1) of the Splashgard concepts.
4. Test P1 and select one design for finalization.
5. Redesign and produce second prototype (P2).
6. Field-test P2.
7. Complete production documentation.
8. Develop marketing plan.
9. Develop quality-control procedure.
10. Prepare patent applications.
11. Establish product appearance.
12. Develop packaging.

TABLE 5.3
Details of Xerox's plan

Xerox phase	Phase in this text (Fig. 4.1)
Pre-concept (33)	Planning for the design process Engineering requirements development
Concept (43)	Conceptual design
Design (37) Development (61) Production planning (23) Launch (24)	Product design
Maintenance (10)	Manufacturing and beyond

TABLE 5.4
Xerox sample C25: issue initial system requirements specification/system operation description

Description	The system requirements specification functionally describes the system and subsystem operation from a user's perspective, as well as describing the electrical, hardware, and software that underlies the functionality.
Purpose	Purpose is to ensure that the product requirements are understood and internalized by the design teams to enable the development of the initial prototype (P1).
Supplier	Engineering
Customer	Program Management, Marketing, Production Planner, Customer and Service Education, Support Groups, Manufacturing
Output	System requirement specification 1. System functional description a. System overview b. Features and functions c. System operation d. System faults and diagnosis 2. Subsystem function a. Input module description b. Copy handling module description c. Xerographic description d. Optics module description e. Fuser module description f. User interface description 3. Terminology and index

For each of these tasks the design team developed objectives (step 2) and estimated the personnel, time, and other resources needed to meet the objectives (step 3). The results of this activity for the first four tasks are given below. The remaining tasks were similarly developed.

Task 1: Complete specification development. (Note new title.)

Objective: Demonstrate problem understanding through generation of the following documentation:

Customer requirements:
Interview at least 20 potential customers in bike shops and at mountain-bike club meetings and rides.

Competition benchmarks:
Evaluate all the current types of fender fastenings relative to customer requirements.

Engineering requirements and targets:
Develop a set of measurable engineering targets for the product.

Personnel requirements:
Design engineer: 100 percent time for half-month.
Product marketing manager: 50 percent time for half-month.
Manufacturing manager: 10 percent time for half-month.

Task 2: Establish two concepts for product development.

Objective: Based on a clear understanding of the functions required, generate sketches of at least seven potential concepts. Evaluate each concept relative to customer requirements. Choose the best two. Document the selection with the generation of decision matrices.

Personnel requirements:
Design engineer: 100 percent time for half-month.
Technician: 100 percent time for half-month.

Note: The technician was to aid the engineer in evaluating concepts.

Task 3: Develop first prototype (P1) of the Splashgard concepts.

Objective: Utilizing the concepts developed in task 2, refine the designs to the point that the first prototype of both Splashgards can be built. This would require the development of the following documents:

Assembly drawings.
Detail drawings of all components.

A parts list, or bill of materials (BOM).
Rough draft of manufacturing and assembly procedure.
Ninety-percent cost estimate (estimate with an accuracy of ± 10 percent).

Personnel requirements:
Design engineer: 100 percent time for one month.
Technician: 50 percent time for one month.
Materials specialist: 20 percent time for one month.
Product marketing manager: 10 percent time for one month.
Drafter: 60 percent time for half-month.

Task 4: Test P1 and select one design for finalization.

Objective: Manufacture four samples of each design developed in task 3. These must be produced using the final materials but not necessarily the final production methods. Test three samples of each design in the laboratory to evaluate their ability to meet the engineering targets. These three samples will be used for testing to give a statistical basis for all results and backups in case of failure. The fourth sample will be tested by a panel of potential customers, including at least five mountain-bike riders who are neither members of the design team nor are company employees. They will be asked to score each design relative to the customer requirements. Any new requirements that arise during this testing will be added to the original list. The results of the testing will be used to select one design for final development.

Personnel requirements:
Design engineer: 50 percent for one month.
Technician: 50 percent for one month.
Product marketing manager: 10 percent for one month.
Machinist: 100 percent for one month.

Developing a sequence for the tasks (step 4) in this project is not difficult because of its simplicity. This step was partially accomplished in doing step 1, in which many of the tasks were listed sequentially as they were identified. The relation between tasks was further refined in writing the objectives (step 2).

The bar chart developed for the Splashgard (Fig. 5.3) was the result of several iterations; yet even after these iterations, two problems were still apparent. In the second month the design engineer was scheduled for more than 100 percent of his or her time. Either more help had to be assigned during these two months or the schedule had to be redrawn to make the total time commitment work out. Also, the schedule was that of a six-month project, while the original goals were for a three-month design effort and production in twelve months. It became apparent after plan development that this was not a very realistic goal. Although the design effort was now projected to take six months, production was still scheduled to fall within the initial goal.

The chart also shows that the designer and the technician were scheduled over a continuous period of time. The other team members were needed only sporadically, which was acceptable if they were working on other projects whose schedules could be interwoven with that of the Splashgard project.

The product development costs of the Splashgard can be rapidly estimated. Say that the average salary for each member of the design team was $30,000 a year. Figure 5.3 shows a total of 14.50 months of labor (1.21 years of effort) in direct salaries, an expenditure of $36,375. With overhead and fringe benefits typically at 75 percent of this figure, the total cost to the company for the design team was $63,656. (Additional costs for facilities, equipment, and supplies are not included in this estimate.)

5.8 SUMMARY

* Planning is an important engineering activity.
* Progressive companies have a generic product development process that serves as a basis for planning each product development activity.
* Design projects commonly fall into one of four types: variation of an existing product, improvement of an existing product, development of a new product for low-volume production, and development of a new product for high-volume production.
* Design teams may have representatives from many different disciplines, and they may be organized in one of five different structures.
* The use of prototypes and models is important to consider during planning.
* There are five planning steps: identify the tasks, state their objectives, estimate the resources needed, develop a sequence, and estimate the cost.

5.9 SOURCES

Larson, E., and D. Gobeli: "Organizing for Product Development Projects," *Journal of Product Innovation Management,* 1988, no. 5, pp. 180–190. The study in Section 5.2.3 on design team structure is from this paper.

Meredith, D. D., K. W. Wong, R. W. Woodhead, and R. H. Wortman: *Design Planning of Engineering Systems,* Prentice Hall, Englewood Cliffs, N.J., 1985. Good basic coverage of mathematical modeling, optimization, and project planning, including CPM and PERT.

MicroSoft Planner. Software that supports the planning activity.

Steward, D. V.: *System Analysis and Management Structure, Strategy and Design,* Petrocelli, New York, 1981. Gives methods for scheduling tasks; includes CPM, PERT, and other, more sophisticated methods.

Ward, A., J. Liker, J. Cristiano, and D. Sobek: "The Second Toyota Paradox: How Delaying Decisions Can Make Better Cars Faster," *Sloan Management Review,* spring 1995, pp. 43–61.

5.10 EXERCISES

5.1. Develop a plan for the original design problem begun in Exercise 4.1.

(a) Identify the participants on the design team.

(b) Identify and state the objective for each needed task.

(c) Justify the use of prototypes.
(d) Estimate the resources needed for each task.
(e) Develop a schedule and a cost estimate for the design project.

5.2. For the redesign problem begun in Exercise 4.2, develop a plan as in Exercise 1.

5.3. Develop a plan for making a breakfast consisting of toast, coffee, a fried egg, and juice. Be sure to state the objective of each task in terms of the results of the activities performed, not in terms of the activities themselves.

CHAPTER 6

UNDERSTANDING THE PROBLEM AND THE DEVELOPMENT OF ENGINEERING SPECIFICATIONS

6.1 INTRODUCTION

A famous design story centers on the *Mariner IV* satellite project. The satellite was to be packaged in a rocket, its solar panels folded against its sides. After launch the satellite was to spin so that the solar panels would unfold by centrifugal force and be locked in a straight-out position. Because these panels were quite large and very fragile, there was concern that they would be damaged when they hit the stops that determined their final position. To address this problem, the major aerospace firm that had the *Mariner* contract initiated a design project to develop a retarder (dampener) to gently slow the motion of the panels as they reached their final position. The constraints on the retarders were quite demanding: they would have to work in the vacuum and cold of space, work with great reliability, and not leak, since any foreign substance on the panels would harm their capability of capturing the sun's energy during the satellite's nine-month mission to Mars. Millions of dollars and thousands of hours were spent to design these retarders, yet after extensive design work, testing, and simulation, no acceptable devices evolved. With time running out, the design team ran a computer simulation of what would happen if the retarders failed completely; to the team's amazement, the simulation showed that the panels would be safely deployed without any dampening at all. In the end, they realized that there was no need for retarders, and *Mariner IV* successfully went to Mars without them.

This story is an example of how a lot of time and money can be wasted designing the wrong product. Surveys show that poor product definition is a factor in 80 percent of all time-to-market delays. Further, getting a product to market late is more costly to a company than being over cost or having less than optimal performance. Finding the "right" problem to be solved may seem a simple task; unfortunately, it often is not.

An even more difficult and expensive problem for most companies is what is often called "creeping specifications," which is a change in the specifications for the product during the design process. It is estimated that fully 35 percent of all product development delays are directly caused by such changes. There are three factors that cause creeping specifications. First, as the design process progresses, more is learned about the product and so more features can be added. Second, since design takes time, new technologies and competitive products become available during the design process. It is a difficult decision whether to ignore these, incorporate them (i.e., change the specifications) or start all over (i.e., decide that the new developments have eliminated the market for what you are designing). Third, since design requires decision making, any specification change causes a readdressing of all the decisions dependent on that specification. Even a seemingly simple specification change can cause redesign of virtually the whole product. The point is that when specification changes become necessary, they should be done in a controlled and informed manner.

In the preceding chapters the importance of the early phases of the design process has been repeatedly emphasized. As pointed out in Chap. 1, careful requirements development is a key feature of concurrent engineering. In this chapter the focus is on understanding the problem that is to be solved. The ability to write a good set of engineering specifications is proof that the design team understands the problem. There are many techniques used to generate engineering specifications. One of the best and currently most popular is called *quality function deployment (QFD).* Regardless of whether this technique or some other is used, the important information that must be developed is the same. It is the information developed that is emphasized in this chapter.

The QFD method was developed in Japan in the mid-1970s and introduced in the United States in the late 1980s. Using this method, Toyota was able to reduce the costs of bringing a new car model to market by over 60 percent and to decrease the time required for its development by one-third. It achieved these results while improving the quality of the product. A recent survey of 150 U.S. companies shows that 69 percent use the QFD method and that 71 percent of these have begun using the method since 1990. A majority of companies use the method with cross-functional teams of 10 or fewer members. Of the companies surveyed, 83 percent felt that the method had increased customer satisfaction and 76 percent indicated that it facilitated rational decisions.

QFD requires a time commitment, but its effectiveness dictates that it be followed from the beginning of all design projects. Before itemizing the steps that compose this technique for understanding a design problem, we list some important points.

1. No matter how well the design team thinks it understands a problem, it should employ the QFD method for all original design or redesign projects requiring new features, for in the process the team will learn what it does not know about the problem.
2. The customers' requirements must be translated into *measurable design targets for identified critical parameters.* You cannot design a car door that is "easy to open" when you do not know the meaning of "easy." Is easiness measured by force, time, or what? If force is a critical parameter, then is "easy" 20 N or 40 N? The answer has to be known before time and resources are invested in the design effort.
3. The QFD method can be applied to the entire problem and any subproblem. (Note that the design of a door mechanism, above, is a subproblem in automobile design.)
4. It is important to first worry about *what* needs to be designed and, only after that is fully understood, to worry about *how* the design will look and work. Our cognitive capabilities generally lead us to try to assimilate the customers' functional requirements (what is to be designed) in terms of form (how it will look); these images then become our favored designs and we get locked onto them. The QFD procedure helps overcome this cognitive limitation.
5. This method takes time to complete. In some design projects, about one-third of the total project time is spent on this activity. Ford spends 3–12 months

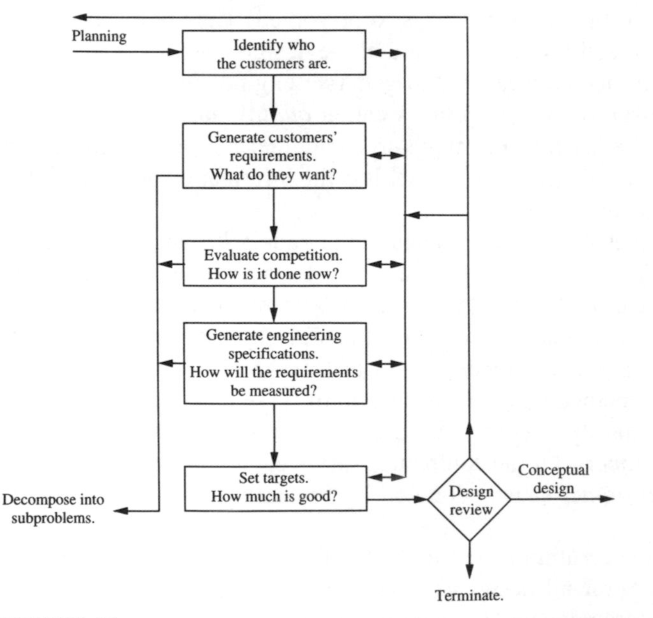

FIGURE 6.1
The engineering specifications development phase of the design process.

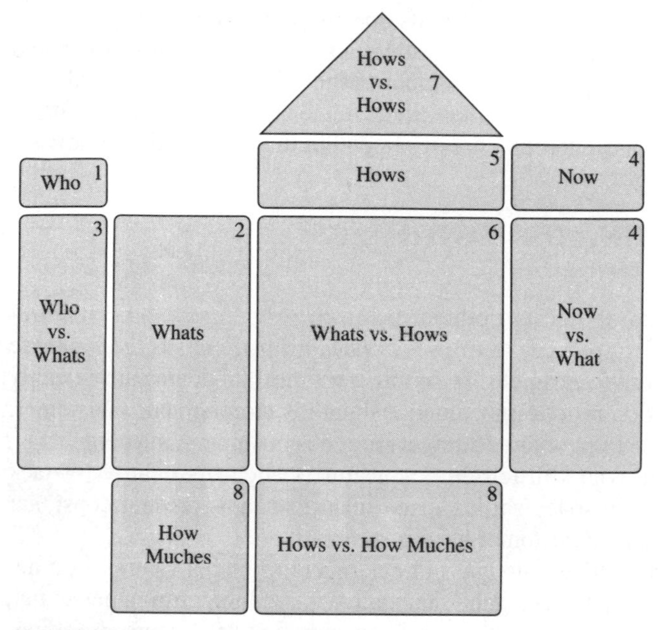

FIGURE 6.2
The house of quality, also known as the QFD diagram.

developing the QFD for a new feature. Time spent here saves time later. Not only does the technique help in understanding the problem, it also helps set the foundation for concept generation.

The QFD method helps generate the information needed in the engineering specifications development phase of the design process (Fig. 4.1) That phase is reproduced in Fig. 6.1. Each block in the diagram is a major section in this chapter and a step in the QFD method.

Applying the QFD steps builds the *house of quality* shown in Fig. 6.2. This is a house of many rooms, each containing valuable information. This figure will be completed in detail for the bicycle Splashgard as a running example. Before we describe each step for filling in Fig. 6.2, a brief description of the figure is helpful. The numbers in the figure refer to the steps that are detailed in the sections below. Developing information begins with identifying *who* (step 1) the customers are and *what* (step 2) it is they want the product to do. In developing this information, we also determine to whom the what is important—*who vs. what* (step 3). Then it is important to identify how the problem is solved *now* (step 4), in other words, what the competition is for the product being designed. This information is compared to what the customers desire—*now vs. what* (step 4 continued)—to find out where there are opportunities for an improved product. Next comes one of the more difficult steps in developing the house, determining *how* (step 5) you are going to measure the product's ability to satisfy the customers' requirements. The hows are the engineering specifications, and their correlation to the customers' requirements is given by *whats vs. hows* (step 6). The engineering specifications can also be compared to each other to find their interrelations—*how vs. how* (step 7). Finally, target information—*how much* (step 8 continued)—is developed in the basement of the house and compared with the engineering specifications—*how vs. how much* (step 8 continued). Details of all these steps and why they are important are developed in the sections that follow.

6.2 STEP 1: IDENTIFY THE CUSTOMERS: WHO ARE THEY?

The goal in understanding the design problem is to *translate customers' requirements into a technical description of what needs to be designed.* Or, as the Japanese say, "Listen to the voice of the customer." To do this, we must first determine exactly who the customers are. For most design situations there is more than one customer; for many products the most important customers are the consumers, the people who will buy the product and who will tell other consumers about its quality (or lack thereof). Some products—a space shuttle or an oil drill head—are not consumer products but still have a broad customer base.

There is usually more than one class of customers to be considered in a design situation. The consumer, the designer's management, manufacturing personnel, sales staff, and service personnel must also be considered as customers. Additionally, standards organizations should be viewed as customers, as they too may set requirements for the product. For many products there are five or more classes of customers whose voices need to be heard.

It was only in the 1980s that we began to regard the people involved in manufacturing and assembly as customers. Prior to this time there was little communication between design and production (manufacturing plus assembly) in most companies. Remember the discussion of the over-the-wall design process in Chap. 1: the inefficiency of this method of design led management to implement efforts to overcome the communication problems. The multidiscipline team concept is one result of those efforts. Such teams have some of the customers as members.

Regardless of the type of product, it is the desire of the customers that drives the development of the product, not the engineer's vision of what the customers *should* want. The mental image created by the designer in trying to understand the problem (recall Section 3.3) may not be an accurate picture of what the customers really want in the product. Many products have been poorly received by consumers simply because the engineer's concept of the customers' desires was inaccurate.

The Splashgard is definitely a consumer product; the main customers are bicycle riders. In thinking about mountain-bike riders in general, the members of the design team concluded that there were three different types of riders: the hard-core mountain cyclist, the fair-weather rider, and the dual-use rider (the one who both rides off-road and commutes). They further realized that their market was primarily with the dual-use rider. Additionally, bicycle shop mechanics were also considered as customers, as they would have to sell and maybe install the Splashgard. Two

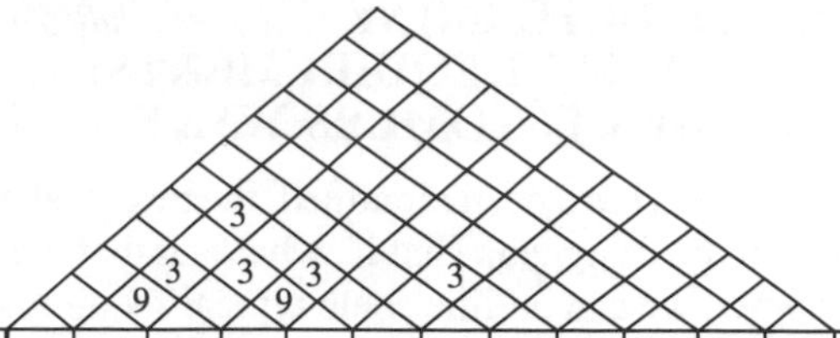

	Mechanic	Marketing	Rider	Water hitting rider (%)	Steps to attach (#)	Time to attach (sec)	Steps to detach (#)	Time to detach (sec)	Number of parts (#)	Weight (g)	Customers finding it visually appealing (%)	Colors available (#)	Bikes that it fits (%)	Upward release force (N)	Sales price ($)	Whale Tale	Norco	Raincoat
Functional performance																		
Keeps water off rider	1	1	7	9										3		1	4	2
Fast to attach	5	4	8		3	9			3	1						1	4	3
Fast to detach	9	5	10				3	9	3	1						2	4	3
Can attach when dirty	7	13	12		3	3										3	3	2
Can detach when dirty	11	12	13				3	9								3	3	2
Human factors																		
Easy to attach	4	6	9		9				3	1						1	3	3
Easy to detach	10	7	11				9		3	1						1	4	3
Looks fast	2	10	2								9					4	2	2
Color matches bike	12	11	5								3	9				3	2	2
Interface with bike																		
Fits bike	3	3	3										9			3	2	4
Does not mar bike	8	8	6											1		3	1	4
Light weight	6	9	4						9							3	3	4
Competitive sales price	13	2	1												9	2	3	1
Whale Tail				25	5	25	2	5	6	130	75	5	94	5	12			
Norco				0	3	5	1	3	2	140	65	1	65	15	15			
Raincoat				30	3	10	3	10	1	100	35	4	100	0	20			
Target				0	1	2	2	3	2	130	85	5	95	5	10			

FIGURE 6.3
House of quality for the Splashgard.

functions in the company were also listed as customers to ensure capture of production and marketing needs. Although large companies might break this down more finely, the team thought this sufficient.

Thus, the customers used in this example are the mechanic, marketing, and the rider, the most important. These customers are listed in the "who" section of Fig. 6.3, the house of quality for the Splashgard.

6.3 STEP 2: DETERMINE THE CUSTOMERS' REQUIREMENTS: WHAT DO THE CUSTOMERS WANT?

Once the customers have been identified, the next goal of the QFD method is to determine *what* is to be designed. That is, what is it that the customers want? Depending on the customer, we can outline some typical requirements.

Typically, as shown by the customer survey in Table 1.2, the *consumers* want a product that works as it should, lasts a long time, is easy to maintain, looks attractive, incorporates the latest technology, and has many features.

Typically, the *production customer* wants a product that is easy to produce (both manufacture and assemble), uses available resources (human skills, equipment, and raw materials), uses standard parts and methods, uses existing facilities, and produces a minimum of scraps and rejected parts.

Typically, the *marketing/sales customer* wants a product that meets consumers' requirements; is easy to package, store, and transport; is attractive; and is suitable for display.

The sections below give some important background information on collecting customers' requirements.

6.3.1 The Kano Model of Customer Satisfaction

Our goal is to find requirements that not only satisfy the customers but also excite them to want to purchase the product and recommend it to others. According to the Kano[1] model of customer satisfaction, there are three different types of product quality that give customer satisfaction: *basic quality, performance quality,* and *excitement quality,* shown in Fig. 6.4.

Basic quality refers to customers' requirements that are not verbalized as they specify assumed functions of the device. The only time a customer will mention them is if they are missing. If they are not fully implemented in the final product, the customer will be disgusted with it. If they are included, the customer will be neutral. This type of requirement should not be included in the QFD. An example is the requirement that a bicycle should have brakes.

Performance quality refers to customers' requirements that are verbalized in the form that the better the performance, the better the product. This type of requirement is a major part of the QFD.

The prime goal in determining the voice of the customer is to find the requirements that result in a product of *excitement quality.* These requirements are often unspoken because customers do not expect them to be met in the product. If they are absent, the customers are neutral. However, if the customers' reaction to the final product is surprise and delight at the additional functions, then the product's chance of success in the market is high. Requirements for excitement-quality features are often called "wow requirements."

Over time, excitement level requirements become performance level requirements and, ultimately, basic requirements. This is true for most features of home

---

[1]The Kano model was developed by Dr. Noriaki Kano in the early 1980s to describe customer satisfaction.

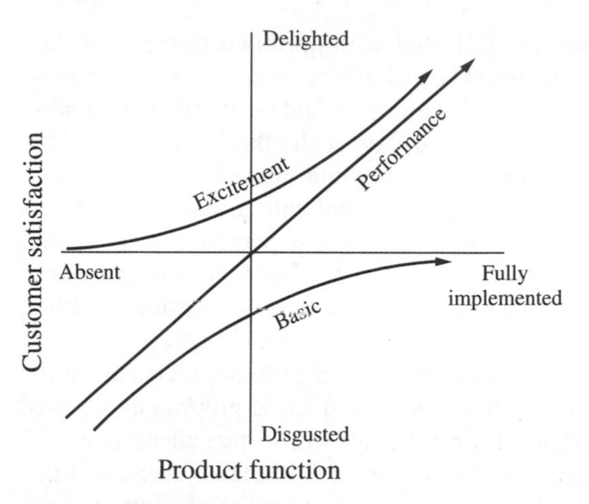

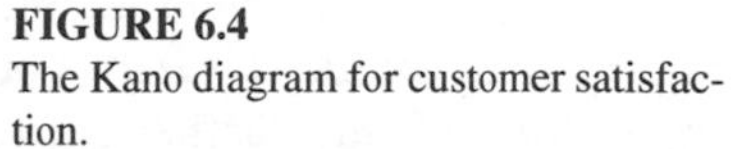
FIGURE 6.4
The Kano diagram for customer satisfaction.

entertainment systems, cars, and other consumer products. When first introduced, a new feature is special on one brand and consumers are surprised and delighted. The next year every brand has the feature and some perform better than others. After a few years the feature is not even mentioned in advertising because it is an expected feature of the product.

6.3.2 Collection Methods for Customers' Requirements

The key to this QFD step is collecting information from customers. There are essentially three methods commonly used: observations, surveys, and focus groups. Each will be introduced, and then steps for collecting the information presented.

Fortunately, most new products are refinements of existing products, so many requirements can be found by observing customers using the existing product. For example, automobile manufacturers send engineers into shopping center parking lots to observe customers putting purchases into cars to better understand one aspect of car door requirements.

Surveys are generally used to gather specific information or ask people's opinions about a well-defined subject. Surveys use questionnaires that are carefully crafted and applied either through the mail, over the telephone, or in face-to-face interviews. Surveys are well suited for collecting requirements on products to be redesigned or on new, well-understood product domains. For original products or to gather the customers' ideas for product improvement, focus groups are best.

The focus-group technique was developed in the 1980s to help capture customers' requirements from a carefully chosen group of potential customers. The method begins by identifying 7–10 potential customers and asking if they will attend a meeting to discuss a new product. One member of the design team acts as moderator and another as note taker. It is also best to electronically record the session. The goal in the meeting is to find out what is wanted in a product that does not yet exist, and so it relies on the customers' imaginations. Initial questions about the participants' use of similar products are followed with questions designed to find performance and excitement requirements. The goal of the moderator is to use questions to guide the discussion, not control it. The group should need little intervention from the moderator, because the participants build on each other's comments. One technique that helps elicit useful requirements during interviews is for the moderator to repeatedly ask "Why?" until the customers respond with information in terms of time, cost, or quality. To elicit good information takes experience, training, and multiple sessions with different participants. Usually the first focus group leads to questions needed for the second group. It often takes as many as six sessions to obtain stable information.

Later in the design process, surveys can be used to gather opinions about the relative merit of different alternatives. Observation and focus groups can be used both to generate ideas that may become alternatives and to evaluate alternatives. All these types of information gathering rely on questions formulated ahead of time. With a survey the questions and the answers must be formalized. Both surveys and observations usually use closed questions (i.e., questions with predetermined answers); focus groups use open-ended questions.

Regardless of the method used, the following steps will help the design team develop useful data:

STEP 2.1: SPECIFY THE INFORMATION NEEDED. Reduce the problem to a single statement describing the information needed. If no single statement represents what is needed, more than one data-collecting effort may be warranted.

STEP 2.2: DETERMINE THE TYPE OF DATA-COLLECTION METHOD TO BE USED. Base the use of focus groups, observations, or surveys on the type of information being collected.

STEP 2.3: DETERMINE THE CONTENT OF INDIVIDUAL QUESTIONS. A clear goal for the results expected from *each question* should be written. Each question should have a single goal. For a focus group or observation, this may not be possible for all questions, but it should be for the initial questions and other key questions.

STEP 2.4: DESIGN THE QUESTIONS. Each question should seek information in an unbiased, unambiguous, clear, and brief manner. Key guidelines are as follows:

Do not assume the customers have more than common knowledge.
Do not use jargon.
Do not lead the customer toward the answer you want.
Do not tangle two questions together.
Do use complete sentences.

Questions can be in one of four forms

- Yes–no–don't know.

- Ordered choices (strongly agree, mildly agree, neither agree nor disagree, mildly disagree, strongly disagree; or 0–4, 5–8, 9–12, greater than 12). Be sure that any ordered list is complete (i.e., that it covers the full range possible and that the choices are unambiguously worded).
- Unordered choices (a, b, and/or c).
- Ranking (a is better than b is better than c).

For focus groups avoid simple yes-no answers.

The best questions ask about attributes, not influences. Attributes express what, where, how, or when. *Why* questions should lead to what, where, how, or when as they describe time, quality, and cost.

STEP 2.5: ORDER THE QUESTIONS. Order the questions to give context. This will help participants in focus groups or surveys follow the logic.

STEP 2.6: TAKE DATA. It usually takes repeated application to generate usable information. The first application of any set of questions should be considered a test or verification experiment.

STEP 2.7: REDUCE THE DATA. A list of customers' requirements should be made in the customers' own words, such as "easy," "fast," "natural," and other abstract terms. A later step of the design process will be to translate these terms into engineering parameters. The list should be in positive terms—what the customers want, not what they do not want. We are not trying to patch a poor design; we are trying to develop a good one. (Even so, as will be seen shortly, negative statements are sometimes needed to convey a requirement.)

Below are some of the questions developed by the Splashgard team to help guide a focus group of mountain bikers they had assembled. These questions were split into two sets: a presession questionnaire and questions to be used in the focus-group session itself.

Here is a sample of presession questions:

Q1. How often do you ride a mountain bike? *(Circle the number of your answer.)*
1. Never
2. Once a month
3. Once a week
4. 2–5 times a week
5. 6–10 times a week
6. More than 10 times a week

Q2. How often do you use your mountain bike off-road? *(Circle the number of your answer.)*
1. Never
2. Once a month
3. Once a week
4. 2–5 times a week
5. 6–10 times a week
6. More than 10 times a week

Q3. What do you use your mountain bike for? *(Circle all that apply.)*
1. Off-road riding
2. Commuting to work or school
3. Touring
4. Shopping
5. Other ______________________

Q4. Suppose you could add a device to your mountain bike for use on the street in wet weather. Which of the following is most important to you? *(Rank 1 through 6.)*

_____ Staying dry
_____ Low weight
_____ Ease of installation and removal
_____ Attractive style
_____ Stiffness

Questions prepared for the focus group included the following:

Q1. Describe your experience using your bike for going to work or school. (Note that this is not a question yet it leads the customers to describe their activities and can lead to many questions that hone in on time, cost, and quality.)

Q2. Why do you ride your bike if you know it will make you wet and muddy?

6.3.3 The Types of Customers' Requirements

A checklist of the major types of requirements is given in Table 6.1. Comparing this list with the list of requirements developed for a product can reveal missing information. It is also useful in developing questions to ask in surveys and focus groups and as a key to information that needs to be found before design begins. The major types of requirements in this list are detailed in the paragraphs below.

Functional performance requirements are those elements of the performance that describe the product's desired behavior. Although the customers may not use technical terms, function is usually described as the flow of energy, information, and materials or as information about the operational steps and their sequence. In the next chapter we develop concepts by building a functional model, based on the flow of energy, information, and materials. It will be seen that *developing functional requirements for the QFD and building a functional model of the product are often iterative*. The more the function is understood, the more complete are the requirements that can be developed.

Any product that is seen, touched, heard, tasted, smelled, or controlled by a human will have *human factors requirements* (see App. D for details on human factors). This includes nearly every product. One frequent customers' requirement is that the product "looks good" or looks as if it has a certain function. These are areas in which a team member with knowledge about industrial design is essential. Other requirements focus on the flow of energy and information between the product and the human. Energy flow is usually in terms of force and motion but can take other forms as well. Information flow requirements apply to the ease of controlling and sensing the state of the product. Thus, human factors requirements are often functional performance requirements.

TABLE 6.1
Types of customer requirements

Functional performance	Life-cycle concerns (continued)
Flow of energy	Diagnosability
Flow of information	Testability
Flow of materials	Repairability
Operational steps	Cleanability
Operation sequence	Installability
Human factors	Retirement
Appearance	Resource concerns
Force and motion control	Time
Ease of controlling and sensing state	Cost
Physical requirements	Capital
Available spatial envelope	Unit
Physical properties	Equipment
Reliability	Standards
Mean time between failures	Environment
Safety (hazard assessment)	Manufacturing requirements
Life-cycle concerns	Materials
Distribution (shipping)	Quantity
Maintainability	Company capabilities

Physical requirements include needed physical properties and spatial restrictions. Some physical properties often used as requirements are weight, density, and conductivity of light, heat, or electricity (i.e., flow of energy). Spatial constraints relate how the product fits with other, existing objects. Almost all new design efforts are greatly affected by the physical interface with other objects that cannot be changed.

In the *Time* magazine survey on quality quoted in Chap. 1, the second most important consumer concern was "Lasts a long time," or the product's *reliability.* It is important to understand what acceptable reliability means to the customer. The product may only have to work once with near-absolute certainty (e.g., a rocket), or it may be a disposable product that does not need much reliability. A measure of reliability is the *mean time between failures.* This topic is discussed in Chap. 12.

A part of reliability involves the questions, What happens when the product does fail? What are the *safety* implications? Product safety and hazard assessment are very important to the understanding of the product, and they are covered in Chap. 8.

An often overlooked class of requirements is the class of those relating the product life cycle other than product use. All life-cycle phases listed in Table 6.1 were taken from Fig. 1.7. In designing the BikeE recumbent bicycle, mentioned in the Preface, one of the design requirements set by sales/marketing was that the bicycle had to be shipped by a commercial parcel service. Such services have weight and size limits, which greatly affected the design of the product. If the advantages of distributing the product by commercial parcel service had not been realized early, extensive redesign might have been necessary. The same applies to the other life-cycle phases listed in Table 6.1 and Fig. 1.7.

A limited resource on every design project is time. *Time requirements* may come from the consumer; more often they originate in the market or in manufacturing needs. In some markets there are built-in time constraints. For example, toys must be ready for the summer buyer shows so that Christmas orders can be taken; new automobile models traditionally appear in the fall. Contracts with other companies might also determine time constraints. Even for a company without an annual or contractual commitment, time requirements are important. As discussed earlier, in the 1960s and 1970s Xerox dominated the copier market, but by 1980 its position had been eroded by domestic and Japanese competition. Xerox discovered that one of the problems was that it took it twice as long as some of its competitors to get a product to market, and Xerox put new time requirements on its engineers. Fortunately, Xerox helped its engineers work smarter, not just faster, by introducing techniques similar to those we talk about here.

Cost requirements concern both the capital costs and the costs per unit of production. Included in capital costs are expenditures for the design of the product. For a Ford automobile, design costs make up 5 percent of the manufacturing cost (Fig. 1.1). Many product ideas never get very far in development because the initial requirements for capital are more than the funds available. (Cost estimating will be covered in detail in Section 12.2.)

Standards spell out current engineering practice in common design situations. The term *code* is often used interchangeably with *standard.* Some standards serve as good sources of information. Other standards are legally binding and must be adhered to—for example, the ASME pressure vessel codes. Although the actual information contained in standards does not enter into the design process in this early phase, knowledge of which standards apply to the current situation are important to requirements and must be noted from the beginning of the project.

Standards that are important to design projects generally fall into three categories: performance, test methods, and codes of practice. There are *performance standards* for many products, such as seat-belt strength, crash-helmet durability, and tape-recorder speeds. The *Product Standards Index* lists U.S. standards that apply to various products; most of those referenced are also covered by ANSI (American National Standards Institute), which does not write standards but is a clearinghouse for standards written by other organizations.

Test method standards for measuring properties such as hardness, strength, and impact toughness are common in mechanical engineering. Many of these are developed and maintained by the American Society for Testing and Materials (ASTM), an organization that publishes over four thousand individual standards covering the properties of materials, specifying equipment to test the properties and outlining the procedures for testing. Another set of testing standards that are important to product design are those developed by the Underwriters Laboratories (UL). This organization's standards are intended to prevent loss of life and property from fire, crime, and casualty. There are over 350 UL standards. Products that have been tested by UL and have met their standards can display the words "Listed UL" and the standard number. The company developing the product must pay for this testing. Consumer products are usually not marketed without UL listing because the liability risk is too high without this proof of safe design.

Codes of practice give parameterized design methods for standard mechanical components, such as pressure vessels, welds, elevators, piping, and heat exchangers.

Standards information is given in the Sources section at the end of the chapter; most technical libraries carry an up-to-date set of ASME codes, ANSI standards, and UL standards.

It is important for the design team to ensure that requirements imposed by *environmental concerns* have been identified. Since the design process must consider the entire life cycle of the product, it is the design engineer's responsibility to establish the impact of the product on the environment during production, operation, and retirement. Thus, requirements for the disposal of wastes produced during manufacture (whether hazardous or not), as well as for the final disposition of the product, are the concern of the design engineer. This topic is further discussed in Chap. 12.

Some of the *manufacturing/assembly requirements* are dictated by the quantity of the design to be produced and the characteristics of the company producing the design. The quantity to be produced often affects the kind of manufacturing processes to be used. If only one unit is to be produced, then custom tooling cannot be amortized across a number of items and off-the-shelf components should be selected when possible (see Chap. 10). Additionally, every company has internal manufacturing resources whose use is preferable to contracting work outside the company. Such factors must be considered from the very beginning.

6.3.4 The Splashgard Requirements

Even though we are only on step 2 of the QFD house, some work on step 4, finding out how the problem is now solved, is helpful now. As shown in Fig. 6.1, there is much iteration in this early design phase.

To gather customer information, the Splashgard design team interviewed dual-use riders and bicycle store mechanics and asked them what features they would want in a removable mountain-bicycle splash guard. The answers were tape-recorded; in reviewing these, the team generated the list shown in Table 6.2, which incorporates the customers' own words.

Note that this list tells *what* was needed, not *how* the final product would look or operate. This is extremely important; if there were the requirement, "The fender should be quick to attach," then it is implied that the Splashgard is a fender of some sort, which was certainly a possibility, but it was too early to commit to any such type of design.

Although the list in Table 6.2 is a good first try, it is hard to tell if it is complete. Therefore the team organized the items on the list into the major areas of concern that appear in Table 6.1. Many of the requirements for functional performance could, alternatively, be considered human factors requirements because the rider or mechanic supplies the energy and control (see App. D). By utilizing the checklist in Table 6.1, the design team could refine customers' requirements for completeness. The original requirements were then transformed into a more uniform listing, and in this way some omissions were brought to light. It is important to identify all requirements as early as possible. Partial results of the refinement are shown in Fig. 6.3, the sample Splashgard QFD diagram.

TABLE 6.2
Preliminary list of customers' requirements for the Splashgard

Riders' and bike shop mechanics' requirements
Keeps water off rider
Is easy to attach
Is easy to detach
Is quick to attach
Is quick to detach
Won't mar bicycle
Won't catch water/mud/debris
Won't rattle
Won't wobble
Won't bend
Has a long life
Won't wear out
Is lightweight
Won't rub on wheel
Is attractive
Fits universally
If permanent piece on bike, then is small
If permanent piece on bike, then is easy to attach
If permanent piece on bike, then is fast to attach
If permanent piece on bike, then is noninterfering
Won't interfere with lights, rack, panniers, or brakes
Company management requirements
Capital expenditure is less than $15,000
Can be developed in 3 months
Can be marketable in 12 months
Manufacturing cost is less than $3
Estimated volume is 200,000 per year for five years

6.4 STEP 3: DETERMINE RELATIVE IMPORTANCE OF THE REQUIREMENTS: WHO VERSUS WHAT

The next step in the QFD technique is to evaluate the importance of each of the customers' requirements. This is accomplished by generating a weighting factor for each requirement and entering it in Fig. 6.2. The weighting will give an idea of how much effort, time, and money to invest in achieving each requirement. Two questions are addressed here: (1) To whom is the requirement important? and (2) How is a measure of importance developed for this diverse group of requirements?

Since a design is "good" only if the customers think it is good, the obvious answer to the first question is, The customer. However, we know that there may be more than one customer. In the case of a piece of production machinery, the desires of the workers who will use the machine and those of management may not be the same. This discrepancy must be resolved at the beginning of the design process, or the requirements may change partway through the job. Sometimes a designer's hardest job is to determine whom to please.

The region of the house of quality labeled "who vs. whats" in Fig. 6.2 is for input of the importance of each requirement. There are many ways to do this rating. What is suggested here is that representatives of each customer group be given a list of the requirements and asked to rank-order them. One way to do this is to write each requirement on a piece of self-stick note paper, put the notes on a wall, and ask each customer to arrange them in order of importance. Another method is to give the customers a list of the requirements and have them label each requirement with a sequential number, starting with "1" as most important. They can leave blank the requirements that have no importance to them. The results of rank-ordering the requirements for the Splashgard are shown in Fig. 6.3 for each of the three customers.

6.5 STEP 4: IDENTIFY AND EVALUATE THE COMPETITION: HOW SATISFIED IS THE CUSTOMER NOW?

The goal here is to determine how the customer perceives the competition's ability to meet each of the requirements. Even though the Splashgard is a new design, there is competition. Figure 6.5 shows two commercially available removable fenders. The purpose for studying existing products is twofold: first, it creates an awareness of what already exists (the "now"), and, second, it reveals opportunities to improve on what already exists. In some companies this process is called *competition benchmarking* and is a major aspect of understanding a design problem. In benchmarking, each competing product must be compared with customers' requirements (now vs. what). Here we are only concerned with a subjective comparison that is based on customer opinion. Later, in step 8, we will do a more objective comparison. For each customers' requirement, we rate the existing design on a scale of 1 to 5:

1. The design does not meet the requirement at all.
2. The design meets the requirement slightly.
3. The design meets the requirement somewhat.
4. The design meets the requirement mostly.
5. The design fulfills the requirement completely.

Though these are not very refined ratings, they do give an indication of how the competition is perceived by the customer.

This step is very important as it shows opportunities for product improvement. If all the competition rank low on one requirement, this is clearly an opportunity. This is especially so if the customers ranked that specific requirement highly important in step 3. If one of the competitors meets the requirement completely, this product should be studied and good ideas used from it (note patent implications, Section 7.4.1). In the 1980s there was a commercial on television for a family van in which the manufacturer bragged that its product was so good that one of its competitors bought and studied it. The commercial shows the competitor's technicians in white coats disassembling the van. What the commercial did not say was that the advertiser

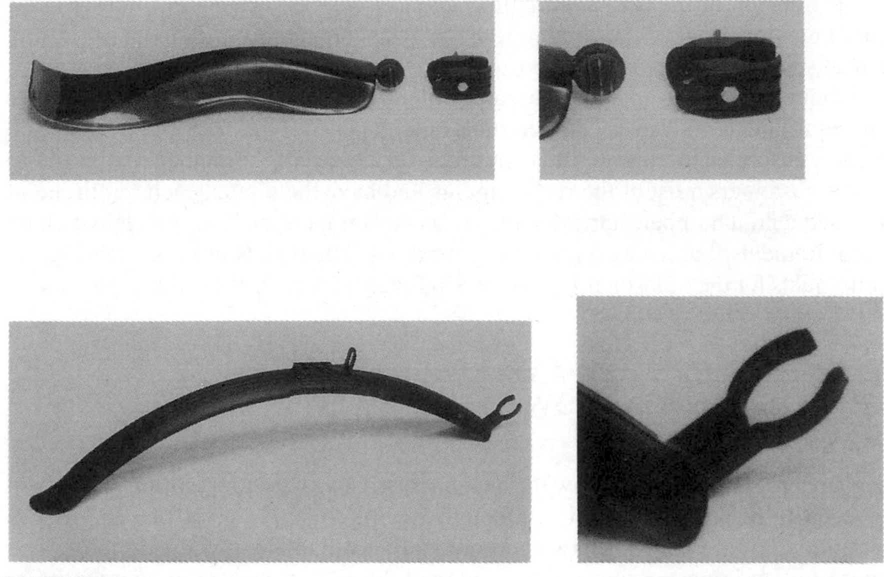

FIGURE 6.5
Existing fender products.

also bought and studied its competitor's product and that this is just good design practice.

Results of this step for the Splashgard are shown in Fig. 6.3. The competition consists of the two fenders shown in Fig. 6.5 and a raincoat. The first fender, sold under the name of the "Whale Tail," is made of injection-molded plastic. It is attached to the bike seat post with a two-piece molded clamp. Tightening the wing nut on the bolt holding the two pieces together not only clamps the seat post but also clamps the ball molded on the end of the fender. This ball allows the fender to be positioned in a wide range of positions. The second fender is also of injection-molded plastic. To mount it, the front C-shaped piece is snapped over the down tube (the tube connecting the seat to the pedals) and the raised section is slid over a metal bracket permanently attached to the bolt that holds the rear brakes on. This use of a part permanently attached to the bike made the design team reexamine the work to date. Namely, they had to determine if the customers' requirements for "easy" and "fast" referred to the Splashgard itself or to permanent attachments. They concluded that the information to date could be used for both but decided to develop targets in step 8 for the Splashgard only—assuming that any permanent attachments were already installed.

To determine how well the three competitors met the requirements, the design team used a focus group to evaluate them. The average results are shown in the "now vs. what" section of the diagram. Note that none of the competitors meets the customers' requirements and thus there is opportunity for a superior product.

6.6 STEP 5: GENERATE ENGINEERING SPECIFICATIONS: HOW WILL THE CUSTOMERS' REQUIREMENTS BE MET?

The goal here is to develop a set of *engineering specifications* from the customers' requirements. These specifications are the restatement of the design problem in terms of parameters that can be measured and have target values. Without such information the engineers cannot know if the system being developed will satisfy the customers. The parameters are developed in this step, and the target values for these parameters are developed in step 8.

In this step, parameters that tell *how* we know if each customers' requirement has been met are developed. These parameters are the measures of the customers' requirements. Some customers' requirements are directly measurable; this step does not apply to them. For example, the requirement that a new device be able to lift 100 kg or that a paper tray hold size A paper is clearly measurable. Abstract requirements for the Splashgard, like "easy to attach," must be refined in order to be measurable. It is toward these more abstract requirements that the following is directed.

We begin by finding as many engineering parameters as possible that indicate a level of achievement for each customers' requirement. For example, the "easy to attach" customers' requirement can be measured by (1) the number of steps needed to attach it, (2) the time to attach it, (3) the number of parts, and (4) the number of standard tools used. Note that a set of units is associated with each of these measures—step count, time, part count, and tool count. *If units for an engineering parameter cannot be found, the parameter is not measurable and must be readdressed.* Each engineering parameter must be measurable and thus must have units of measure. However, "time to attach" may not be a reliable measure as it will be dependent on the skill and training of the customer. Either the customer skill level needs to be defined or this parameter eliminated.

An important point here is that every effort must be made to find as many ways as possible to measure each customers' requirement. If there are no measurable engineering parameters for a specific customers' requirement, then the customers' requirement is not well understood. Possible solutions are to break the requirement into finer independent parts or to redo step 2 with specific attention to that specific requirement.

When developing the engineering specifications, carefully check each entry to see what nouns have been used. Each noun refers to an object that is part of the product or its environment and should be considered to see if new objects are being assumed. For example, if one specification in the Splashgard problem was for "weight of fender," then a fender has been assumed as part of the solution. If the design team has made a decision that the Splashgard is to be a fender, this is acceptable. However, if no such assumption has been made, the product solution has been unknowingly limited. Paying attention to the objects that are part of the product is a major topic in concept generation.

A sample of Splashgard engineering specifications is shown in Fig. 6.3.

6.7 STEP 6: RELATE CUSTOMERS' REQUIREMENTS TO ENGINEERING SPECIFICATIONS: HOWS MEASURE WHATS?

To complete this step, we fill in the center portion of the house of quality. Each cell of the form represents how an engineering parameter relates to a customers' requirement. Many parameters will measure more than one customers' requirement. The strength of this relationship can vary, with some engineering parameters providing strong measures for a customers' requirement and others providing no measure at all. The relation is conveyed through numerical values; we will use four:

9 = strong relationship

3 = medium relationship

1 = weak relationship

Blank = no relationship at all

Sometimes symbols are used for the correlation, and often the numbers 3,2,1 are used instead of 9,3,1. Fig. 6.3 shows the results of this step for the Splashgard.

6.8 STEP 7: IDENTIFY RELATIONSHIPS BETWEEN ENGINEERING REQUIREMENTS: HOW ARE THE HOWS DEPENDENT ON EACH OTHER?

Engineering specifications may be dependent on each other. It is best to realize these dependencies early in the design process. Thus, the roof is added to show that as you work to meet one specification, you may be having a positive or negative affect on others.

In Fig. 6.3 the roof for the Splashgard QFD shows diagonal lines connecting the engineering specifications. If two specifications are interdependent, a number is noted at the intersection. Here the 9,3,1 system has been used to note strong relationship, medium relationship, or weak relationship. A blank denotes no relationship. The roof shows that the average time to attach is strongly related to the number of tools to attach and weakly related to the number of steps to attach. These ratings were developed by the design team.

6.9 STEP 8: SET ENGINEERING TARGETS: HOW MUCH IS GOOD ENOUGH?

The last step in the QFD technique is to determine a target value for each engineering measure. As the product evolves, these target values are used to evaluate the product's ability to satisfy customers' requirements. There are really two actions here. The first is to ascertain how the competition, examined in step 4, meets the engineering specifications, and the second is to establish the target for the new product.

In step 4 competition products were compared to customers' requirements. In this step they will be measured relative to engineering specifications. This ensures that both knowledge and equipment exist for evaluation of any new products developed in the project. Also, the values obtained by measuring the competition give a basis for establishing the targets. This usually means obtaining actual samples of the competition's product and making measurements on them in the same way that measurements will be made on the product being designed.

Setting targets early in the design process is important; targets set near the end of the process are easy to meet but have no meaning as they always match what has been designed. However, setting targets too tightly may eliminate new ideas. Some companies refine their targets throughout concept development and then make them firm. Their initial targets, set here, may have ± 30 percent tolerance on them.

Some customers' requirements will have ready-made targets—the requirement that a device lift 100 kg, for instance, is measurable and provides a specific target. For other requirements, realistic targets must be set. These values define an ideal product and must be based on what is physically realizable, which is why it is essential to examine the competing products.

The best targets are set for a specific value. Less precise, but still usable, are those set within some defined range. A third type of target is a value made to be as large or as small as possible. Although measurable, these extremes are poor targets, since they give no information that tells when the performance of the new product is acceptable. In all three cases, evaluation of the competition can be used to give an achievable target value or range of values.

Here is a final comment on target setting. If a target is much different than the values achieved by the competition, it should be questioned. Specifically, what do you know that the competition does not know? Do you have a new technology, do you know of new concepts, or are you just smarter than your competition?

For the Splashgard, the targets were developed after studying the competition. The team thought they could do better than the competition in the number of steps to attach the product and so set their target lower than the others. They did this because it appeared to them to offer an opportunity. One member of the team was experienced in human factors and felt that one-step attachment was possible. For other specifications, the targets were set to be the best value shown by the competition. The data shown in Fig. 6.3 are for the Splashgard only—not for any permanent attachments to the bike.

6.10 FURTHER COMMENTS ON QFD

The QFD technique ensures that the problem is well understood. It is useful with all types of design problems and results in a clear set of customers' requirements and associated engineering measures. It may appear to slow the design process, but in actuality it does not, as time spent developing information now is returned in time saved later in the process.

Even though this technique is presented as a method for understanding the design requirements, it forces such in-depth thinking about the problem that many good design solutions develop from it. No matter how hard we try to stay focused on the requirements for the product, product concepts are invariably generated. This is one situation where a design notebook is important. Ideas recorded as brief notes or sketches during the problem understanding phase may be useful later; however, it is important not to lose sight of the goals of the technique and drift off to one favorite design idea.

The QFD technique automatically documents this phase of the design process. Diagrams like Fig. 6.3 serve as a design record and also make an excellent communication tool. Specifically, the structure of the house of quality makes explaining this phase to others very easy. In one project a member of the sponsoring organization was blind. A verbal description of the structure helped him understand the project.

Often, when working to understand and develop a clear set of requirements for the problem, the design team will realize that it can be decomposed into a set of loosely related subproblems, each of which may be treated as an individual design problem, as is shown in Fig. 6.1. Thus, a number of independent houses may be developed.

The QFD technique can also be applied during later phases of the design process. Instead of developing customers' requirements, we may use it to develop a better measure for functions, assemblies, or components in terms of cost, failure modes, or other characteristics. To accomplish this, review the steps, replacing customers' requirements with what is to be measured and engineering requirements with any other measuring criteria.

A final comment about developing the engineering specifications. Since much learning occurs during the design process, the QFD is considered a working document that is reviewed and updated as needed. The formality and complexity of the technique forces any change to be carefully considered and thus keeps the project moving toward completion. Without a system like QFD, changes in specifications can occur at the whim of a manager or without the design team even realizing it. These changes lead to a failure to meet the schedule and a potentially poor product.

6.11 SUMMARY

* Understanding the design problem is best accomplished through a technique called quality function deployment (QFD). This method transforms customers' requirements into targets for measurable engineering requirements.
* Important information to be developed at the beginning of the problem includes customers' requirements, competition benchmarks, and engineering specifications complete with measurable benchmarks.
* Time spent completing the QFD is more than recovered later in the design process.
* There are many customers for most design problems.
* Studying the competition during problem understanding gives valuable insight into market opportunities and reasonable targets.

6.12 SOURCES

Adams, J. L.: *Conceptual Blockbusting,* Norton, New York, 1976. The story about the *Mariner IV* satellite comes from this book on creativity.

Clausing, D.: *Total Quality Development,* ASME Press, New York, 1994.

Cristiano, J. J., J. K. Liker, and C. C. White: "An Investigation into Quality Function Deployment (QFD) Usage in the U.S.," in *Transactions for the 7th Symposium on Quality Function Deployment,* June 1995, American Supplier Institute, Detroit, 1995. Statistics on QFD usage were taken from the study in this paper.

Hauser, J. R., and D. Clausing: "The House of Quality," *Harvard Business Review,* May–June 1988, pp. 63–73. A basic paper on the QFD technique.

Index of Federal Specifications and Standards, U.S. Government Printing Office, Washington, D.C. A sourcebook for federal standards.

Krueger, R. A.: *Focus Groups: A Practical Guide for Applied Research,* Sage Publishing, Newbury Park, CA 1988. A small book with direct help for getting good information from focus groups.

Roberts, V. L.: *Products Standards Index,* Pergamon, New York, 1986. A sourcebook for standards.

Salant, P., and D. Dillman: *How to Conduct Your Own Survey,* John Wiley & Sons, New York, 1994. A very complete book on how to do surveys to collect opinions.

Software packages

QFD/CAPTURE, version 2.2.1, International TechneGroup Inc. (ITI), Milford, Ohio. Software support for QFD diagrams. Fig. 6.3 was developed on this software.

QFD Designer, QS Software Corporation, Rochester Hills, Mich.

6.13 EXERCISES

6.1. For the original design problem (Exercise 4.1), develop a house of quality and supporting information for it. This must include the results of each step developed in this chapter. Make sure you have at least three types of customers and three benchmarks. Also, make a list of the ideas for your product that were generated during this exercise.

6.2. For the features of the redesign problem (Exercise 4.2) to be changed, develop a QFD matrix to assist in developing the engineering specifications. Use the current design as a benchmark. Are there other benchmarks? Be careful to identify the features needing change before spending too much time on this. The methods in Chap. 7 may be used iteratively to help refine the problem.

6.3. Develop a house of quality for the following:

(a) The controls on an electric mixer

(b) An all-terrain bicycle

(c) An attachment for electric drills to cut equilateral-triangle holes in wood. The wood can be up to 50 mm thick, and the holes must be adjustable from 20 mm to 60 mm per side.

(d) A tamper-proof fastener as used in public toilet facilities

CHAPTER 7

CONCEPT GENERATION

7.1 INTRODUCTION

In Chap. 6 we went to great lengths to understand the design problem and to develop its specifications and requirements. Now our goal is to use this understanding as a basis for generating concepts that will lead to a quality product. In doing this, we apply a simple philosophy: *Structure,* or *form, follows function.*

In some companies, design begins with a concept to be developed into a product. This is a weak philosophy and does not generally lead to quality designs. On the average, industry spends about 15 percent of design time developing concepts. Based on comparison of the companies in Fig. 1.4, this should be 20–25 percent to minimize changes later.

A concept is an idea that is sufficiently developed to evaluate the physical principles that govern its behavior. Confirming that the proposed product will operate as anticipated and that, with reasonable further development, it will meet the targets set is a primary goal in concept development. Concepts must also be refined enough to evaluate the technologies needed to realize them, to evaluate the basic architecture (i.e., form) of them, and, to some limited degree, to evaluate their manufacturability. Concepts can be represented in a rough sketch or flow diagram, a proof-of-concept prototype, a set of calculations, or textual notes—an abstraction of what might someday be a product. However a concept is represented, the key point is that enough detail must be developed to model performance so that the functionality of the idea can be ensured.

Some concepts are naturally generated during the engineering requirements development phase, since in order to understand the problem, we have to associate it with things we already know (see Chap. 3). There is a great tendency for designers to take their first idea and start to refine it toward a product design. This is a very weak methodology, best expressed by the adage, *If you generate one idea, it will probably be a poor idea; if you generate twenty ideas, you might have one good idea.* This statement and the methods in this chapter support one of the key features of concurrent engineering: generate multiple concepts. The main goal of this chapter, then, is to present techniques for the generation of many concepts.

The flow of conceptual design is shown in Fig. 7.1. Here, as with all problem solving, the generation of concepts is iterative with their evaluation. Also part of the iterative loop, as shown in the figure, is the communication of design information, the updating of the plans, and the decomposition of the problem into subproblems.

In line with our basic philosophy, the techniques we will look at here for generating design concepts encourage the consideration of the function of the device being designed. These techniques aid in decomposing the problem in a way that affords the greatest understanding of it and the greatest opportunity for creative solutions to it.

We will focus on techniques to help with *functional decomposition* and *concept variant generation.* These are based on the fact that many important customer requirements are concerned with the functional performance desired in the product. These requirements become the basis for the concept generation techniques. Functional decomposition is designed to further refine the functional requirements; concept variant generation aids in transforming the functions to concepts.

The techniques support a divergent-convergent design philosophy. This philosophy requires that a design problem be expanded into many solutions before it is narrowed to one final solution. The consideration of multiple configurations is part of the concurrent engineering method described in Chap. 1.

Before continuing, note that the techniques presented here are useful during the development of an entire system and also for each subsystem, component, and

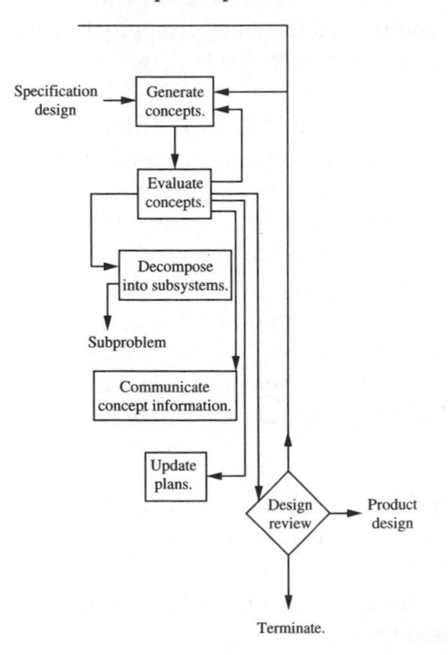

FIGURE 7.1
The conceptual design phase of the design process.

feature. This is not to say that the level of detail presented here needs to be undertaken for each flange, rib, or other detail; however, it is useful for difficult features.

The Splashgard problem will be used once again throughout the chapter to demonstrate various steps in the techniques. The results shown for this problem are actual concepts from the design team's notebooks and their final project documentation.

7.2 A TECHNIQUE FOR FUNCTIONAL DECOMPOSITION

In Chap. 2 we discussed how some design problems are naturally decomposed into the design of functionally independent subsystems. Although some problems can be easily decomposed, most require a structured effort. In this section we will develop a technique for decomposing any problem into smaller, more easily managed parts by working to understand the functions required of the device and treating each function as a separate subsystem.

In reading this section, it is important to remember that *function* tells *what* the product must do, whereas its *form,* or *structure,* conveys *how* the product will do it. The effort in this chapter is to develop the *what* and then map the *how.* This is similar to the QFD in Chap. 6, where *what* the customer required was mapped into *how* the requirements were to be measured.

Function can be described in terms of the logical flow of energy (including static forces), material, or information. For example, in order to attach any component to another, a person or mechanical assembler must *grasp* the component, *position* it, and *attach* it in place. These functions must be completed in a logical order: grasp, position, and then attach. In undertaking these actions, the human provides information and energy in controlling the movement of the component and in applying force to it. The three flows—energy, material, and information—are rarely independent of each other. For instance, the control and the energy supplied by the human cannot be separated. However, it is important to note that both are occurring and that both are supplied by the human.

The functions associated with the flow of energy can be classified both by the type of energy and by its action in the system. The types of energy normally identified with electromechanical systems are mechanical, electrical, fluid, and thermal. As these types of energies flow through the system, they are transformed, stored, transferred (conducted), supplied, and dissipated. These are the "actions" of the components or assemblies in the system. Thus, all terms used to describe the flow of energy are action words; this is characteristic of all descriptions of function. Also considered as part of the flow of energy is the flow of forces even when they do not result in motion. This concern for force flows is further developed in Section 10.2.

The functions associated with the flow of materials can be divided into three main types:

1. *Through-flow,* or material-conserving processes. Material is manipulated to change its position or shape. Some terms normally associated with through-flow are *position, lift, hold, support, move, translate, rotate,* and *guide.*
2. *Diverging flow,* or dividing the material into two or more bodies. Terms that describe diverging flow are *disassemble* and *separate.*

3. *Converging flow,* or assembling or joining materials. Terms that describe converging flow are *mix, attach,* and *position relative to.*

The functions associated with information flow can be in the form of mechanical signals, electrical signals, or software. Generally the information is used as part of an automatic control system or to interface with a human operator. For example, if you install a component with screws, after you tighten the screws you wiggle the component to see if it is really attached. Effectively you ask the question, Is the component attached? and the simple test confirms that it is. This is a common type of information flow. Software is used to modify information that flows through an electronic circuit—a computer chip—designed to be controlled by the code. Thus electrical signals transport information to and from the chip and the software transforms the information.

The goal of the functional modeling technique is to decompose the problem in terms of the flow of energy, material, and information. This forces a detailed understanding, at the beginning of the design project, of *what* the product-to-be is to do. The functional decomposition technique is very useful in the development of new products and in understanding already existing products. It can be used in benchmarking and redesign as well as original design problems.

There are four basic steps in applying the technique and several guidelines for successful decomposition. These steps are used iteratively and may be reordered as needed. This technique can be used iteratively with the QFD to help understand the problem. In the following discussion the usefulness of the technique will be demonstrated with the Splashgard to show its use with an original design problem and with the space shuttle aft field joint to show its use with a more complex redesign problem.

7.2.1 Step 1: Find the Overall Function That Needs to Be Accomplished

This is a good first step toward understanding the function. The goal here is to generate a single statement of the overall function on the basis of the customer requirements. All design problems have one "most important" function. This must be reduced to a simple clause and put in a *black box.* The inputs to this box are all the energy, material, and information that flow into the boundary of the system. The outputs are what flows out of the system. Thinking about these flows forces definition of the boundary of the product being designed. In the diagrams that follow, energy flow is noted by a thin line, ________, material flow by a thick line, ▬▬▬▬ and information flow by a dotted line, ...

Guideline: System boundaries must be clearly identified. It is important to establish at the beginning exactly what the system boundary is to be. There is a tendency to limit the problem too tightly and say that other functions are "not my problem." The other extreme is the desire to design too much, to fix all the problems in the larger system.

Guideline: Energy must be conserved. Whatever energy goes into the system must come out or be stored in the system.

Guideline: Material must be conserved. Materials that pass through the system boundary must, like energy, be conserved.

Guideline: All interfacing objects and known, fixed parts of the system must be identified. It is important to list all the objects that interact, or interface, with the system. Here, objects include all features, components, assemblies, humans, or elements of nature that exchange energy, material, or information with the system being designed. These objects may also constrain the system's size, shape, weight, color, etc. Further, there are some objects that are part of the system being designed that cannot be changed or modified. These too must be listed at the beginning of the design process.

Guideline: When adding information flows to the diagram, ask the question, How will I know if the system is performing? Answers to this question will help identify signals that are important.

Finding the overall function: The Splashgard example. For the bicycle Splashgard, the "most important" function is to "protect rider from water and dirt off the rear wheel," so this is entered in the box shown in Fig. 7.2. It is not clear yet what energy may be used by the system being designed; however, the kinetic energy of the water and the bike, gravity, and the rider energy are available. These are noted next to thin lines entering the system.

The material flow into the system is the water potentially mixed with dirt. The water and dirt also have to flow out of the system unless they are stored somewhere, an interesting yet unlikely idea. Also shown on the top of the box are other objects that affect the Splashgard: the rider and the bike. These two objects interact with the system, rather than flow through it, and thus are shown having an influence on it.

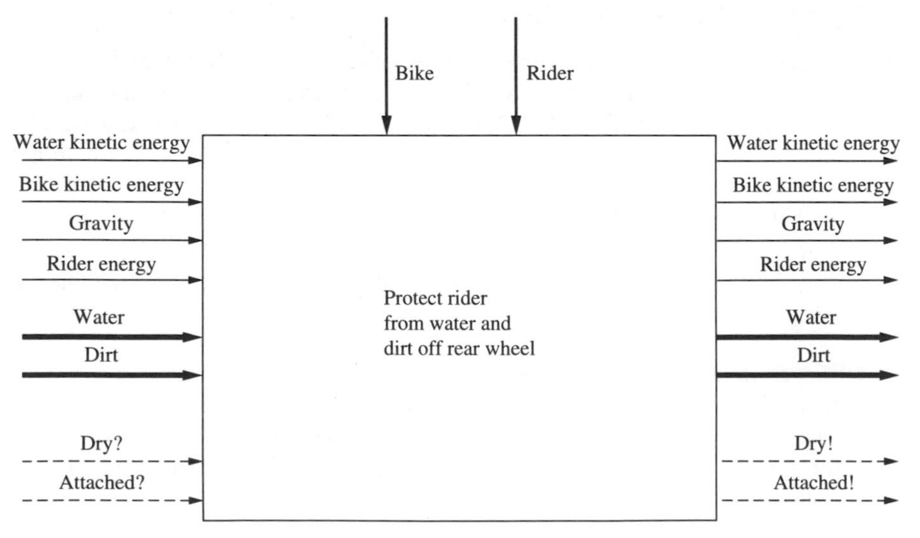

FIGURE 7.2
Splashgard black box.

Finally, the information modified by the system is identified. Many products actively modify the information that passes through them; the Splashgard does not. However, reviewing the customer's requirements developed in the previous chapter reveals information about the Splashgard that the user senses: whether the Splashgard keeps the rider dry, and whether it is attached. The question marks after "dry" and "attached" on the input side of Fig. 7.2 signify questions that must be answered for the system to meet the design requirements. They are shown answered on the output side of the diagram.

Finding the overall function: The space shuttle *Challenger* aft field joint example. Here the problem is to redesign the joint between the aft mid segment and aft segment (see Fig. 4.4). It will be shown through this example that for the most part, *function occurs at the interfaces between components.* The main function of the joint is to seal the segments together. It is the subfunction decomposition of this main function that adds to an understanding of the joint.

The objects involved in this example are, as seen in Fig. 4.9, the aft segment tang, aft midsection clevis, front O-ring, rear O-ring, steel pin, insulation, putty, and propellant. Some of these, like the O-rings, the clevis and the tang, may be redesigned; others, like the propellant, surely will not.

7.2.2 Step 2: Create Subfunction Descriptions

The goal of this step is to decompose the overall function in the black box. When this step is complete, we will have simple descriptions of the subfunctions that need to be accomplished for the product to meet the requirements. The subfunctions will be in terms of verb-noun-modifier combinations. Note that the overall function in the black box in Fig. 7.2 is of this form: "protect [verb] rider [noun] from water and dirt off the rear wheel [modifier]." This step focuses on identifying the subfunctions needed, and the next step concerns their organization.

There are three reasons for decomposing the overall function: First, the resulting decomposition controls the search for solutions to the design problem. Since concepts follow function and products follow concepts, we must fully understand the function before wasting time generating products that solve the wrong problem.

Second, the division into finer functional detail leads to a better understanding of the design problem. Although all this detail work sounds counter to creativity, most good ideas come from fully understanding the functional needs of the design problem. Since it improves understanding, sometimes it is useful to begin this process before the QFD process in Chap. 6 is complete and use the functional development to help determine the engineering specifications.

Finally, breaking down the functions of the design may lead to the realization that there are some already existing components that can provide some of the functionality required.

The following guidelines for accomplishing the decomposition will help ensure usable results. Each subfunction developed will ultimately have the following components:

- A box
- An action verb

- The object(s) on which the verb acts
- Possibly a modifier giving details of the function
- Known flows of materials, energy, and control

It may take several iterations to finalize all this information.

The subfunction boxes can be drawn as block diagrams as in step 1 on a single sheet of paper, or better yet, it is suggested that each subfunction be written on a piece of self-stick removable note paper. The reason for this will become clear. After the guidelines below, the examples will help clarify this very important step.

Guideline: Consider *what,* not *how.* It is imperative that only *what* needs to happen—the function—be considered. Detailed, structure-oriented *how* considerations must be suppressed as they add detail too soon. Even though we remember functions by their physical embodiments, it is important that we try to abstract this information. If, in a specific problem solution, it is not possible to proceed without some basic assumptions about the form or structure of the device, then document the assumptions. A way to control this is to use only nouns previously used (e.g., in the QFD or in step 1 above) to describe the material flow or interfacing objects. If any other nouns are used during this step, either something is missing in the first step (go back to step 1 and reformulate the overall function) or a design decision to add another object to the system has been made (consider very carefully, as in the Splashgard example below).

Guideline: Break the function down as finely as possible. This is best done by starting with the overall function of the design and breaking it into the separate functions. Let each function represent a change or transformation in the flow of material, energy, or information. Action verbs often used in this activity are given in Table 7.1. Some are shown as opposing pairs. Obviously, many other action verbs can be used in developing the functional breakdown; the list contains only those most commonly used.

TABLE 7.1
Typical mechanical design functions

Absorb/remove	Dissipate	Release
Actuate	Drive	Rectify
Amplify	Hold or fasten	Rotate
Assemble/disassemble	Increase/decrease	Secure
Change	Interrupt	Shield
Channel or guide	Join/separate	Start/stop
Clear or avoid	Lift	Steer
Collect	Limit	Store
Conduct	Locate	Supply
Control	Move	Support
Convert	Orient	Transform
Couple/interrupt	Position	Translate
Direct	Protect	Verify

Guideline: List all alternative functions. This step helps develop alternative functions. Some of these are just different words to describe the same activity; others are truly different actions. Condense those that are the same and note those that are different.

Guideline: Include all input and output energy, materials, and information. For each function show the input energy, materials, and information that are affected by the function or are needed to enable it. This is similar to the technique used in Section 5.4 to support planning. It is best if the flows that are affected are shown as input on the left side and output on the right side. The state of each of the flows should also be noted. For example, if water flows through a function that is defined as "absorb energy" then the temperature of the input T_i and that of the output T_o should be labeled because the temperature is a measure of the change of the state of the flowing material. The enablers and other resources are drawn as inputs to the function from above and below.

Guideline: Consider all operational sequences. A product may have more than one use (see Fig. 1.7), so there may be many operating sequences. The functions of the device may be different during each of these. Additionally, prior to the actual *use* there may be some *preparation* that must be modeled, and similarly, after use there may be some *conclusion*. It is often effective to think of each function in terms of its preparation, use, and conclusion.

Guideline: Use standard notation when possible. For some types of systems there are well-established methods for building functional block diagrams. Common notation schemes exist for electrical circuits and piping systems, and block diagrams are used to represent transfer functions in systems dynamics and control. Use these notations if possible.

Guideline: Developing subfunction ideas can be aided by reviewing customer requirements, visualizing the flows, reviewing the list of verbs, or acting the role of the product. The goal is to note as many functions as possible so each of these can later be used to generate concepts. Visualizing flows and role playing will be further discussed in the next step.

Guideline: For redesign problems or to understand benchmarks, disassemble a sample to find subfunctions. If working with a sample of an existing device, try the following:

1. Disassemble the product, one component at a time.
2. As each component is removed, note what objects—features, components, assemblies, humans, or other—it interfaces to and list the flow of materials, energy, and information between them. Each component must conserve energy and materials and be in force equilibrium.
3. For each change in energy, materials, or information, draw a function box showing the inputs and outputs. This box should also show the components that interact to provide the function.

Creating subfunction description: The Splashgard example. The logic of the design team on this step is as follows. First, they decided that they had to consider the functions involved during three operational steps: when the Splashgard is being put onto the bike (preparation), when it is protecting the rider (use), and when it is being taken off the bike (conclusion).

Second, they then thought of all the functions they could in a brainstorming session (see techniques for generating ideas in Section 7.4). These are shown on the self-stick removable notes in Fig. 7.3. During this activity there was much discussion among the team members and much simulation that they carried out by waving their hands in the air (i.e., air-design). Notice that all the functions are noun/verb pairs with some use of modifiers. Some of the verbs are from the list in Table 7.1, and some are not; all are action verbs. Notice that the nouns are all from Fig. 7.2 with the exception of "Splashgard" and "box." The "Splashgard" is assumed to be an object that is to be designed. The "box" is a new object, and if not eliminated, its inclusion in this problem shows that the team now assume that the Splashgard will come in a box.

Creating subfunction description: The space shuttle *Challenger* aft field joint example. In this example the decomposition of the function of the joint greatly aids in its understanding. Here it is easiest to decompose the function in a tree-like structure. Often, for redesign problems, a tree is more effective than the sticky note paper used for the Splashgard. Part of the resulting tree for the aft field joint is shown in Fig. 7.4. Note that the goal is to decompose the function of the joint as finely as possible using action verbs to relate interactions among the objects that were identified in step 1.

The first level of decomposition takes into account the preparation, use, and conclusion of the function of the joint. For preparation the verb *enable* has been used. This is not a strong action verb, nor is *allow,* as neither of them can be related to transfer or

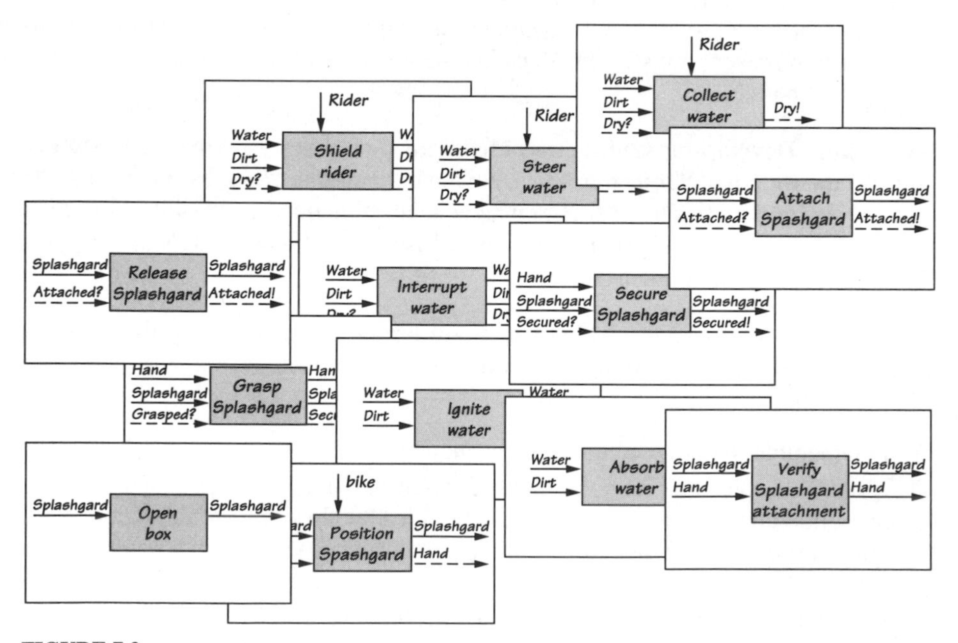

FIGURE 7.3
Functions for the Splashgard.

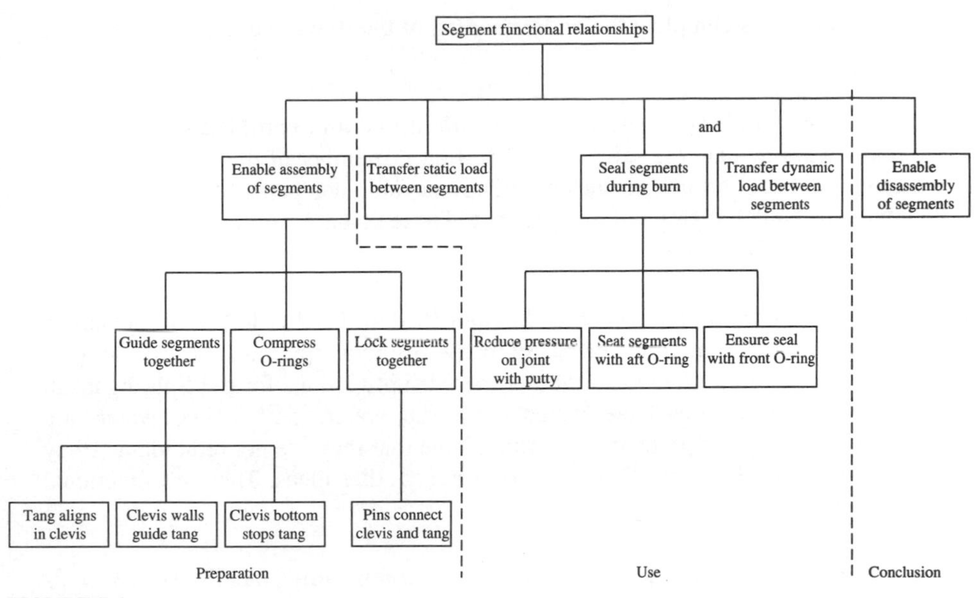

FIGURE 7.4
The space shuttle aft field joint functional decomposition.

modification of energy, material, or information. It would have been better to just say, "Assemble the segments."

After the segments are assembled, the joint must support the static load (effectively a transfer of energy). By thinking of the sequence of events in the life of the joint, we can identify nonobvious, important functions. Focusing on the sequence is covered in step 3; as in all the techniques, there is much iteration and interdependence between techniques and the steps within them.

When the rocket is fired, the joint must seal and transfer dynamic loads. Note that the parallel nature of these two functions is shown in the diagram.

These first-level functions can be further decomposed, as shown in the figure and discussed in step 4.

7.2.3 Step 3: Order the Subfunctions

The goal is to add order to the functions generated in the previous step. For many redesign problems this occurs simultaneously with their identification in step 2, but for some processing systems this is a major step. The goal here is to order the functions found in step 2 to accomplish the overall function in step 1. The guidelines and examples below should help with this step.

Guideline: The flows must be in logical or temporal order. The operation of the system being designed must happen in a logical manner or in a time sequence. This sequence can be determined by rearranging the subfunctions. First arrange them in independent groups (preparation, uses, and conclusion). Then arrange them within each group so that the output of one function is the input of another. This helps complete the understanding of the flows and helps find missing functions.

Guideline: Redundant functions must be identified and combined. Often there are many ways to state the same function. If each member of the design team has written his or her subfunctions on self-stick removable notepaper, all the pieces can be put on the wall and grouped by similarity. Those that are similar need to be combined into one subfunction.

Guideline: Functional choices must be identified and either left as an option or a decision made. Sometimes there are many different subfunctions that can provide the same overall function. In the proposed subfunctions for the Splashgard in Fig. 7.3, the subfunctions "steer water," "interrupt water," and "collect water" are three different ways of protecting the rider. Note that they are not redundant—they still tell *what,* not *how,* and they do lead to many other ideas. They are functional options for the product.

Guideline: Functions not within the system boundary must be eliminated. This step helps the team come to mutual agreement on the exact system boundaries; it is often not as simple as it sounds.

Guideline: Energy and material must be conserved as they flow through the system. Match inputs and outputs to the functional decomposition. Inputs to each function must match the outputs of the previous function. The inputs and outputs represent energy, material, or information. Thus the flow between functions conveys the energy, material, or information without change or transformation.

Guideline: For redesign or benchmarking problems the flow can be established during the disassembly in the previous step.

Ordering the subfunctions: The Splashgard example. When the team rearranged the pieces of paper with the functions written on them (Fig. 7.3), Fig. 7.5 was developed. Their logic for this arrangement was as follows:

1. "Open box to get Splashgard" is not within the system boundaries and was eliminated. Also, "box" is not one of the original objects.
2. The functions "shield rider" and "interrupt water" are the same, and "shield rider" seems to best describe the needed function.
3. "Shield rider," "steer water," and "collect water" are alternatives for the same function. Realizing this, the team thought of other alternatives for "what can you do with water." This is part of concept generation, the topic of the next section of the book, and will be discussed in detail there. For now, the example will proceed with "shield rider" for this subfunction.
4. Three independent flows—attachment (preparation), use, and detachment (conclusion)—are treated separately.

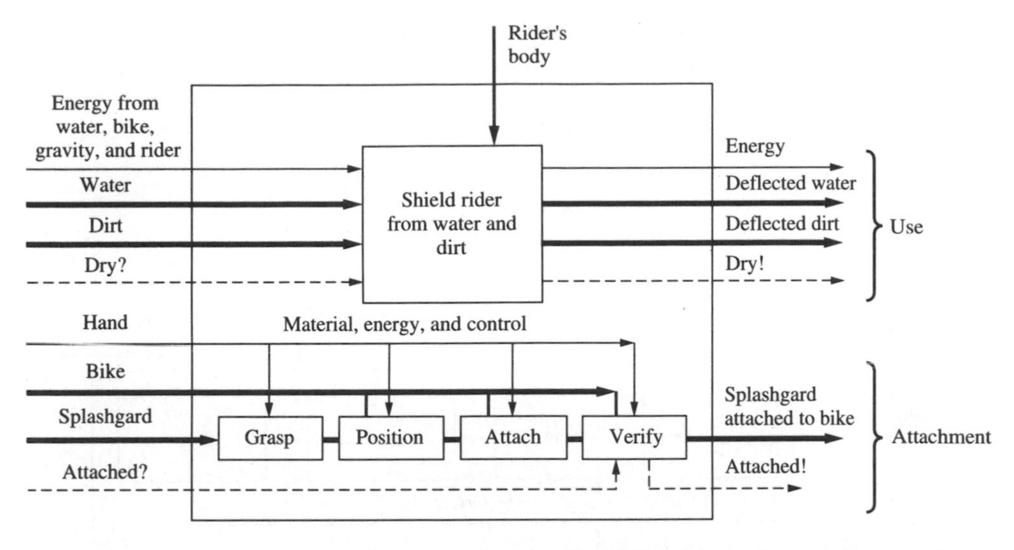

FIGURE 7.5
Refined functional diagram for the Splashgard.

Notice that the description in Fig. 7.5 could apply to a raincoat, a fender, or a number of other products. The team have made no commitment to *how* the Splashgard does the job; they only worked to understand *what* it needs to accomplish.

Ordering the subfunctions: The space shuttle *Challenger* aft field joint example. The tree diagram developed in the previous step (Fig. 7.4) shows the ordering.

7.2.4 Step 4: Refine Subfunctions

The goal is to decompose the subfunction structure as finely as possible. This means examining each subfunction to see if it can be further divided into sub-subfunctions. This decomposition is continued until one of two things happens: "atomic" functions are developed or new objects are needed for further refinement. The term atomic implies that the function can be fulfilled by existing objects. However, if new objects are needed, then you want to stop refining because new objects require commitment to how the function will be achieved, not refinement of what the function is to be. Each noun used represents an object or a feature of an object.

Further refinement of the subfunctions: The Splashgard example. As an example, the function "position" from Fig. 7.5 is further refined as shown in Fig. 7.6. In (*a*) the function is shown in more detail and in (*b*) it is further refined. The hand is shown as material flow because it must move with the Splashgard to position it. The hand also provides energy and control for positioning. These are assumptions that will now affect many downstream decisions. Further, information about whether or not the Splashgard is positioned has been added as a refinement.

What does it mean to "position" the Splashgard? As shown in (*b*), there are, according to the design team, four steps to positioning. These four steps are not the

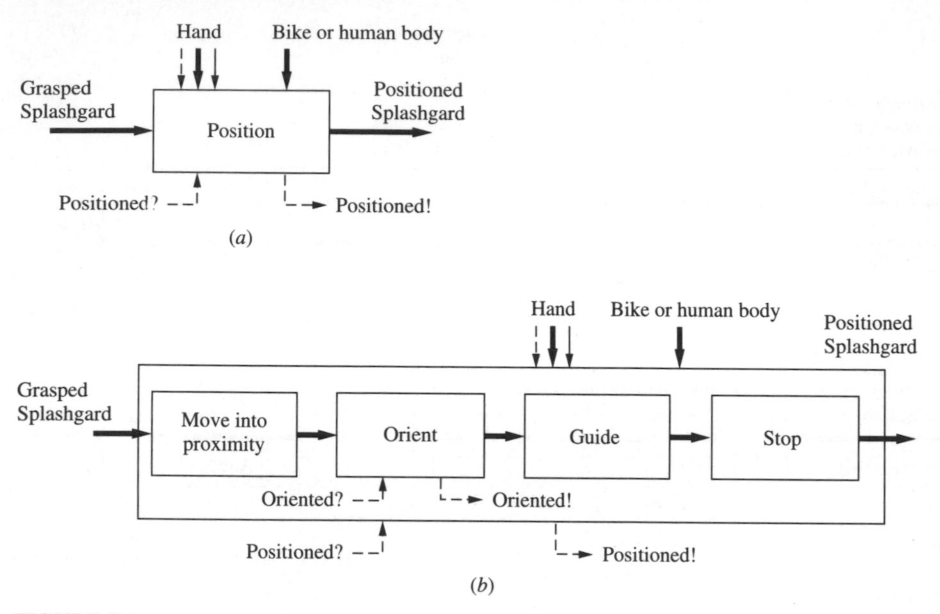

FIGURE 7.6
Further refinement of the position function.

only decomposition possible, nor are the verbs used the only ones possible. This figure is just what the design team developed.

Although decomposition this fine may seem like it is going too far, the breakdown forces thinking about what makes an object easy to position relative to another. Obviously the two objects must be moved into proximity with each other. Then they must be oriented or aligned (an alternative verb for the function). Will the user be able to easily move and align the Splashgard? How will the user know when these functions are accomplished? The realization that these questions exist at this point in the design process is very important, because the design can proceed with knowledge that the product must provide these functions.

Quality designs have built-in guides and stops. Thus, part of positioning the Splashgard ready for attachment is to guide it into place and stop it when it gets there.

Further refining the subfunctions: The space shuttle *Challenger* aft field joint example. As shown in Fig. 7.4, two of the functions are further refined. An approach to decomposition is to ask, What has to happen to "enable assembly of segments?" Continuing to ask, What has to happen? until new objects are needed to answer the question will help functional decomposition. This is the same philosophy as asking Why? in developing the customers' requirements.

As shown in the figure, at least two levels of decomposition are easily realizable, as can be seen by considering the functions of the features of the clevis and tang that help "align," "guide," and "stop" the segments. After the decomposition is finished, it could be used as a basis for developing ideas to replace the clevis and tang (see Section 7.3). The function of transferring the loads between segments is further refined in Section 10.2.

It must be realized that the function decomposition cannot be generated in one pass and that it is a struggle to develop the suggested diagrams. However, it is a fact that the design can be only as good as the understanding of the functions required by the problem. This exercise is both the first step in developing ideas for solutions and another step in understanding the problem. The functional decomposition diagrams are intended to be updated and refined as the design progresses.

7.3 A TECHNIQUE FOR GENERATING CONCEPTS FROM FUNCTIONS

Once the functions of the product are understood, the next goal is to generate concepts that satisfy them. *Concepts are the means of providing function.* Concepts can be represented as sketches, block diagrams, textual descriptions, clay or paper models, or other forms that give some indication of the manner in which the functions are achieved with the least commitment.

The technique presented here uses the functions identified above to foster ideas. There are two steps to this technique. The goal of the first is to find as many concepts as possible that can provide each function identified in the decomposition. The second is to combine these individual concepts into overall concepts that meet all the functional requirements. The design engineer's knowledge and creativity are crucial here, as the ideas generated are the basis for the remainder of the design evolution. This technique is often called the "morphological technique," and the resulting table a "morphology," which means "a study of form or structure."

7.3.1 Step 1: Developing Concepts for Each Function

The goal of this first step is to generate as many concepts as possible for each of the functions identified in the decomposition. There are two activities here that are similar to each other. First, for each function develop as many alternative functions as possible. This was begun for the Splashgard example with "interrupt water," "collect water," and "steer water" all being optional ways of providing the needed function. Second, for each subfunction the goal is to develop as many means of accomplishing the function as possible. For example, in Fig. 7.6 one of the functions that must be performed in attaching the Splashgard is to guide it onto the bike or human body. Common ways of mechanically guiding one object onto another involve the use of a pin, a wall, a rail, a slot, or a hole. Two points can be made about this list. First, there are other ways to guide objects. Wild, impractical ideas often lead to good ideas (see Section 7.4.4). We could use electrical or magnetic fields or a laser beam with an optical sensor and an active control system. These ideas are not practical for this product, so they do not appear on the list of possibilities. Second, the concepts on the list are all abstract in that they have no specific geometry. Rough sketches of these concepts would be simple blocks and cylinders.

If there is a function for which there is only one conceptual idea, this function should be reexamined. There are few functions that can be fulfilled by only one concept. The following situations explain the lack of more concepts.

First, the designer has made a fundamental assumption without realizing it. For example, one function that has to occur in positioning the Splashgard is "move into proximity" (Fig. 7.7). In developing the concepts, it was realized that the only logical way to move the Splashgard into position was by hand. (Robots and cranes seemed extreme.) Here an assumption was made about the operation of attaching the Splashgard. It is reasonable to make an assumption such as this, provided there is an awareness that an assumption has been made.

Second, the function is directed at *how*, not *what*. If one idea gets built into a function, the true function of the design must then be reconsidered. For example, if "guide" in Fig. 7.6 had been stated as "move along rail," it should come as no surprise that the only concept for this function would be to have the Splashgard's direction controlled by a rail.

Finally, domain knowledge is limited. In this case, help is needed to develop other ideas. (See Section 7.4.)

It is a good idea to keep the concepts as abstract as possible and at the same level of abstraction. Suppose one of the functions is to move some object. Moving requires a force applied in a certain direction. The force can be provided by a hydraulic piston, a linear electric motor, the impact of another object, or magnetic repulsion. The problem with this list of concepts is that they are at different levels of abstraction. The first two refer to fairly refined mechanical components. (They could be even more refined if we had specific dimensions or manufacturers' model numbers.) The last two are basic physical principles. It would be difficult to compare these concepts because of this difference in level of abstraction. We could begin to correct this situation by abstracting the first item, the hydraulic piston. We could cite instead the use of fluid pressure, a more general concept. Then again, air might be better than hydraulic fluid for the purpose, and we would also have to consider the other forms of fluid components that might give more usable forces than a piston. We could refine the "impact of another object" by developing how it will provide the impact force and what the object is that is providing the force. Regardless of what is changed, it is important that all concepts be equally refined.

Developing concepts for each function: The Splashgard example. In Table 7.2 concepts for each function of the bicycle Splashgard are listed. The ideas listed in this table came from the design team's knowledge and creativity. Overall concepts for the form of the Splashgard later came from this concept list.

7.3.2 Step 2: Combining Concepts

The result of applying step 1 is a list of concepts generated for each of the functions. Now we need to combine the individual concepts into complete conceptual designs. The method here is to select one concept for each function and combine those selected into a single design. So, for example, in the Splashgard design we may choose to have a device between the rider and the wheel that is to be grasped in two hands, oriented using a wall, guided into place using a pin, and secured with a suction cup. There are pitfalls to this method, however.

First, if followed literally, this method generates too many ideas. The Splashgard design team generated five concepts for the guide function and nine concepts

TABLE 7.2
Morphology for the Splashgard

Attach						
Grasp	Contact	One finger	Thumb and finger	Fist	Two hands	
	Hold	Hole	Friction	Post	Grip around device	
Position	Move	Assume all by hand(s) of user				
	Orient	Pin	Wall(s)	Depression		
	Guide	Pin	Wall(s)	Rail	Slot	Hole
	Stop	Interfere	Bottom out	Jam		
Secure		Screw	Suction cup	Snap	Spring clip	Over-center linkage
		Velcro	Pin	Magnet	Elastic rope	
Verify		Resistance to pull	Noise	Color	Visual gap	
Hold		*See* Secure				
Detach	Grasp	*See* Attach				
	Release	*See* Attach				
	Guide	*See* Attach				
	Move	*See* Attach				
Deflect water		Device on rider	Device between rider and wheel	Device next to wheel, like fender	Scrape water from wheel	Keep water from getting on wheel

for the secure function. These two functions alone could combine to yield 45 possible designs for guiding and securing. For the entire example there would be thousands of possibilities.

The second problem with this method is that it erroneously assumes that each function of the design is independent and that each concept satisfies only one function. Generally, this is not the case. For example, if a pin is used to orient the Splashgard, then it is at least reasonable, for a first try, to consider using the same pin to guide the Splashgard into place. These two functions are not really mutually dependent, but as overall ideas are generated, often the same concept can satisfy more than one function.

Third, the results may not make any sense. Although the method is a technique for generating ideas, it also encourages a coarse ongoing evaluation of the ideas. Still, care must be taken not to eliminate concepts too readily; a good idea could conceivably be prematurely lost in a cursory evaluation. A goal here is to do only a coarse evaluation and generate all the ideas that are reasonably possible. In the next chapter we will evaluate the concepts and decide between them.

Even though the concepts developed here may be quite abstract, this is the time that back-of-the-envelope sketches begin to be useful. Prior to this time, most of the design effort has been in terms of text, not graphics. Now the design is developing to the point that rough sketches can be drawn. Sketches of even the most abstract concepts are increasingly useful from this point on: (1) As discussed in Chap. 3, we remember functions by their forms; thus our index to function is form. (2) The only way to design an object with any complexity is to use sketches to extend the short-term memory. (3) Sketches made in the design notebook provide a clear record of the development of the concept and the product.

Keep in mind that the goal is only to develop concepts and that effort must not be wasted worrying about details. Often a single-view sketch is satisfactory; if a three-view drawing is needed, a single isometric view may be sufficient.

Combining concepts: The Splashgard example. Fig. 7.7 shows selections from the notebooks of the members of the Splashgard design team. The seven concepts shown are a good sample of the many ideas developed.

- Concept I is a simple poncho or raincoat. Even though this idea is included as a benchmark, the design team also wanted to include it in concept development.
- Concept II is a device between the rider and the wheel that attaches to the seat post or seat. The team realized during conceptual design that the water-deflection function could be treated virtually independently from the attach-to-bike function.

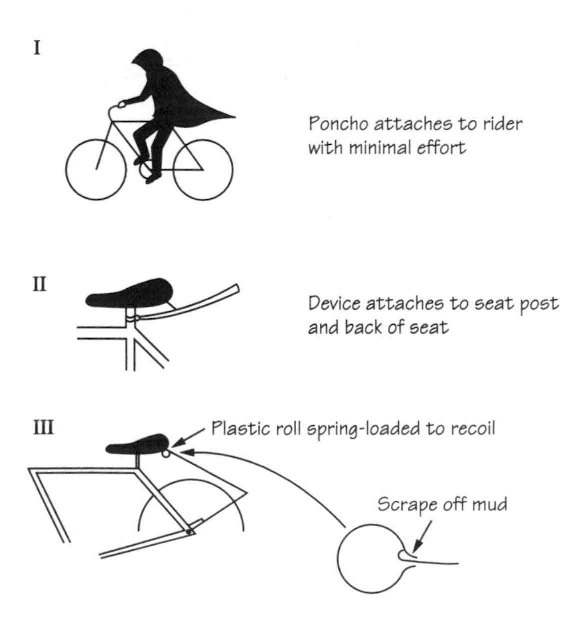

FIGURE 7.7
Concepts sketched in the Splashgard design team's notebooks.

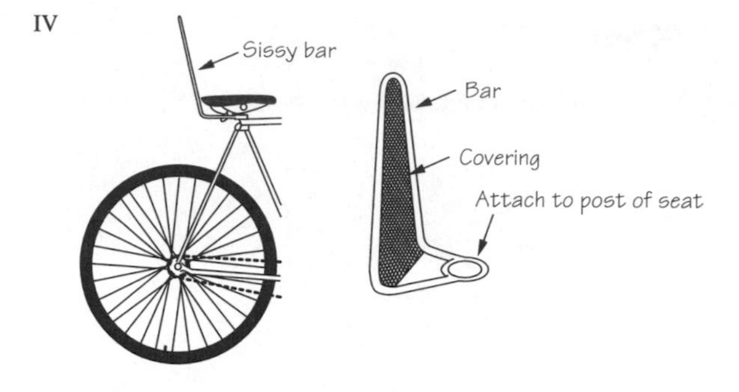

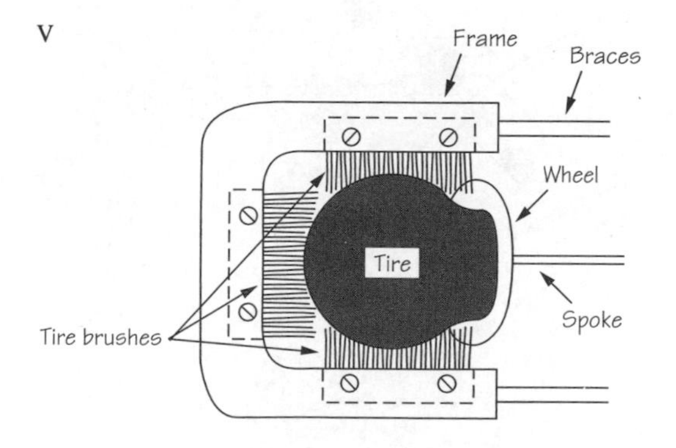

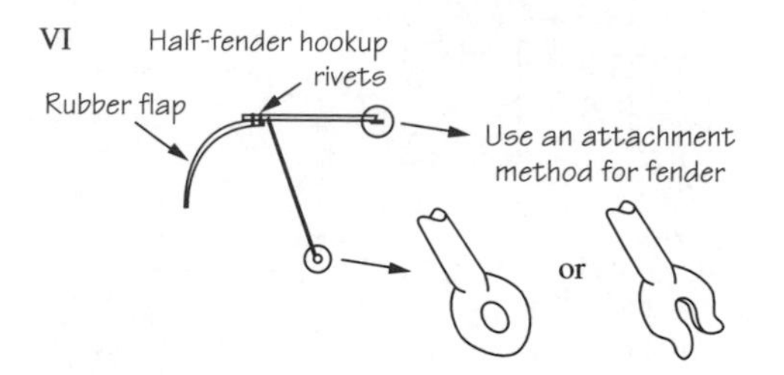

FIGURE 7.7 (continued)
Concepts sketched in the Splashgard design team's notebooks.

This allowed the problem to be decomposed into two separate subproblems. If the project were a larger one, a new design project might then have been initiated for each subproblem. (The attachment subproblem is further developed in the sketches for concept VII.)

- Concept III is a plastic window-shade device, spring-loaded to roll up in a tube when not in use. The tube scrapes off the mud and water as the Splashgard is rolled up inside it.
- Concept IV, the sissy bar, is another device between the rider and the wheel.
- Concept V is a device to scrape or brush the water from the wheel.
- Concept VI combines a rack for carrying baggage and a fender.

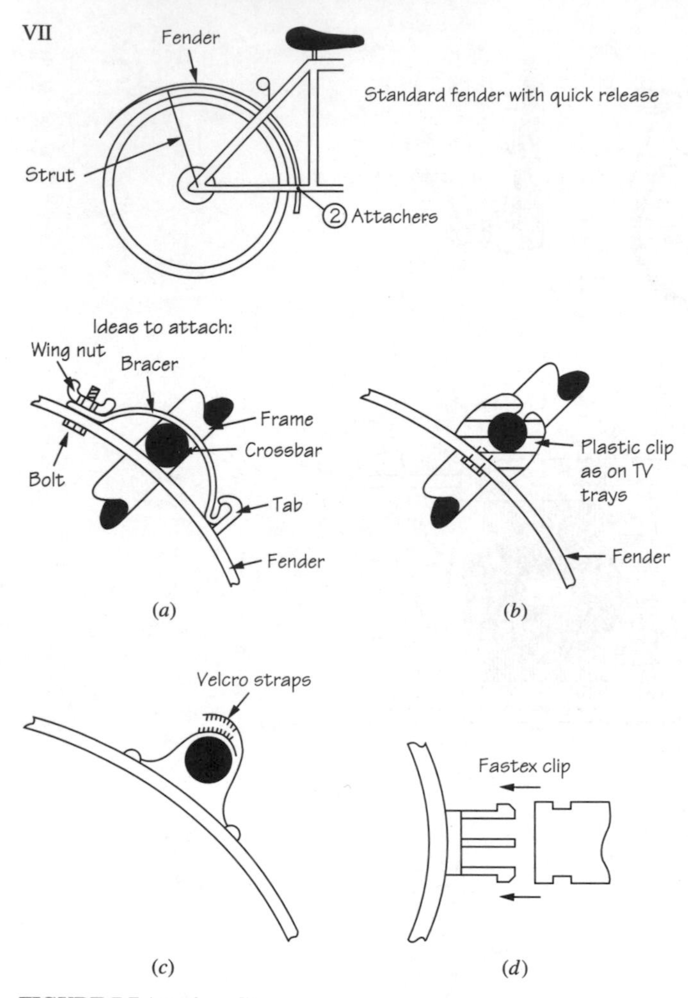

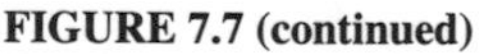

FIGURE 7.7 (continued)
Concepts sketched in the Splashgard design team's notebooks.

- Concept VII is a standard fender. All variations of the concept—(*a*) through (*d*)—pertain to the method by which the fender could be attached to the bike. Note that concepts VIIb and VIId have a very distinct method for orienting, guiding, and stopping the attachment; the verification would be through a "tug" or a "snapping" sound. Concepts VIIa and VIIc have no designed-in method for guiding or stopping the Splashgard. If these methods had been chosen for further refinement, these two functions would have had to be added.

In reconsidering this list, the design team saw that concepts II, IV, V, and VII, could all be decomposed into a water-deflection system and an attachment system. Having already developed a good understanding of each of these subsystems through problem appraisal and functional modeling, they did not need to repeat these techniques for each of these new problems.

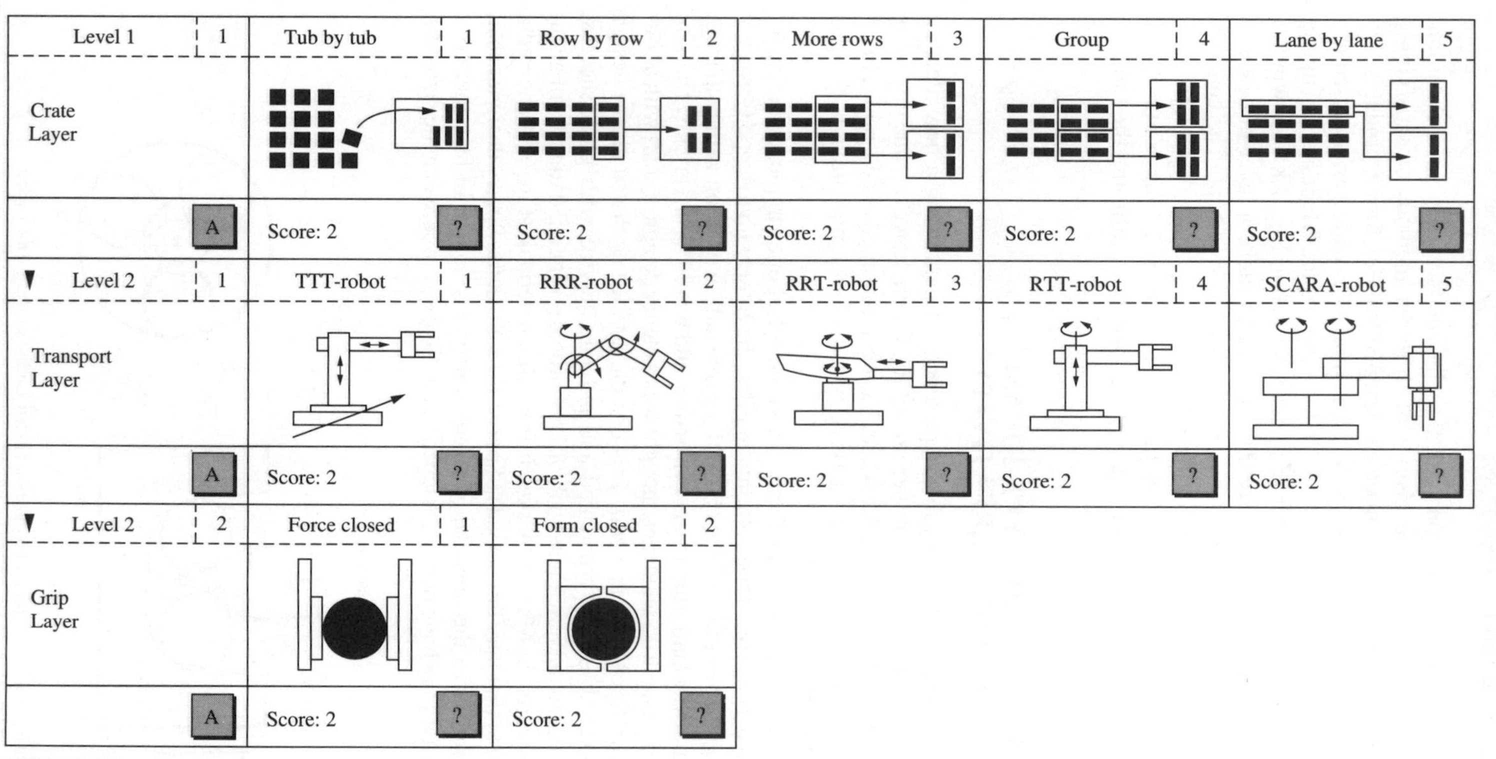

FIGURE 7.8
Example of a morphology.

Although simplistic, this morphological technique is widely used. Additionally, it can also be used to keep a history of different methods of satisfying a function. This history serves as a source of design ideas for future products. For example, Unilever Ltd., a multinational manufacturer of a wide variety of consumer and grocery products has a computerized system, called Modessa, to support engineers. A sample window from this program is shown in Fig. 7.8 This window supports an engineer developing a system to create and move a layer of boxes of ice cream bars prior to putting them into a larger box for shipment. The concepts for each function are ones that have been used in previous assembly line products developed at Unilever. Selecting an icon such as "form closed," for example, will yield information on previous uses of this type of concept in fulfilling the function "grip layer."

7.4 SOURCES FOR CONCEPT IDEAS: TECHNIQUES TO PRIME CREATIVITY

Dreaming up new ideas is an enjoyable experience. In generating a new idea, an engineer can take great pride in his or her creativity. Often, however, one discovers that the idea is not original after all; the wheel has been invented once again. This is not surprising, since there are no good indexing systems for finding previous designs categorized by function.

It is impossible to know about all previous concepts. For example, in the 1920s, while designing a gyroscope for an automatic pilot, the Sperry Gyroscope Company needed a bearing that would hold the end of the gyro shaft in position with both axial and lateral accuracy, would support the gyroscope, and would have very low friction. The designers came up with what they thought was a clever design, a shaft with a conical end riding on three balls in a cup. This one clever idea achieved all the design functions; it was subsequently patented and put into service with great success. In 1965 a previously unknown notebook of Leonardo da Vinci's, dating from about 1500, was discovered in Madrid. A sketch from this notebook (Fig. 7.9) showed a design identical to that later developed at Sperry. Of course, the Sperry designers had no way of knowing that the idea had been previously developed. In fact, there is a good chance that it may have been developed many times over between the sixteenth

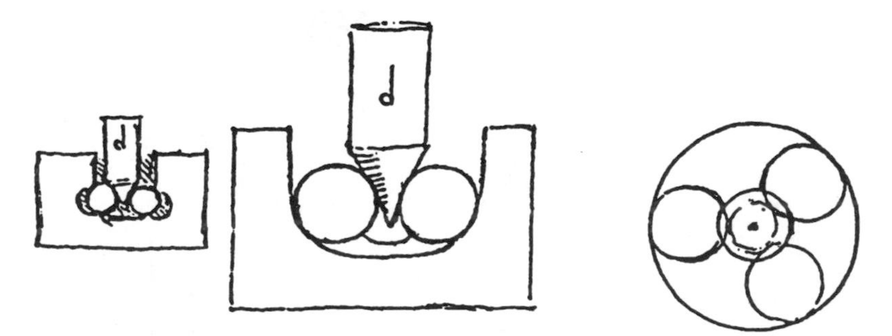

FIGURE 7.9
A low-friction bearing from da Vinci's notebooks. (From Ladislao Reti, *The Unknown Leonardo,* McGraw-Hill, New York, 1974, p. 285.)

and twentieth centuries and not recorded in any fashion. The point is that every effort must be made to find design ideas that have been previously developed; the problem is that most previous ideas are not recorded by function during conceptual design.

What follows in this section is a list of useful sources of design information that might keep a designer from reinventing the bearing. A majority of these sources refer to products that are already embodied and can therefore influence the concepts being generated. Unfortunately, there is no good way to adequately represent design concepts other than graphically, by their form. A good designer will abstract the function of a design from the form, utilize the important aspects, and discard the rest.

7.4.1 Using Patents as an Idea Source

Patent literature is a good source of ideas. It is relatively easy to find patents on just about any subject imaginable. Some of the problems in using it are that it is hard to find exactly what you want in the literature; it is easy to find other, interesting, distracting things not related to the problem at hand; and patents are not very easy to read.

There are two main types of patents: *utility patents* and *design patents.* The term *utility* is effectively synonymous with *function,* so the claims in a utility patent are about how an idea operates or is used. Design patents only cover the look or form of the idea, so here the term *design* is used in the visual sense. Design patents are not very strong, as a slight change in the form of a device that makes it look different is considered to bring about a different product. Utility patents are very powerful, because they cover how the device works, not how it looks.

There are over five million utility patents, each with many diagrams and each having diverse claims. To cull these to a reasonable number, a *patent search* must be performed. That is, all the patents that relate to a certain utility must be found. Any individual can do this, but it is best accomplished by a professional familiar with the literature.

The easiest way to do patent searches is on CASSIS (Classification and Search Support Information System), a computer index to the patent numbers. This system can help find the classification numbers for a specific area of interest and search for all patents issued under these numbers. It will output the title and, sometimes, a brief abstract of each patent. Most technical libraries have CASSIS available.

Whether using paper indices or the computer system, a patent search requires the following three steps.

STEP 1: IDENTIFY THE CLASS OR SUBCLASS OF THE AREA OF INTEREST. All patents are organized by their class and subclass numbers. This step entails finding the class or subclass that best represents the area.

The classes and subclasses are listed alphabetically in the *Index to Classification* available in most technical libraries. Unfortunately, use of this document usually turns up a number of potential classes of interest, since the classification system is not precise and a given device might be classified in any number of ways. For example, in studying the *Index to Classification* to find the class for the Splash-gard, the design team found the following:

	Class	Subclass
Bicycle	280	200
⋮		
Wheel	301	5R
Guards	280	160.1
Scrapers and cleaners .	280	158.1

The 280 class with a 160.1 subclass seemed the most promising.

To help refine the classes, the *Manual of Classification* or the *Classification Definitions* can help. The *Manual* is a numerical list of the classes and subclasses and gives slightly more detail than the *Index.* To find a full definition of a class or subclass, the *Classification Definitions* is needed. Looking in the *Manual of Classification* under class 280 yielded the following for the Splashgard team (out of more than 700 subclasses in class 280):

CLASS 280 LAND VEHICLES
WHEELED
. Attachments
⋮

Subclass
847 . . Dust and mud guards
152.05 . . . With wheel or tire carrying means
152.1 . . . Velocipede type
152.2 . . . . Combined with wheel guards
152.3 . . . . Flexible or sectional
⋮
848 . . . Body attached
154 . . . . Securing device
⋮
160 . . Wheel guards
160.1 . . . Velocipede type

The 280/154 class looked even more promising than 160.1 for ideas for attaching the Splashgard to the bike.

STEP 2: FIND THE NUMBERS OF THE SPECIFIC PATENTS THAT HAVE BEEN FILED IN THE CLASSES IDENTIFIED. Two patents from CASSIS found during the search for ideas for the Splashgard are shown in Fig. 7.10. The first patent, classified under 280/154, had a complete abstract. Upon reading, it became obvious to the team that the patent was not relevant to the problem. The second patent did not have an abstract; however, if the title sounded interesting to the design team, it could be

```
Patent Number  4591178
Issue Year     986
Assignee Code  182700
State/Country  IA
Classification 280/154 280/851
Title          QUICK ATTACH FENDER AND METHOD FOR USING SAME
Abstract       The quick attach fender of the present invention
               comprises a stationary bracket adapted to be
               mounted to a vehicle frame. The bracket has at
               least two horizonally extending shafts which
               extend from the vehicle frame toward the wheel of
               the vehicle.  Detachably mounted to the bracket is
               a shield assembly comprising a flat sheet member
               and a shield frame attached to the sheet member.
               The shield frame has at least two sleeve members
               sized and positioned to telescopically slide over
               the shafts of the bracket so as to support the
               shield in at least partial covering relation over
               the wheel.  Bolts threadably engage the sleeve
               members and the horizontal shafts to permit
               selective attachment and detachment of the sleeve
               members to the shafts.  The fender can be quickly
               removed by loosening the bolts and sliding the
               sleeves off of the shafts.  Similarly, it can be
               reattached merely by sliding the sleeves onto the
               shafts and by tightening the bolts.

Patent Number  4706980
Issue Year     986
Assignee Code  0
State/Country  DEX
Classification 280/154 24/456 24/535 24/568 81/487 267/74 267/158
               269/254R
Title          VEHICLE QUARTER FENDER AND ASSEMBLY FOR MOUNTING
               THE SAME
```

FIGURE 7.10
CASSIS listing for two patents.

further searched as detailed in step 3. (Note that this patent is classified under many other numbers; searching these may lead to other useful classes or subclasses. Care must be taken, or the search will quickly get out of hand.)

STEP 3: FOR A SPECIFIC PATENT NUMBER, SEARCH EITHER CASSIS, THE *OFFICIAL GAZETTE*, OR THE PATENTS THEMSELVES. The *Official Gazette* is a weekly publication that lists, in numerical order, an abstract of all patents issued in the previous week. (Most technical libraries subscribe to the *Gazette;* its monthly and yearly indices give the same information as found in CASSIS.) The abstract in the *Gazette* comprises one figure from the patent and the first claim in the patent. A full patent usually contains many figures and many claims about the uniqueness

4,706,980
VEHICLE QUARTER FENDER AND ASSEMBLY FOR MOUNTING THE SAME

Timothy R. Hawes, Muskegon; Steven A. Antekeier, North Shores; Glenn R. Cryderman; David I. Munger, both of Muskegon, Leonard A. Gould, Fruitport, and Louis E. Eklund, Jr., Muskegon, all of Mich., assignors to Fleet Engineers Inc., Muskegon, Mich.

Filed Mar. 19, 1986, Ser. No. 841,280
Int. Cl.^{4} B62B *9/16*

U.S. Cl. 280–154 20 Claims

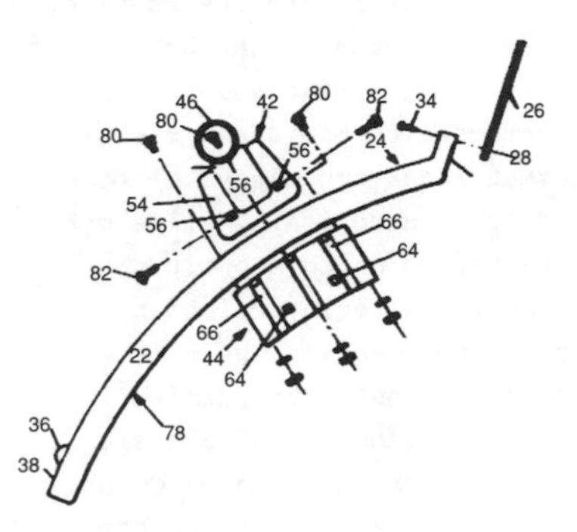

1. The combination of a fender and an assembly for mounting said fender to a vehicle frame, said fender having top and bottom surfaces, a width and a flange on an inside longitudinal edge of said fender and said assembly comprising a rod secured to and positioned substantially transversely of a longitudinal axis of said frame, wherein said assembly further comprises:

a mounting means extending transversely over less than one-half of said width of said top surface of said fender, engaging said flange and adapted to engage said rod to mount said fender to said frame; and

a rigidifying means extending transversely over more than one-half of said full width of said bottom surface of said fender, engaging said flange, secured to said mounting means and for providing rigidity to said fender when mounted to said frame.

FIGURE 7.11
Official Gazette listing for patent 4,706,980.

of the concept. The first claim and figure are usually the most general and are all that appear in the *Gazette*. The *Gazette* listing for patent 4,706,980 from the March 19, 1986, issue is shown in Fig. 7.11.

Actual patents are housed at about 50 patent depositories in the United States. Some patents are on paper; most are on microfiche. Copies of the full patents can be requested from the depositories. As seen in the single claim in Fig. 7.11, the wording in a patent makes for slow reading, so a limited patent search is necessary.

Patents will be discussed again in Chap. 13, where the process for application is discussed.

7.4.2 Finding Ideas in Reference Books and Trade Journals

Most reference books give analytical techniques that are not very useful in the early stages of a design project. In some you will find a few abstract ideas that are useful at this stage—usually in design areas that are quite mature and with ideas so decomposed that their form has specific function. A prime example is the area of linkage design. Even though a linkage is mostly geometric in nature, most linkages can be classified by function. For example, there are many geometries that can be classified by their function of generating a straight line along part of their cycle. (The function is to move in a straight line.) These straight-line mechanisms can be grouped by function. Two such mechanisms are shown in Fig. 7.12.

Many good ideas are published in trade journals that are oriented toward a specific discipline. Some, however, are targeted at designers and thus contain information from many fields. A listing of design-oriented trade journals is given in Sources at the end of this chapter (Section 7.7).

7.4.3 Using Experts to Help Generate Concepts

If designing in a new domain, one in which we are not experienced, we have two choices in gaining the knowledge sufficient to generate concepts. We either find someone with expertise in that domain or spend time gaining experience on our own. It is not always easy to find an expert; the domain may even be one that has no experts. With the Splashgard, for example, there were no experts knowledgeable in keeping riders dry; it is too broad an area.

How do we become an expert in an area that is new or unique? How do we become expert when we cannot find or afford the existing experts? Evidence of expertise can be found in any good designer's office. The best designers work long and hard in a domain, performing many calculations and experiments themselves to find out what works and what does not. Their offices also contain many reference books, periodicals, and sketches of concept ideas.

A good source of information is manufacturers' catalogs and, even better, manufacturers' representatives. A competent designer usually spends a great deal of time on the telephone with these representatives, trying to find sources for specific items or trying to find "another way to do it." One way to find manufacturers is through indexes such as the *Thomas Register,* a gold mine of ideas. All technical libraries subscribe to the 23 annually updated volumes, which list over a million producers of components and systems usable in mechanical design. Beyond a limited selection of reprints of manufacturers' catalogs, the *Thomas Register* does not give information directly but points to manufacturers that can be of assistance. The hard part of using the *Register* is finding the correct heading, which can take as much time as the patent search.

7.4.4 Brainstorming as a Source of Ideas

Brainstorming, initially developed as group-oriented technique, can also be used by an individual designer. What makes brainstorming especially good for group efforts

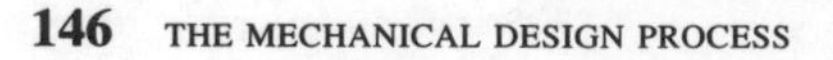

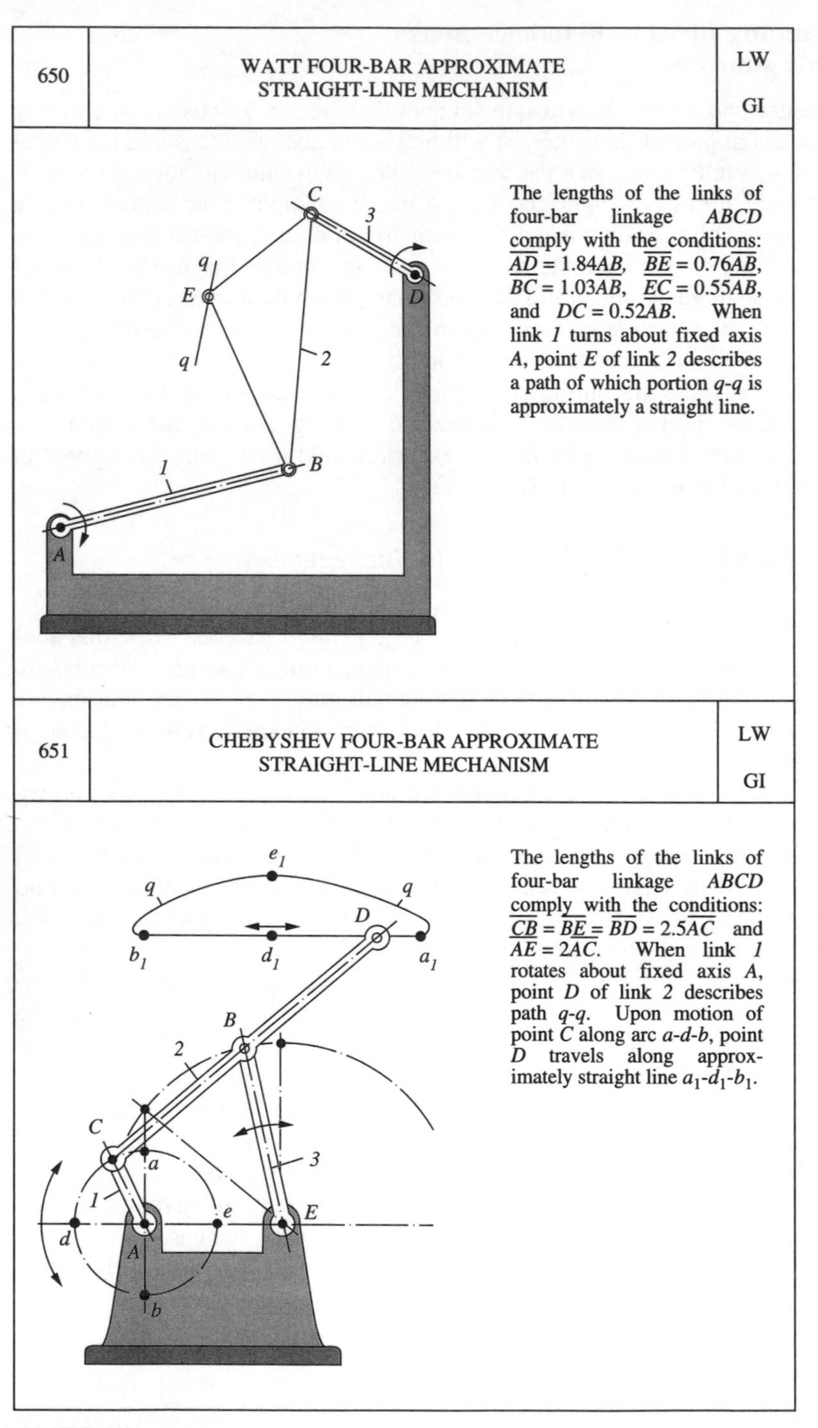

650
WATT FOUR-BAR APPROXIMATE STRAIGHT-LINE MECHANISM
LW
GI
C
3
q
E
D
q
2
1
B
A
The lengths of the links of four-bar linkage $ABCD$ comply with the conditions: $\overline{AD} = 1.84\overline{AB}$, $\overline{BE} = 0.76\overline{AB}$, $\overline{BC} = 1.03\overline{AB}$, $\overline{EC} = 0.55\overline{AB}$, and $\overline{DC} = 0.52\overline{AB}$. When link 1 turns about fixed axis A, point E of link 2 describes a path of which portion q-q is approximately a straight line.
651
CHEBYSHEV FOUR-BAR APPROXIMATE STRAIGHT-LINE MECHANISM
LW
GI
e_1
q
q
D
b_1
d_1
a_1
B
2
C
a
1
3
d
e
E
A
b
The lengths of the links of four-bar linkage $ABCD$ comply with the conditions: $\overline{CB} = \overline{BE} = \overline{BD} = 2.5\overline{AC}$ and $\overline{AE} = 2\overline{AC}$. When link 1 rotates about fixed axis A, point D of link 2 describes path q-q. Upon motion of point C along arc a-d-b, point D travels along approximately straight line a_1-d_1-b_1.

FIGURE 7.12
Straight-line mechanisms. (From I. I. Artobolevsky, *Mechanisms in Modern Engineering Design,* MIR Publishers, Moscow, 1975.)

is that each member of the group contributes ideas from his or her own viewpoint. The rules for brainstorming are quite simple:

1. Record all the ideas generated. Appoint someone as secretary at the beginning: this person should also be a contributor.
2. Generate as many ideas as possible, and then verbalize these ideas.
3. Think wild. Silly, impossible ideas sometimes lead to useful ideas.
4. Do not allow evaluation of the ideas; just the generation of them. This is very important. Ignore any evaluation, judgment, or other comments on the value of an idea and chastise the source.

In using this method, there is usually an initial rush of obvious ideas, followed by a period when ideas will come more slowly with periodic rushes. In groups, one member's idea will trigger ideas from the other team members. A brainstorming session should be focused on one specific function and allowed to run through at least three periods during which no ideas are being generated. It is important to encourage humor during brainstorming sessions as even wild, funny ideas can spark useful concepts. This is a proven technique that is useful when new ideas are needed.

7.4.5 Using the 6-3-5 Method as a Source of Ideas

A drawback to brainstorming is that it can be dominated by one or a few team members (see Section 3.5.3). The 6-3-5 method forces equal participation by all. This method is effectively brainstorming on paper and is called *brainwriting* by some. The method is similar to that shown in Fig. 7.13.

To perform the 6-3-5 method, arrange the team members around a table. The optimal number of participants is the "6" in the method's name. In practice, there can be as few as 3 participants or as many as 8. Each takes a clean sheet of paper and divides it into three columns by drawing lines down its length. Next, each team member writes 3 ideas for how to fulfill a specific agreed-upon function, one at the top of each column. The number of ideas is the "3" in the method's name. These ideas can be sketched or written as text. They must be clear enough that others can understand the important aspects of the concept.

After 5 minutes of work on the concepts, the sheets of paper are passed to the right. The time is the "5" in the method's name. The team members now have another 5 minutes to add 3 more ideas to the sheet. This should only be done after studying the previous ideas. They can be built on or ignored as seen fit. As the papers are passed in 5-minute intervals, each team member gets to see the input of each of the other members, and the ideas that develop are some amalgam of the best. After the papers have circulated to all the participants, the team can discuss the results to find the best possibilities.

There should be no verbal communication in this technique until the end. This rule forces interpretation of the previous ideas solely from what is on the paper, possibly leading to new insight and also eliminating evaluation.

"It's our new assembly line. When the person at the end of the line has an idea, he puts it on the conveyor belt, and as it passes each of us, we mull it over and try to add to it."

FIGURE 7.13
Automated brainwriting. (From S. Harris, *Einstein Simplified: Cartoons on Science,* Rutgers University Press, New Brunswick, N.J.,1989.)

7.4.6 Using Existing Products and Concepts as Idea Sources

One of the best sources of ideas is to look at existing products. Many hundreds of engineering hours have been spent developing the features of existing products, and to ignore this work is foolish. The QFD method, featured in Chap. 6, encourages the study of existing product benchmarks. Studying the ideas of others helps avoid the "NIH complex": since the idea was *N*ot *I*nvented *H*ere, we won't use it.

The functional decomposition method introduced in this chapter helps develop good habits when studying other products. Specifically, how do energy, materials, and information flow through the system? How does the interface between components create the functionality? How does an existing product account for preparation, use, and conclusion?

Another technique for generating ideas is to begin with a product or concept and modify it. There are five types of modifications to be considered at this conceptual stage.

Geometric modifications are centered around changing the length, width, height, or other geometric feature to understand, enhance, or change the manner in which the device functions. An example is to reconsider Concept III (Fig. 7.7) by modifying the geometry of a common window shade.

Energy-flow modifications can be of two forms: a change in the path or a change in the form of energy. In Chap. 10 the flow of forces through a structure will be used to assist in product generation. This type of consideration is also useful during concept generation. Specifically, are there alternative paths for the energy other than those in current products or concepts? Brainstorming may help find these paths. Additionally, it is useful to consider the usefulness of changing the form of energy. Can a product that now has a mechanical link be modified by using electromechanical components, hydraulic components, or just electrical signals? The same question can be asked about modifications in information and material flows.

Finally, although considerations about material selection will not be presented until Chap. 10, it is often worthwhile to develop concepts by modifying materials used in a product.

7.5 COMMUNICATION DURING CONCEPT GENERATION

The techniques outlined in this chapter have focused on generating potential concepts. In performing these techniques, the following documents are produced to support communication to others and archiving the design process: functional decomposition diagrams, literature and patent search results, function-concept mapping, and sketches of overall concepts.

7.6 SUMMARY

* The two techniques presented here—functional decomposition and generating concepts from functions—force the generation of many conceptual ideas to meet the functional requirements.
* Functional decomposition encourages breaking down the needed function of a device as finely as possible, with as few assumptions about the form as possible.
* The functional decomposition of existing products is the best method to understanding them.
* Listing concepts for each function helps generate ideas; this list is often called a *morphology.*
* Sources for conceptual ideas come primarily from the designer's own expertise; this expertise can be enhanced through the use of patent searches, reference books, experts, brainstorming, and the 6-3-5 method.

7.7 SOURCES

Artobolevsky, I. I.: *Mechanisms in Modern Engineering Design,* MIR Publishers, Moscow, 1975. This five-volume set of books is a good source for literally thousands of different mechanisms, many indexed by function.

Chironis, N. P.: *Machine Devices and Instrumentation,* McGraw-Hill, New York, 1966. Similar to Greenwood's *Product Engineering Design Manual.*

———: *Mechanism, Linkages and Mechanical Controls,* McGraw-Hill, New York, 1965. Similar to the above.

Damon, A., H. W. Stoudt, and R. A. McFarland: *The Human Body in Equipment Design,* Harvard University Press, Cambridge, Mass., 1966. This book has a broad range of anthropometric and biomechanical tables.

Design News, Cahners Publishing, Boston. Similar to *Machine Design.*

Edwards, B.: *Drawing on the Right Side of the Brain,* Tarcher, Los Angeles, 1982. Although not oriented specifically toward mechanical objects, this is the best book available for learning how to sketch.

Greenwood, D. C.: *Product Engineering Design Manual,* Krieger, Malabar, Fla., 1982. A compendium of concepts for the design of many common items, loosely organized by function.

———: *Engineering Data for Product Design,* McGraw-Hill, New York, 1961. Similar to the above.

Human Engineering Design Criteria for Military Systems, Equipment, and Facilities, MIL-STD 1472, U.S. Government Printing Office, Washington, D.C. This Standard contains four hundred pages of human factors information.

Machine Design, Penton Publishing, Cleveland, Ohio. One of the best mechanical design magazines published, it contains a mix of conceptual and product ideas along with technical articles. It is published twice a month.

Norman, D.: *The Psychology of Everyday Things,* Basic Books, New York, 1988. This book is light reading focused on guidance for designing good human interfaces.

Plastics Design Forum, Advanstar Communications Inc., Cleveland, Ohio. A monthly magazine for designers of plastic products and components.

Product Design and Development, Chilton, Radnor, Pa. Another good design trade journal.

Thomas Register of American Manufacturers, Thomas Publishing, Detroit, Mich. This 23-volume set is an index of manufacturers and is published annually.

7.8 EXERCISES

7.1. For the original design problem (Exercise 4.1), develop a functional model by
   (a) Stating the overall function.
   (b) Decomposing the overall function into subfunctions. If assumptions are needed to refine this below the first level, state the assumptions. Are there alternative decompositions that should be considered?
   (c) Identifying all the objects (nouns) used and defending their inclusion in the functional model.

7.2. For the redesign problem (Exercise 4.2), apply items a–c from Exercise 1 and also study the existing device(s) to establish the following:
   (a) Which subfunction(s) must remain unchanged during redesign?
   (b) Which subfunctions (if any) must be changed to meet new requirements?
   (c) Which subfunctions may cease to exist?

7.3. For the functional decomposition developed in Exercise 1,
   (a) Develop a morphology as in Fig. 7.8 to aid in generating concepts.
   (b) Combine concepts to develop at least 10 complete conceptual designs.

7.4. For the redesign problem functions that have changed in Exercise 2,
   (a) Generate a morphology of new concepts as in Fig. 7.8.
   (b) Combine concepts to develop at least five complete conceptual designs.

7.5. Find at least five patents that are similar to an idea that you have for the following:
   (a) The original design problem begun in Exercise 4.1.
   (b) The redesign problem begun in Exercise 4.2.
   (c) A perpetual motion machine. In recent times the patent office has refused to consider such devices. However, the older patent literature has many machines that violate the basic energy conservation laws.

7.6. Use brainstorming to develop at least 25 ideas for the following:
(a) A way to fasten together loose sheets of paper.
(b) A device to keep water off a mountain-bike rider (Splashgard).
(c) A way to convert human energy to power a boat.
(d) A method to teach the design process.

7.7. Use brainwriting to develop at least 25 ideas for the following:
(a) A device to leap tall buildings in a single bound.
(b) A way to fasten a gear to a shaft and transmit 500 watts.

CHAPTER 8

CONCEPT EVALUATION

8.1 INTRODUCTION

In the previous chapter we developed techniques for generating promising conceptual solutions for a design problem. In this chapter we explore techniques for choosing the best of these concepts for development into products. The goal is to expend the least amount of resources on deciding which concepts have the highest potential for becoming a quality product. The difficulty in concept evaluation is that we must choose which concepts to spend time developing when we still have very limited knowledge and data on which to base this selection.

There is the question of how soon to narrow down to a single concept. Ideally, enough information about each concept should be known at this point to make a choice and put all resources into developing this one concept. On the other hand it is less risky to refine a number of concepts before committing to one of them. However, this requires that instead of being concentrated, resources may be spread thin. Many companies generate only one concept and then spend time developing it. Others develop many concepts in parallel, eliminating the weaker ones along the way. Again, every company has its own culture for product development and there is no one "correct" number of concepts to select. Design is learning, and resources are limited.

How can a rough conceptual idea be evaluated? Concepts are abstract, have little detail, and cannot be measured. Should time be spent refining them, giving them structure, making them measurable so that they can be compared with the engineering targets developed during problem specifications development? Or should the concept that seems like the best one be developed into a product design in the hope that it will become a quality product? In this chapter a method will be developed that will help in making a knowledgeable decision with limited information. This method uses four different techniques to reduce the many concepts generated to the few concepts most promising for development into quality products. The techniques, presented in the first four boxes in Fig. 8.1, are the topics of the sections of this chapter. The fifth box, "comparison with engineering requirements," is covered in Chap. 11. Before discussing the techniques, however, we will define the term *evaluation.*

Evaluation, as used in this text, implies both *comparison* and *decision making.* These are tightly interrelated actions; it is the comparisons between alternative concepts and the requirements they must meet that gives the information necessary to make a decision. This is true for all design decisions, and thus the techniques in this chapter and in the chapters that focus on product evaluation (Chaps. 11 and 12) will emphasize (1) itemizing the alternatives and the criteria for their evaluation and (2) comparison of the alternatives to the criteria and to each other.

There are two possible types of comparisons. The first type is absolute in that each alternative concept is directly (i.e., absolutely) compared with some set of criteria. The second type of comparison is relative in that alternative concepts are compared with each other using measures defined by the criteria. As shown in Fig. 8.1, the first three comparison techniques, all absolute, are used as a filter for the relative comparison technique, called a decision matrix. These four techniques together

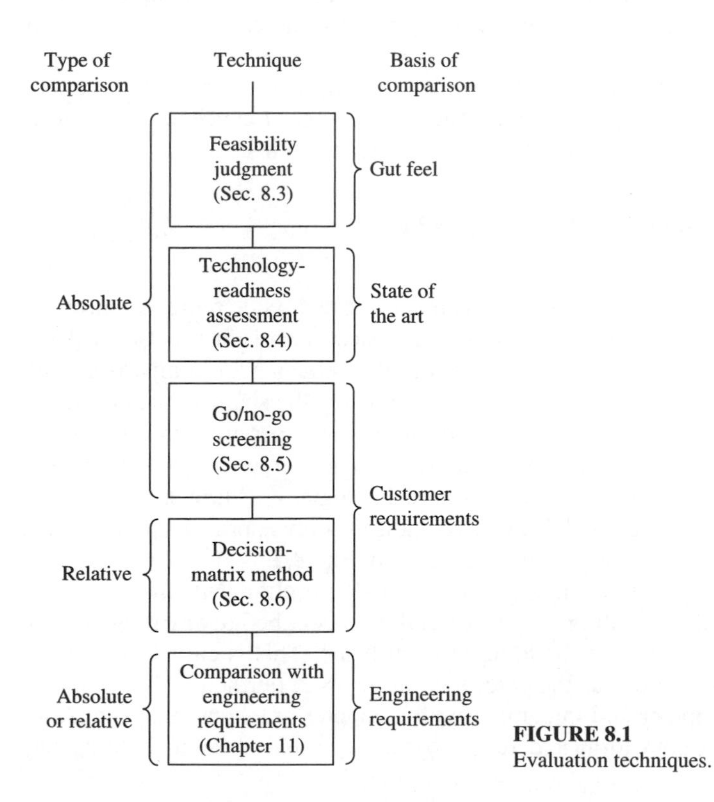

FIGURE 8.1
Evaluation techniques.

are the best tools for the evaluation of concepts. Each technique can be helpful in deciding which concepts are worth continued effort. The last technique, comparison with the engineering requirements, is used mainly with products—when values for parameters are known for comparison. However, some concepts are refined sufficiently to use this level of evaluation.

Before continuing, some important points about concept evaluation will be raised. In order to be compared, alternatives and criteria must be in the same language and they must exist at the same level of abstraction. Consider, for example, the spatial requirement that a product fit in an area 2.000 ± 0.005 in. long. An unrefined concept for this product may be described as "short." It is impossible to compare "2.000 ± 0.005 in." to "short" because the concepts are in different languages—a number versus a word—and they are at different levels of abstraction—very concrete versus very abstract. It is simply not possible to make a comparison between the "short" concept and the requirement of fitting a 2.000 ± 0.005 in. slot.

An additional problem with concept evaluation is that abstract concepts are fuzzy; as they are refined, their behavior can differ from that initially anticipated. The greater the knowledge about a concept, the fewer the surprises. However, even in a well-known area, as the concept is refined to the product, unanticipated factors arise.

Finally, there is the question of how soon to narrow down to a single concept. Ideally, a designer would know enough about each concept at this point to choose one and put all possible resources into developing this one concept. On the other hand it might be wise to refine a number of concepts before committing to one of them. Designers at Toyota follow what they call a "parallel set narrowing process," in which they continue parallel development of a number of concepts. As more is learned, they slowly eliminate those that show the least promise.

8.2 INFORMATION REPRESENTATION IN CONCEPT EVALUATION

In planning for the project, we identified the models to be used to represent information during concept development (Table 5.1). Physical models or proof-of-concept prototypes support evaluation by demonstrating the behavior for comparison with the functional requirements (see Fig. 2.4) or by showing the shape of the design for comparison with form constraints. Sometimes these prototypes are very crude—just cardboard, wire, and other minimal materials thrown together to see if the idea makes sense. Often, when one is designing with new technologies or complex known technologies, building a physical model and testing it is the only approach possible. This *design-build-test cycle* is shown as the inner loop in Fig. 8.2.

The time and expense of building physical models is eliminated by developing analytical models and simulating (i.e., testing) the concept before anything is built. All the iteration occurs without building any hardware. This is called the *design-test-build cycle* and is shown as the outer loop in Fig. 8.2. Further, if the analytical models are on a computer and integrated with computer graphical representations of the concept, then both form and function can be tested without building any

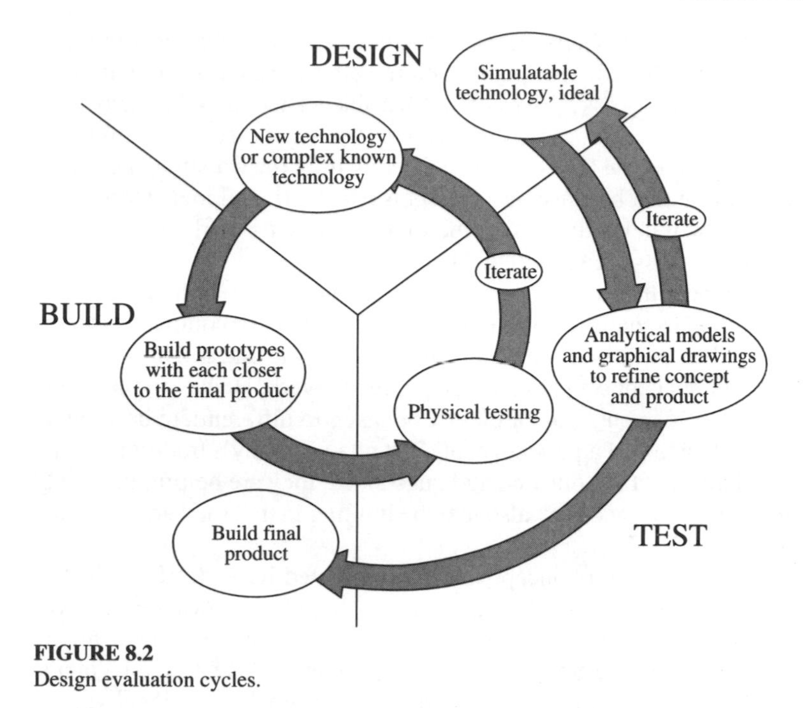

FIGURE 8.2
Design evaluation cycles.

hardware. This is obviously ideal as it has the potential for minimizing time and expense. This is the promise of virtual reality, the simulation of form and function in a way that richly supports concept and product evaluation. However, analysis can only be performed on systems that are understood and can be modeled mathematically. New and existing technologies, complex beyond the ability of analytical models, must be explored with physical models.

Regardless of which representation is chosen, the techniques shown in Fig. 8.1 are used either formally or informally to evaluate concepts. Each of the first four techniques is developed in detail in one of the following sections. The use of analytical models to support comparison with the engineering requirements is covered in Chap. 11.

8.3 EVALUATION BASED ON FEASIBILITY JUDGMENT

As a conceptual design is generated, the designer has one of three immediate reactions: (1) it is not feasible, it will never work; (2) it might work if something else happens; (3) it is worth considering. These judgments about a concept's feasibility are based on "gut feel," a comparison made with prior experience stored as design knowledge. The more design experience, the more reliable an engineer's knowledge and the decision at this point. Let us consider the implications of each of the possible initial reactions more closely.

IT IS NOT FEASIBLE. If a concept seems infeasible, or unworkable, it should be considered briefly from different viewpoints before being rejected. Before an idea is discarded, it is important to ask, Why is it not feasible? There may be many reasons. It may be obviously technologically infeasible. It may not meet the customer's requirements. It may just be that the concept is different from the way things are normally done. Or it may be that because the concept is not an original idea, there is no enthusiasm for it. We will delay discussing the first two reasons until the following sections, and we will discuss the second two here.

As for the judgment that a concept is "different," humans have a natural tendency to prefer tradition to change. Thus an individual designer or company is more likely to reject new ideas in favor of ones that are already established. This is not all bad because the traditional concepts have been proven to work. However, this view can block product improvement, and care must be taken to differentiate between a potentially positive change and a poor concept. Part of a company's tradition lies in its standards. Standards must be followed and questioned; they are helpful in giving current engineering practice, and they also may be limiting in that they are based on dated information.

As for the judgment that a concept was "not invented here" (NIH): It is always more ego-satisfying to individuals and companies to use their own ideas. However, ideas borrowed from others are sometimes better. Remembering the example from the previous chapter of the Sperry engineers reinventing Leonardo da Vinci's bearing, we can make the case that very few ideas are original. Additionally, part of the technique presented in Chap. 6 for understanding the design problem involved benchmarking the competition. One of the reasons for doing this was to learn as much as possible about existing products to aid in the development of new products.

A final reason to further consider ideas that at first do not seem feasible is that they may give new insight to the problem. Part of the brainstorming technique introduced in the last chapter was to build from the wild ideas that were generated. Before discarding a concept, see if new ideas can be generated from it, effectively iterating from evaluation back to concept generation.

IT IS CONDITIONAL. The initial reaction might be to judge a concept workable *if something else happens.* Typical of other factors involved are the readiness of technology, the possibility of obtaining currently unavailable information, or the development of some other part of the product. Consider the space shuttle project. The concept of using recycled booster segments was predicated on being able to design a reliable joint.

IT IS WORTH CONSIDERING. The hardest concept to evaluate is one that is not obviously a good idea or a bad one, but looks worth considering. Engineering knowledge and experience are essential in the evaluation of such a concept. If sufficient knowledge is not immediately available for the evaluation, it must be developed. This is accomplished by developing models or prototypes that are easily evaluated.

8.4 EVALUATION BASED ON TECHNOLOGY-READINESS ASSESSMENT

The second evaluation technique shown in Fig. 8.1 is to determine the readiness of the technologies that may be used in the concept. This technique refines the evaluation by forcing an absolute comparison with state-of-the-art capabilities. If a technology is to be used in a product, it must be mature enough that its use is a design issue, not a research issue. The vast majority of technologies used in products are mature, and the measures discussed below are readily met. However, in a competitive environment, there are high incentives to include new technologies in products. Recall the *Time* survey in Chap. 1, which showed that a majority of people think that including the latest technology in a product is a sign of quality. Care must be taken to ensure that the technology is *ready* to be included in the product.

Consider the technologies listed in Table 8.1. Each of these technologies required many years from inception to the realization of a physical product. Besides these major technologies, each product utilizes many other lesser technologies. An attempt to design a product before the necessary technologies are ready leads either to a low-quality product or to a project that is canceled before a product reaches the market because it is behind schedule and over cost. How, then, can the maturity of a technology be measured? Six measures can be applied to determine a technology's maturity:

CAN THE TECHNOLOGY BE MANUFACTURED WITH KNOWN PROCESSES? If reliable manufacturing processes have not been refined for the technology, then either the technology should not be used or there must be a separate program for developing the manufacturing capability. There is a risk in the latter alternative, as the separate program could fail, jeopardizing the entire project.

ARE THE CRITICAL PARAMETERS THAT CONTROL THE FUNCTION IDENTIFIED? Every design concept has certain parameters that are critical to its proper operation. It has been estimated that only about 10 to 15 percent of the dimensions on a finished component are critical to the operation of the product. It is important to know which parameters (e.g., dimensions, material properties, or other features)

TABLE 8.1
A time line for technology readiness

Technology	Development time, years
Powered human flight	403 (1500–1903)
Photographic cameras	112 (1727–1839)
Radio	35 (1867–1902)
Television	12 (1922–1934)
Radar	15 (1925–1940)
Xerography	17 (1938–1955)
Atomic bomb	6 (1939–1945)
Transistor	5 (1948–1953)
High-temperature superconductor	? (1987–)

are critical to the function of the device. For a simple cantilever spring, the critical parameters are its length, its moment of inertia about the neutral axis, the distance from the neutral axis to the most highly stressed material, the modulus of elasticity, and the maximum allowable yield stress. These parameters allow for the calculation of the spring stiffness and the failure potential for a given force. The first three parameters are dependent on the geometry; the last two are dependent on the material properties.

ARE THE SAFE OPERATING LATITUDE AND SENSITIVITY OF THE PARAMETERS KNOWN? In refining a concept into a product, the actual values of the parameters may have to be varied to achieve the desired performance or to improve manufacturability. It is essential to know the limits on these parameters and the sensitivity of the product's operation to them. This information is only known in a rough way during the early design phases; during the product evaluation, it will become extremely important.

HAVE THE FAILURE MODES BEEN IDENTIFIED? Every type of system has characteristic failure modes. It is in general a useful design technique to continuously evaluate the different ways a product might fail. This is expanded on in Chap. 13.

DOES HARDWARE EXIST THAT DEMONSTRATES POSITIVE ANSWERS TO THE ABOVE FOUR QUESTIONS? The most crucial measure of a technology's readiness is its prior use in a laboratory model or another product. If the technology has not been demonstrated as mature enough for use in a product, the designer should be very wary of assurances that it will be ready in time for production.

IS THE TECHNOLOGY CONTROLLABLE THROUGHOUT THE PRODUCT'S LIFE CYCLE? This question addresses the later stages of the product's life cycle: its manufacture, service, and retirement. It also raises other questions. What manufacturing by-products come from using this technology? Can the by-products be safely disposed of? How will this product be retired? Will it degrade safely? Answers to these questions are the responsibility of the design engineer.

Often, if these questions are not answered in the positive, a consultant or vendor is added to the team. This is especially true for manufacturing technologies in which the design engineer cannot possibly know all the methods needed to manufacture a product.

8.5 EVALUATION BASED ON GO/NO-GO SCREENING

Once it has been established that the technologies used in a concept are mature, the basis of comparison moves to criteria defined by the customer requirements developed in Chap. 6. Each alternative concept must be compared with the requirements in an absolute fashion. In other words, each customer requirement must be transformed into a question to be addressed to each concept. The questions should be answerable as either yes or maybe (go), or no (no-go). This type of evaluation not only weeds out designs that should not be further considered, but also helps generate new ideas. If a concept has only a few no-go responses, it may be worth modifying rather than being eliminated. This evaluation rapidly points out the weak areas in a concept so that they can be "patched"—changed or modified to fix the problem. During patching, the functional decomposition and morphology should be referred to and possibly updated as more is learned through the evaluation.

Go/no-go screening: The Splashgard example. Consider concept IV for the Splashgard, the sissy bar shown in Fig. 7.7. Was this concept worth pursuing? The team began by comparing it with the customers' requirements from Fig. 6.3 and Table 6.2:

Question: Is concept IV easy to attach?

Answer: Maybe (go).

Question: Will concept IV fit most bicycles?

Answer: Yes (go).

They then proceeded in this manner through the remainder of the requirements.

Since each response for this concept was a yes or a maybe, concept IV was kept in consideration as a design option. All other concepts shown in Fig. 7.9 passed this evaluation with the exception of concept V, which would have brushed the water from the tire rather than coming between the tire and the rider, as did all the others. Since it was not clear if concept V could pass this screening, the design team decided to build a model, a proof-of-concept prototype, to determine experimentally if it was worth further investment. Two pages from a team member's notebook are shown in Fig. 8.3. Note that the team tried various types of brushes and that, contrary to the sketch in Fig. 7.7, they used a brush only on the tread of the tire, not on the sides.

The results of this crude test proved that the brush did remove a good percentage of the road water. You can see in the notes the estimate of 95 percent removal. This was enough evidence for the team to pass the concept in the go/no-go screening.

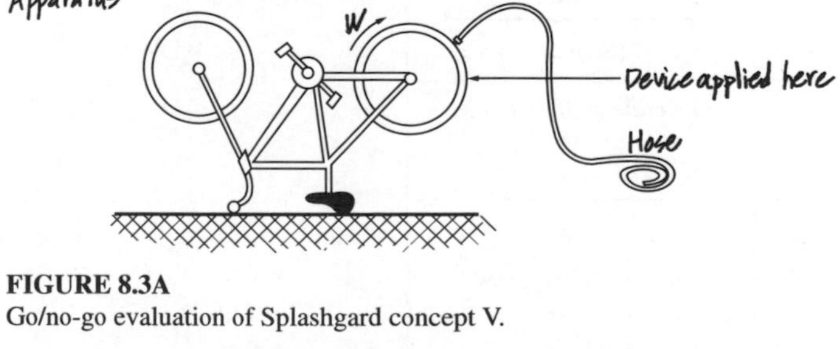

Testing Roller/Brush Theory

Date: 5/11/90 Location: 934 NW 21st Corvallis

All members present, Jeff P.: Supply Bicycle

Purchased:
- 2.5 inch paint roller with handle (1.5" dia)
- 2 inch olefin paint brush (cheap)
- 2 inch sponge-type paint brush (cheaper)

Apparatus

W/

Device applied here

Hose

FIGURE 8.3A

Go/no-go evaluation of Splashgard concept V.

• As one member supported and cranked the rear tire another member applied water from a hose at various locations. The conditions were found to exceed that of actual rain coming off the tire; therefore it was an adequate test. A third member of the group then applied the three devices to the tire while observing the effectiveness of each. The results are as follows.

Device	Water % removed	Water displaced	Effectiveness 10 = dry	Comments
Roller	80	Spray in all directions	5	Would require building a separate shield to cover spray
Brush	95	Pulled off by gravity	9	Minimal friction with good results
Sponge	90	Pulled off by gravity	7	Would tend to wear out quickly

• Of course all of the values and opinions are based on our idea of the customer requirements

FIGURE 8.3B
Go/no-go evaluation of Splashgard concept V.

8.6 EVALUATION BASED ON A DECISION MATRIX

The *decision-matrix method,* or *Pugh's method,* which is fairly simple, has proven effective for comparing alternative concepts that are not refined enough for direct comparison with the engineering specifications discussed in Chap. 6. The basic form for the method is shown in Table 8.2; in essence, the method provides a means of scoring each alternative concept relative to another in its ability to meet criteria set by the customer requirements. Comparison of the scores thus developed then gives insight to the best alternatives and useful information for making decisions. (In actuality this technique is very flexible and is easily used in other, nondesign situations—such as which job offer to accept or which car to buy.)

The decision-matrix method is an iterative evaluation method that tests the completeness and understanding of requirements, rapidly identifies the strongest

TABLE 8.2
Decision-matrix form

		Concepts
Criterion	**Wt**	***(Step 2)***
(Step 1) ⋮	⋮	*Generate score (step 3)* ⋮
Total + Total − Overall total Weighted total		*Generate totals (step 4)*

alternatives, and helps foster new alternatives. This method is most effective if each member of the design team performs it independently and the individual results are then compared. The results of the comparison lead to a repetition of the technique, with the iteration continuing until the team is satisfied with the results. As shown in Table 8.2, there are four steps to this method.

Decision matrices can be easily managed on the computer using a common spreadsheet program. Using a spreadsheet allows for easy iteration and comparison of team members' evaluations.

STEP 1: CHOOSE THE CRITERIA FOR COMPARISON. First it is necessary to know the basis on which the alternatives are to be compared with each other. Using the QFD method in Chap. 6, an effort was made to develop a full set of customer requirements for a design. These were then used to generate a set of engineering requirements and targets that will be used to ensure that the resulting product will meet the customer requirements. However, the concepts developed in the previous chapter might not be refined enough to compare with the engineering targets for evaluation. If they are not, we have a mismatch in the level of abstraction and use of the engineering targets must wait until the concept is refined to the point that actual measurements can be made on the product designs. Usually the basis for comparing the design concepts must be the customer requirements, which, like the concepts, are abstract and thus suitable as a basis for comparison.

If the customer requirements have not been developed, this evaluation technique can still be used. The first step should be to develop criteria for comparison. The methods discussed in Chap. 6 should help with this task.

STEP 2: SELECT THE ALTERNATIVES TO BE COMPARED. The alternatives to be compared are the different ideas developed during concept generation. It is important that all the concepts to be compared be at the same level of abstraction and in the same language. This means it is best to represent all the concepts in the same way. Generally, a simple sketch is best. In making the sketches, ensure that knowledge about the functionality, structure, technologies needed, and manufacturability is at a comparable level in every figure.

STEP 3: GENERATE SCORES. By this time in the design process, every designer has a favorite alternative, one that he or she thinks is the best of the concepts that have yet to be developed. This concept is used as a *datum,* all other designs being compared with it as measured by each of the customer requirements. If the problem is for the redesign of an existing product, then this product, abstracted to the same level as the concepts, can be used as the datum.

For each comparison, the concept being evaluated is judged to be either better than, about the same as, or worse than the datum. If it is better than the datum, the concept is given a + score. If it is judged to be about the same as the datum or if there is some ambivalence, an S ("same") is used. If the concept does not meet the criterion as well as the datum does, it is given a − score. If the decision matrix is on a spreadsheet use +1, 0, −1 for scoring.

Note that if it is impossible to make a comparison to a design requirement, more information must be developed. This may require more analysis, further experimentation, or just better visualization. It may even be necessary to refine the design, through the methods to be described in the following chapters, and then return to make the comparison.

STEP 4: COMPUTE THE TOTAL SCORE. After a concept is compared with the datum for each criterion, four scores are generated: the number of plus scores, the number of minus scores, the overall total, and the weighted total. The overall total is the difference between the number of plus scores and the number of minus scores. The weighted total can also be computed. It is the sum of each score multiplied by the importance weighting, in which an S counts as 0, a + as +1, and a − as −1. Both the weighted and the unweighted scores must not be treated as absolute measures of the concept's value; they are for guidance only. The scores can be interpreted in a number of ways:

- If a concept or group of similar concepts has a good overall total score or a high + total score, it is important to notice what strengths they exhibit, that is, which criteria they meet better than the datum. Likewise, groupings of − scores will show which requirements are especially hard to meet.
- If most concepts get the same score on a certain criterion, examine that criterion closely. It may be necessary to develop more knowledge in the area of the criterion in order to generate better concepts. Or it may be that the criterion is ambiguous, is interpreted differently by different members of the team, or is unevenly interpreted from concept to concept. If the criterion has a low importance weighting, then do not spend much time clarifying it. However, if it is an important criterion, effort is needed either to generate better concepts or to clarify the criterion.
- To learn even more, redo the comparisons, with the highest-scoring concept used as the new datum. This iteration should be redone until a clearly "best" concept or concepts emerge.

After each team member has completed this procedure, the entire team should compare their individual results. The results can vary widely, since neither the concepts nor the requirements may be refined. Discussion among the members of the group should result in a few concepts to refine. If it does not, the group should clarify the criteria or generate more concepts for evaluation.

A variation that may be of use in situations where enough information is available is use of a finer scoring system than the three-level system, for instance, a seven-level scale:

+3 meets criterion in a way vastly superior to the datum
+2 meets criterion much better than the datum
+1 meets criterion better than the datum
0 meets criterion as well as the datum
−1 meets criterion not as well as the datum
−2 meets criterion much worse than the datum
−3 meets criterion far worse than the datum

Using the decision matrix: The Splashgard example. In generating the design concepts for the Splashgard (Fig. 7.7), the team realized that the problem could be decomposed into a water-deflection system and an attaching system for most of the concepts. Taking advantage of this decomposition, they made independent decisions for each of the subsystems.

The evaluation for the water-deflection system is shown in Table 8.3. All the customer requirements are listed. Initially, each member of the design team did the evaluation independently. Table 8.3 shows the consensus of their opinions. Although

TABLE 8.3
Initial decision matrix for Splashgard water-deflection subsystem

		Concepts						
Criterion	**Wt**	**I**	**II**	**III**	**IV**	**V**	**VI**	**VII**
Easy attach	7	+	+	+	+	+	S	D
Easy detach	4	−	+	+	+	+	S	A
Fast attach	3	+	+	+	+	+	S	T
Fast detach	1	+	+	+	+	+	S	U
Attach when dirty	3	+	+	+	+	S	S	M
Detach when dirty	1	−	+	−	+	S	+	
Not mar	10	+	+	+	+	S	S	
Not catch water	7	−	+	−	S	S	S	
Not rattle	8	−	−	−	−	S	S	
Not wobble	7	−	−	−	S	S	S	
Not bend	4	−	−	−	S	−	S	
Long life	11	−	S	−	S	−	S	
Lightweight	7	+	S	S	−	S	S	
Not release accidently	10	+	S	S	S	S	S	
Fits most bikes	7	+	S	S	S	S	S	
Streamlined	5	−	S	−	−	+	S	
Total +		8	8	6	7	5	1	0
Total −		8	3	7	3	2	0	0
Overall total		0	5	−1	4	3	1	0
Weighted total		1	17	−15	9	5	1	0

this is a very subjective process, it did give a clear direction. At this point a fender (concept VII) or fenderlike device (concept VI) was not favored because it imposed difficulty in attaching and detaching, regardless of the actual attaching/detaching device. It was also clear that the poncho (concept I) and the window-shade idea (concept III) were not well thought of because of their flimsiness. This left concepts II, IV, and V in consideration; both their weighted and their unweighted scores were similar. The decision matrix was then repeated, with concept II as the new datum (Table 8.4).

TABLE 8.4
Second decision matrix for Splashgard water-deflection subsystem

		Concepts		
Criterion	**Wt**	**II**	**IV**	**V**
Easy attach	7	D	S	S
Easy detach	4	A	S	S
Fast attach	3	T	S	S
Fast detach	1	U	S	S
Attach when dirty	3	M	S	–
Detach when dirty	1		S	S
Not mar	10		S	–
Not catch water	7		S	S
Not rattle	8		S	S
Not wobble	7		S	–
Not bend	4		S	S
Long life	11		S	–
Lightweight	7		S	–
Not release accidently	10		S	–
Fits most bikes	7		S	–
Streamlined	5		–	+
Total +		0	0	1
Total –		0	2	5
Overall total		0	–2	–4
Weighted total		0	–12	–33

The results of this second evaluation showed that concept II was best, with concept IV a strong second choice. The brush design, concept V, was down-rated because it attached to a dirty part of the bike and would also have to be attached to the painted frame, and so might mar the bike. Because of the similarity of concepts II and IV, it was decided to eliminate concept IV. Concept II, a device between the rider and the wheel that would attach to the seat post or the back of the seat, was found to be the best concept for development.

The comparison for the attachment subsystem, parts a to d of concept VII in Fig. 7.7, is shown in Table 8.5. In this evaluation, concept a will be assumed to represent any screw-type fastener, concept b any type of one-piece snap fastener, concept c a Velcro fastener, and concept d any two-piece snap fastener. Concept d assumed that a part called a *keeper* would be permanently attached to the bike seat post. The keeper must initially be attached to the bike and then the clip (a second component) snapped into it during use. The added effort of attaching a permanent part to the bike had to be evaluated. The design team prepared a questionnaire to establish the relative importance of the necessity of installing the keeper. This iteration back to problem assessment resulted in determining that the extra, one-time effort was of little importance to the customers and therefore mounting the keeper was not included in the decision matrix of Table 8.5. On the basis of this decision matrix, the two-piece snap fastener was clearly the best choice. Its exact configuration was not yet known; it might or might not be like that sketched originally.

Concept evaluation for the Splashgard revealed that the product ought to be a device that could be snapped onto the bike with a two-part fastener and would block the water between the wheel and rider.

8.7 PRODUCT SAFETY AND LIABILITY

One area of product understanding that is often overlooked until late in the project is product safety. It is valuable to consider both safety and the engineer's responsibility for liability, as they are an integral part of human-product interaction and greatly affect the perceived quality of the product. These topics are best thought of early in the design process and thus are covered here. Formal failure analysis will be discussed in Chap. 12.

8.7.1 Product Safety

Design for safety means ensuring that the product will not cause injury or loss. There are two issues that must be considered in designing a safe product. First, who or what is to be protected from injury or loss during the operation of the product? Second, how is the protection actually implemented in the product?

The main consideration in design for safety is the protection of people from injury by the product. As addressed later in this section, there are many ways of providing this protection. However, design for safety extends beyond protection against human injury and includes concern for the loss of other property affected by the product and the product's impact on the environment in case of failure. Neglect in ensuring the safety of any of these objects may lead to a dangerous and potentially litigious situation. Concern for affected property means considering the effect the product can have on other devices, either during normal operation or upon failure.

TABLE 8.5
Initial decision matrix for Splashgard attachment subsystem

		Concepts			
Criterion	**Wt**	**a**	**b**	**c**	**d**
Easy attach	7	D	+	+	+
Easy detach	4	A	+	+	+
Fast attach	3	T	+	+	+
Fast detach	1	U	+	+	+
Attach when dirty	3	M	S	−	+
Detach when dirty	1		+	+	+
Not mar	10		−	+	+
Not catch water	7		S	S	+
Not rattle	8		+	S	S
Not wobble	7		−	−	+
Not bend	4		S	S	+
Long life	11		S	−	+
Lightweight	7		+	+	S
Not release accidently	10		−	−	S
Fits most bikes	7		−	+	S
Streamlined	5		+	+	+
Total +		0	8	9	12
Total −		0	4	4	0
Overall total		0	4	5	12
Weighted total		0	2	14	63

For example, the manufacturer of a fuse or circuit breaker that fails to cut the current flow to a device may be liable because the fuse did not perform as designed and caused loss of or injury to another product. There must also be concern for the product's effect on the environment.

There are three ways to institute product safety. The first way is to design safety directly into the product. This means that the device poses no inherent danger during normal operation or in case of failure. If inherent safety is impossible, as it is with

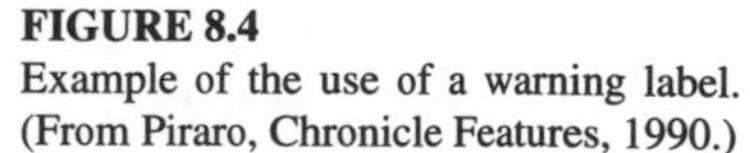

FIGURE 8.4
Example of the use of a warning label. (From Piraro, Chronicle Features, 1990.)

most rotating machinery and vehicles, then the second way to design in safety is to add protective devices to the product. Examples of added safety devices are shields around rotating parts, crash-protective structures (as in automobile body design), and automatic cut-off switches, which automatically turn a device off (or on) if there is no human contact. The third, and weakest, form of design for safety is a warning of the dangers inherent in the use of a product (Fig. 8.4). Typical warnings are labels, loud sounds, or flashing lights.

It is always advisable to design in safety. It is difficult to design protective shields that are foolproof, and warning labels do not absolve the designer of liability in case of an accident. The only truly safe product is one with safety designed into it.

8.7.2 Products Liability

Products liability is the the special branch of law dealing with alleged personal injury or property or environmental damage resulting from a defect in a product. It is important that design engineers know the extent of their responsibility in the design of a product. If, for example, a worker is injured while using a device, the designers of the device and the manufacturer may be sued to compensate the worker and the employer for the losses incurred.

A products liability suit is a common legal action. Essentially, there are two sides in such a case, the plaintiff (the party alleging injury and suing to recover damages) and the defense (the party being sued).

Technical experts, professional engineers licensed by the state, are retained by both plaintiff and defense to testify about the operation of the product that allegedly caused the loss. Usually the first testimony developed by the experts is a technical report supplied to the respective attorney. These reports contain the engineer's expert opinion about the operation of the device and the cause of the situation resulting in the lawsuit. The report may be based on an on-site investigation, on computer or laboratory simulations, or on an evaluation of design records. If this report does not support the case of the lawyer who retained the technical expert, the suit may be dropped or settled out of court. If the investigations support the case, a trial will likely ensue and the technical expert may then be called as an expert witness.

During the trial, the plaintiff's attorney will try to show that the design was defective and that the designer and the designer's company were negligent in allowing the product to be put on the market. Conversely, the defense attorney will try to show that the product was safe and was designed and marketed with "reasonable care."

Three different charges of negligence can be brought against designers in products liability cases:

The product was defectively designed. One typical charge is the failure to use state-of-the-art design considerations. Other typical charges are that improper calculations were made, poor materials were used, insufficient testing was carried out, and commonly accepted standards were not followed. In order to protect themselves from these charges, designers must:

- Keep good records to show all that was considered during the design process. These include records of calculations made, standards considered, results of tests, and all other information that demonstrates how the product evolved.
- Use commonly accepted standards when available. "Standards" are either voluntary or mandatory requirements for the product or the workplace; they often provide significant guidance during the design process.
- Use state-of-the-art evaluation techniques for proving the quality of the design before it goes into production.
- Follow a rational design process (such as that outlined in this book) so that the reasoning behind design decisions can be defended.

The design did not include proper safety devices. As previously discussed, safety is either inherent in the product, added to the product, or provided by some form of warning to the user. The first alternative is definitely the best, the second sometimes a necessity, and the third the least advisable. A warning sign is not sufficient in most products liability cases, especially when it is evident that the design could have been made inherently safe or shielding could have been added to the product to make it safe. Thus it is essential that the design engineers foresee all reasonable safety-compromising aspects of the product during the design process.

The designer did not foresee possible alternative uses of the product. If a man uses his gas-powered lawn mower to trim his hedge and is injured in doing so, is the designer of the mower negligent? Engineering lore claims that a case such as this was found in favor of the plaintiff. If so, was there any way the designer could have foreseen that someone was actually going to pick up a running power mower and turn it on its side for trimming the hedge? Probably not. However, a mower should not continue to run when tilted more than 30° from the horizontal because, even with its four wheels on the ground, it may tip over at that angle. Thus the fact that a mower continues to run while tilted 90° certainly implies poor design. Additionally, this example also shows us that not all trial results are logical and that products must be "idiot-proof."

Other charges of negligence that can result in litigation that are not directly under the control of the design engineer are that the product was defectively manufactured, the product was improperly advertised, and instructions for safe use of the product were not given.

Because safety is such an important concern in military operations, the armed services have a standard—MIL-STD 882B, *System Safety Program Requirements*—focused specifically on ensuring safety in military equipment and facilities. This document gives a simple method for dealing with any *hazard,* which is defined as a situation that, if not corrected, might result in death, injury, or illness to personnel or damage to, or loss of, equipment. MIL-STD 882B defines two measures of a hazard: the likelihood or frequency of its occurrence and the consequence if it does occur. Five levels of *frequency of occurrence* are given in Table 8.6 ranging from "improbable" to "frequent." Table 8.7 lists four categories of the *consequence of occurrence.* These categories are based on the results expected if the hazard does occur. Finally, in Table 8.8 frequency and consequence of recurrence are combined in a hazard assessment matrix. By considering the level of the frequency and the category of the consequence, a hazard-risk index is found. This index gives guidance for how to deal with the hazard. For example, say that during the design of the power lawn mower, the possibility of using the mower as a hedge trimmer was indeed considered. Now, what action should be taken? First, using Table 8.6, we decide that the frequency of occurrence is either remote (D) or improbable (E). Most likely, it is improbable. Next, using Table 8.7, we rate the consequence of the occurrence as critical, category II, because severe injury may occur. Then, using the hazard assessment matrix, Table 8.8, we find an index of 10 or 15. This value implies that the risk of this hazard is acceptable, with review. Thus the possibility of the hazard should not be dismissed without review by others with design responsibility. If the potential for seriousness of injury had been less, the hazard could have been dismissed without further concern. The very fact that the hazard was considered, an analysis was performed according to accepted standards, and the concern was documented might sway the results of a products liability suit.

8.8 COMMUNICATION DURING CONCEPT EVALUATION

When the evaluation techniques discussed here have been completed, there should be a few clearly identified concepts for development into products. Also, some specific documentation should have been generated: documentation to support technology readiness, decision matrices to determine the best concepts, and data from models—analytical, experimental, and graphical—that support the evaluation.

TABLE 8.6
The hazard frequency of occurrence

			MIL-STD 882B
Description	**Level**	**Individual item**	**Inventory**
Frequent	A	Likely to occur frequently.	Continuously experienced.
Probable	B	Will occur several times in life of an item.	Will occur frequently.
Occasional	C	Likely to occur sometime in life of an item.	Will occur several times.
Remote	D	Unlikely, but possible to occur in life of an item.	Unlikely, but can reasonably be expected to occur.
Improbable	E	So unlikely that it can be assumed that occurrence may not be experienced.	Unlikey to occur, but possible.

TABLE 8.7
The hazard consequence of occurrence

Description	Category	Mishap definition
Catastrophic	I	Death or system loss
Critical	II	Severe injury, minor occupational illness, or major system damage
Marginal	III	Minor injury, minor occupational illnes, or minor system damage
Neglible	IV	Less than minor injury, occupational illness, or system damage

TABLE 8.8
The hazard-assessment matrix

	Hazard category			
Frequency of occurrence	**I Catastrophic**	**II Critical**	**II Marginal**	**IV Negligible**
A. Frequent	1	3	7	13
B. Probable	2	5	9	16
C. Occasional	4	6	11	18
D. Remote	8	10	14	19
E. Improbable	12	15	17	20

Hazard-risk index	Criterion
1–5	Unacceptable
6–9	Undesirable
10–17	Acceptable with review
18–20	Acceptable without review

Source: MIL-STD 882B.

With this information, the project is ready for a design review by management. The product potential is clear, and a decision as to whether to continue the project can be intelligently made.

8.9 SUMMARY

* The feasibility of a concept is based on the design engineer's knowledge. Often it is necessary to augment this knowledge with the development of simple models.
* In order for a technology to be used in a product, it must be ready. Six measures of technology readiness can be applied.
* A go/no-go screening based on customers' requirements helps filter the concepts.
* The decision-matrix method provides means of comparing and evaluating concepts. The comparison is between each concept and a datum relative to the customers' requirements. The matrix gives insight into strong and weak areas of the concepts. The decision-matrix method can be used for subsystems of the original problem.
* Product safety implies concern for injury to humans and for damage to the device itself, other equipment, or the environment.
* Safety can be designed into a product, it can be added on, or the hazard can be warned against. The first of these is best.
* A hazard assessment is easy to accomplish and gives good guidance.

8.10 SOURCES

Fox, J.: *Quality through Design,* McGraw-Hill, London, 1993. This book is the source for technology readiness assessment.

Jones, J. V.: *Engineering Design: Reliability, Maintainability and Testability,* TAB Professional and Reference books, Blue Ridge Summit, Pa., 1988. This book considers engineering design from the point of view of military procurement and relies strongly on military specifications and handbooks.

Love, S. F.: *Planning and Creating Successful Engineered Designs: Managing the Design Process,* Advanced Professional Development, Los Angeles, 1986. This is a very readable, low-level book on the design process and was influential in the design of this text.

Pugh, S.: *Total Design: Integrated Methods for Successful Product Engineering,* Addison-Wesley, Wokingham, England, 1991. Gives a good overview of the design process and many examples of the use of decision matrices.

Sunar, D. G.: *The Expert Witness Handbook: A Guide for Engineers,* Professional Publications, San Carlos, Calif., 1985. A paperback, it has details on being an expert witness for products liability litigation.

System Safety Program Requirements, MIL-STD 882B, U.S. Government Printing Office, Washington, D.C., 1982. The hazard assessment is from this standard.

8.11 EXERCISES

8.1 Assess your knowledge of the following technologies by applying the six measures given in Section 8.4.

(a) Chrome plating
(b) Rubber vibration isolators

(c) Fastening wood together with nails
(d) Laser positioning systems

8.2 Use a decision matrix or a series of matrices to evaluate the following:
(a) Concepts for the original design problem (Exercise 4.1)
(b) Concepts for the redesign problem (Exercise 4.2)
(c) The alternatives for a new car
(d) The alternatives between various girlfriends or boyfriends (real or imagined)
(e) The alternatives for a job
Note that for the last three the difficulty is choosing the criteria for comparison.

8.3 Perform a hazard assessment on the following items. If you were an engineer on a project to develop each of the items below, what would you do in reaction to your assessment? Further, for hazardous items, what has industry or federal regulation done to lower the hazard?
(a) A manual can opener
(b) An automobile (with you driving)
(c) A lawn mower
(d) A space shuttle rocket engine
(e) An elevator drive system

CHAPTER 9

THE PRODUCT DESIGN PHASE

9.1 INTRODUCTION

The remainder of this book concerns the product design phase, where the goal is to refine the concepts already generated into quality products. This transformation process could be called hardware design, shape design, or embodiment design, all of which imply giving flesh to what was the skeleton of an idea. As shown in Fig. 9.1, this refinement is an iterative process of generating product designs and evaluating them to verify their ability to meet the requirements. Based on the result of the evaluation, the product is patched and refined (further generation), then reevaluated in an iterative loop. Also, as part of the product generation procedure, the evolving product is decomposed into assemblies and individual components. Each of these assemblies and components requires the same evolutionary steps as the overall product. In product design, generation and evaluation are more closely intertwined than in concept design. Thus, the steps suggested for product generation in Chap. 10 include evaluation, which is covered in Chaps. 11 and 12. In these chapters the product designs are evaluated for their quality and cost. Quality will be measured by the product's ability to meet the engineering requirements and the ease with which it can be manufactured and assembled.

The knowledge gained making the transformation from concept to product can be used to iterate back to the concept phase and possibly generate new concepts. The drawback, of course, is that going back takes time. The natural inclination to iterate back and change the concept must be balanced by the schedule established in the design plan.

In two situations design engineers begin at the product design phase in the design process. In the first of these situations, the concept may have been generated in a corporate research lab and then handed off to the design engineers to "productize."

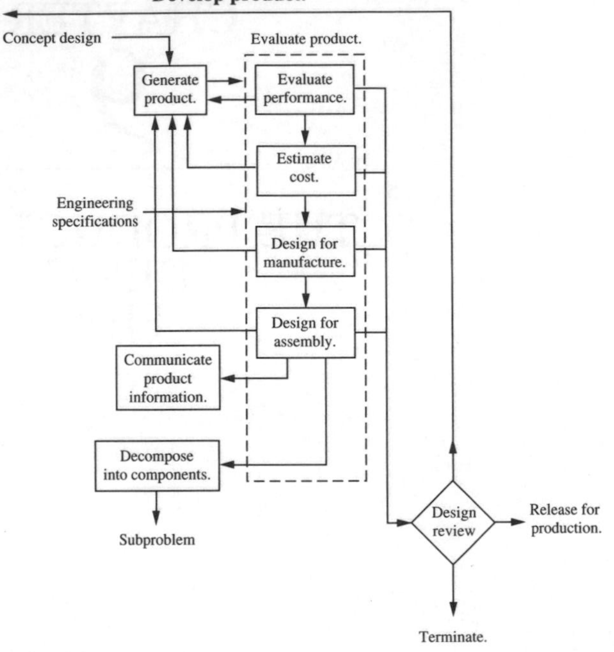

FIGURE 9.1
The product design phase of the design process.

It could be assumed, since the research lab had to develop working models to confirm the readiness of the technology, that the design was well on its way to being a successful product at this point. However, the goal of the researchers is to demonstrate the viability of the technology; their working models are generally handcrafted, possibly held together with duct tape and bubble gum. They probably incorporate very poor product design. This approach is very weak and violates the philosophy of concurrent design. Design engineers, manufacturing engineers, and other stakeholders should have been involved in the process long before the concept was developed to this level of refinement.

In the second situation, the project involves a redesign problem; many problems begin with an existing product that needs only to be redesigned to meet some new requirements. Only "minor modifications" are required, often leading to unexpected, extensive rework resulting in poor-quality products. Consider the history of the field joint on the space shuttle (Chap. 4). The engineers had "only" to modify the Titan rocket joint for use on the space shuttle booster. However, this redesign of a successful product resulted in a joint that left much to be desired.

In either situation, whether the concept comes from a corporate research lab or the project involves only a simple redesign problem, the techniques described in Chaps. 5, 6, 7, and 8 should be applied before the product design phase is ever begun. Only in that way will a good-quality product result.

Before describing the process of refining the concept to hardware, we need to emphasize that we will develop here only enough detail on materials, manufacturing methods, economics, and the engineering sciences to support techniques and examples of the design process. Each area would require a book of its own to develop fully.

Because progress in product development is often measured in the information documented, we introduce the product design phase with a discussion of drawings and bills of materials, the primary records of the product design effort.

9.2 THE IMPORTANCE OF DRAWINGS

Drawings are not only the preferred form of data communication for the designer; they are a necessary part of the design process. Sketching as a form of drawing is an extension of the short-term memory needed for idea generation (see Chap. 3). As the shape of components and assemblies evolve, more formal drawings are used to keep the information organized and easily communicated to others. Thus, a well-trained engineer has standard drafting skills and the ability to represent concepts that are more abstract and best represented as sketches. Specifically, drawings are used to:

1. Archive the geometric form of the design.
2. Communicate ideas between designers and between designers and manufacturing personnel.
3. Support analysis. Missing dimensions and tolerances are calculated on the drawing as it is developed.
4. Simulate the operation of the product.
5. Check completeness. As sketches or other drawings are being made, the details left to be designed become apparent to the designer. This, in effect, helps establish an agenda of design tasks left to accomplish.
6. Act as an extension of the designer's short-term memory. Designers often use sketches to store information they might otherwise forget.
7. Act as a synthesis tool. Sketches and formal drawings allow the piecing together of unconnected ideas to form new concepts.

In making drawings care must be taken because, as a General Electric survey showed, 60 percent of all manufactured components are not made exactly as represented in the drawings. The reasons vary: (1) the drawings are incomplete, (2) the components cannot be made as specified, (3) the drawings are ambiguous, and (4) the components cannot be assembled if manufactured as drawn. Modern solid-modeling CAD tools help to overcome most of these problems.

9.3 DRAWINGS PRODUCED DURING PRODUCT DESIGN

During the design process many types of drawings are generated. Sketches that were encouraged during conceptualization must evolve to final drawings that give enough detail to support production. This evolution usually begins with a layout drawing of the entire product to help define the geometry of the developing assemblies and components. The details of the components and assemblies are partially specified by the information developed on the layouts. As the product is refined, this information is transferred to detail and assembly drawings.

The development of the drawings is synergistic with the evolution of the product geometry and further refinement of its function. As drawings are produced, more knowledge about the product is developed. Some of the major characteristics of the different types of drawings produced during product design—layout, detail, and assembly—and their role in the design process are itemized below.

9.3.1 Layout Drawings

A layout drawing is a working document that supports the development of the major components and their relationships. A typical layout drawing is shown in Fig. 9.2. Consider the characteristics of a layout drawing:

- A layout drawing is a working drawing and as such is frequently changed during the design process. Because these changes are seldom documented, information can be lost. Good records in the design notebook can compensate for this loss.
- A layout drawing is made to scale.
- Only the important dimensions are shown on a layout drawing. In Chap. 10 we see that starting with the spatial constraints sets the stage for developing individual

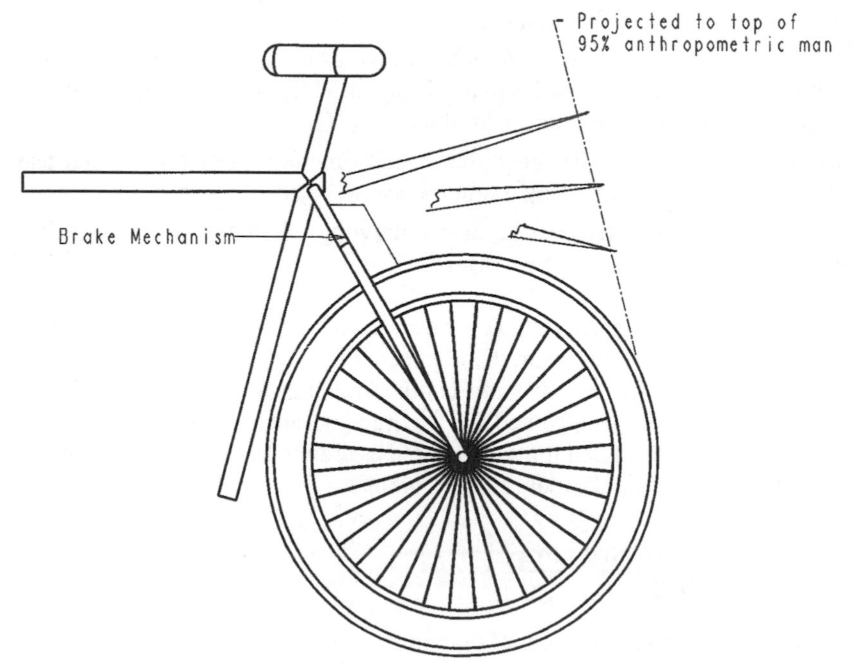

FIGURE 9.2
Typical layout drawing.

components in the product generation process. These constraints are best shown on a layout drawing.

- Tolerances are usually not shown, unless they are critical.
- Notes on the layout drawing are used to explain a design feature or the function of the product.
- A layout drawing often becomes obsolete. As detail drawings and assembly drawings are developed, the layout drawing becomes less useful. However, if the product is being developed on a CAD system, the layout drawing's data file becomes the basis for the detail and assembly drawings.

9.3.2 Detail Drawings

As the product evolves on the layout drawing, the detail of individual components develops. These are documented on detail drawings. A typical detail drawing is shown in Fig. 9.3. Important characteristics of a detail include the following:

- All dimensions must be toleranced. In Fig. 9.3 many of the dimensions are made with unstated company-standard tolerances. Most companies have standard tolerances for all but the most critical dimensions. The upper and lower limits of the critical dimensions in Fig. 9.3 are given.
- Materials and manufacturing detail must be in clear and specific language. Special processing must be spelled out clearly.
- Drawing standards such as those given in ANSI Y14.5M-1994, *Dimensions and Tolerancing*, and in DOD-STD-100, *Engineering Drawing Practices,* or company standards should be followed.
- Since the detail drawings are a final representation of the design effort and will be used to communicate the product to manufacturing, each drawing must be approved by management. A signature block is therefore a standard part of a detail drawing.

9.3.3 Assembly Drawings

The goal in an assembly drawing is to show how the components fit together. There are many types of drawing styles that can be used to show this; one type is an exploded view, as seen in Fig. 9.4. Assembly drawings are similar to layout drawings except that their purpose, and thus the information highlighted on them, is different. An assembly drawing has these specific characteristics:

- Each component is identified with a number or letter keyed to the bill of materials. Some companies put their bill of materials on the assembly drawings; others use a separate document. (The contents of the bill of materials are discussed in the next section.)
- References can be made to other drawings and specific assembly instructions for additional needed information.

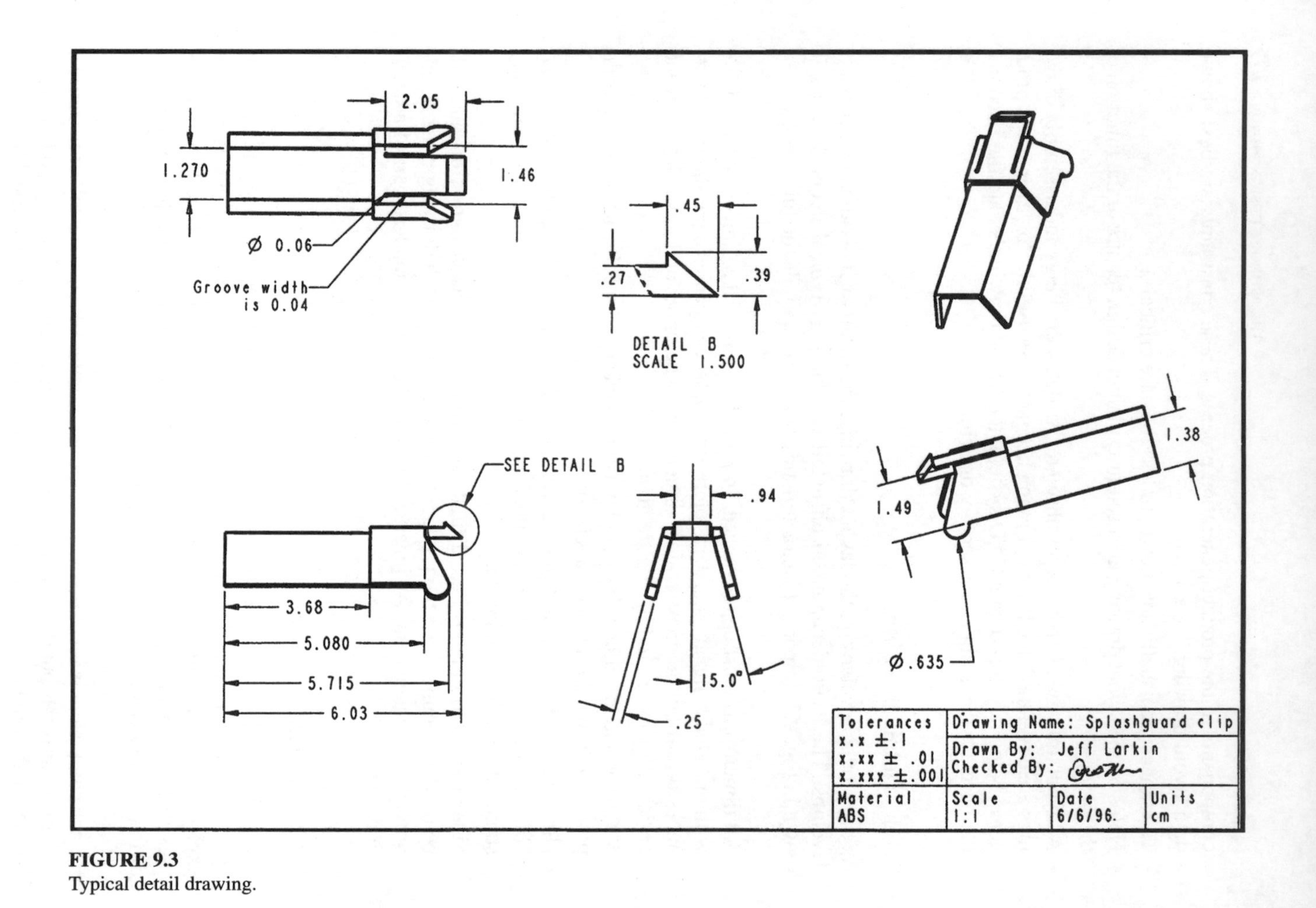

FIGURE 9.3
Typical detail drawing.

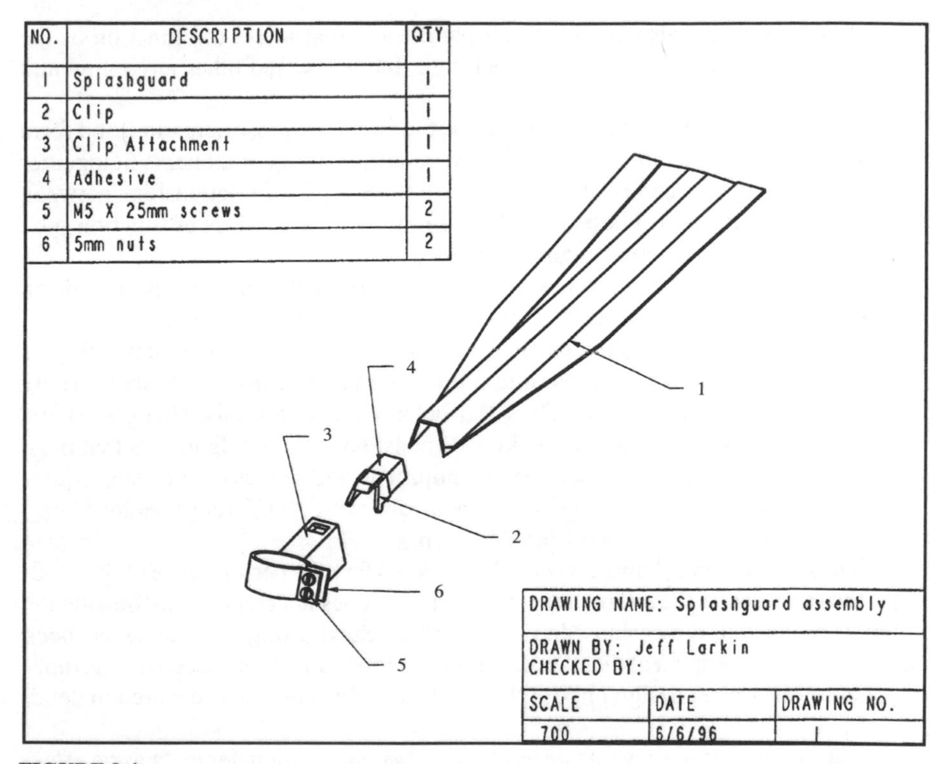

NO.
DESCRIPTION
QTY
1 Splashguard 1
2 Clip 1
3 Clip Attachment 1
4 Adhesive 1
5 M5 X 25mm screws 2
6 5mm nuts 2
1
2
3
4
5
6
DRAWING NAME: Splashguard assembly
DRAWN BY: Jeff Larkin
CHECKED BY:
SCALE
DATE
DRAWING NO.
700
6/6/96
1

FIGURE 9.4
Typical assembly drawing.

- Necessary detailed views are included to convey information not clear in the major views.
- As with detail drawings, assembly drawings require a signature block.

9.4 THE INFORMATION ON DRAWINGS: DIMENSIONING AND TOLERANCING

Drawings, whether on paper or in a CAD system, are the main representation of product information. Many of industry's problems stem from faulty communications using drawings. In fact, one survey showed that most drawings contain errors: components cannot be made as drawn (i.e., they are physically impossible to make), components made as drawn will not fit together, or components that are assembled will not operate as designed. It is essential to make drawings that cannot possibly be misunderstood. The inclusion of manufacturing experts on design teams helps ensure that components can be made. The use of solid models in CAD systems help ensure that components will fit together. However, it is still up to the design engineer to ensure that each system operates as designed, not only initially but under all foreseeable environmental conditions and over its entire operating life.

This section explains one characteristic of the information to be developed in the rest of the book—*robust information*. Although the discussion below is oriented toward dimensions, it also applies to all parameters that affect the function of the product (e.g., material properties, temperature, humidity, and other environmental factors).

Dimensions are information to describe the size, location, orientation, and form of components and assemblies. These in turn determine how components fit together and operate to provide and function. There are three types of relationships between dimensions and function. These are shown on Fig. 9.5, where loss of function (i.e., change in behavior) is plotted against dimensional value.

The top curve represents *sensitive dimensions.* If the dimension is right on target—manufactured to the exact nominal dimension—then the product will perform as designed. As the dimension varies from the target, the functionality is degraded, so the minimum and maximum limits on the dimensions are set to where the function is still acceptable. Thus the behavior continuously changes as the dimension changes. Examples of sensitive dimensions are shaft diameters that must fit in bearings, orifices in flow-measuring equipment, and guides on precision equipment. Some companies label sensitive dimensions with "S/C" (Ford Motor Corp.) or ◇ (General Motors Corp. and Chrysler Corp.).

The middle curve shows a *robust dimension.* Here the performance of the product is the same across the entire range, and behavior does not change until outside the range of acceptable dimensions. In other words, behavior (e.g., function) has been decoupled from component size, location, orientation, and form. Decoupling function, part of the robust design philosophy, is ideal. Robust design is covered in detail in Chap. 11.

The bottom curve shows an *unrelated dimension*, a parameter that has no effect on function regardless of its value. On a typical component, up to 80 percent of the dimensions are unrelated to function.

Dimensions and the tolerances on them should be assigned on the basis of their effect on function, and only unrelated dimensions should be assigned on the basis of the manufacturing process. Dimensions should be clear, concise, consistent, and unambiguous. They should specify the largest possible tolerance the system function will permit.

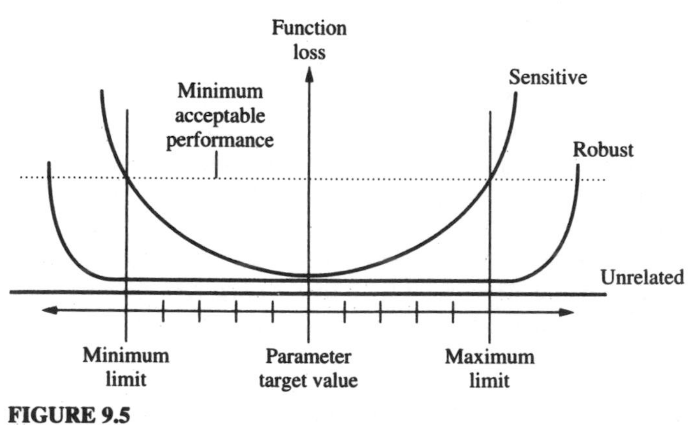

FIGURE 9.5
Function loss curves.

TABLE 9.1
Typical bill of materials

Item	Part	Quantity	Name	Material	Source
1	G-9042-1	1	Governor body	Cast aluminum	Fred's Fine Foundry
2	G-9138-3	1	Governor flange	Cast aluminum	Fred's Fine Foundry
			• • •		
9	X-1784	4	Cover bolt	Plated steel	Phefeer Fasteners

9.5 BILLS OF MATERIALS

The *bill of materials (BOM),* or *parts list,* is like an index to the product. It is good practice to generate the bill of materials as the product evolves. This is commonly done on a spreadsheet, which is easy to update. A typical bill of materials is shown in Table 9.1. To keep lists to a reasonable length, a separate list is usually kept for each assembly. There are six pieces of information on a bill of materials:

1. *The item number or letter.* This is a key to the components on the assembly drawing.
2. *The part number.* This is a number used throughout the purchasing, manufacturing, inventory control, and assembly systems to identify the component. Where the item number is a specific index to the assembly drawing, the part number is an index to the company system.
3. *The quantity needed* in the assembly.
4. *The name or description of the component.* This must be a brief, descriptive title for the component.
5. *The material from which the component is made.* If the item is a subassembly, then this does not appear in the BOM.
6. *The source of the component.* If the component is purchased, the name of the company is listed. If the component is made in-house, this line can be left blank.

9.6 PRODUCT DATA MANAGEMENT

The management of design information is a major undertaking in a company. In fact, this intellectual property is one of the most valuable assets of a company. In past times drawings were stored in large flat drawers or on microfiche and other data stored in file cabinets. Indexing and finding information in this system were usually difficult and often impossible. As product information has become more computer-based, so have methods to manage the information. Generally, computer systems designed to manage both product and process information are referred to as *product data management systems (PDMs).*

Product data management systems are database programs used to support the management of documents and files, product structure, and processes. Within the database there are records with details about the documents and files generated during development of the product. These include CAD files (representing layout, detail and assembly drawings), text documents (including meeting notes and contracts), spread sheets (including QFD, decision matrices, and other analysis), database reports, parts libraries, vendor data, and engineering change orders (ECOs; see Chap. 13). Beyond storing information about these documents and files, a PDM system also allows indexing the information on the basis of the product structure. This can take many forms. For example, a PDM database can be searched to find what components are in an assembly, which components are made of a certain material, or which components are supplied by a certain vendor. In other words, the database in the PDM system allows the product data to be searched and organized according to the immediate information need.

Additionally, PDM systems allow management of the way people create and modify data. Thus, beyond managing product information, a PDM system can also help manage the processes involved in designing and manufacturing a product. Imagine that many people are working on a product currently under development. The information on this product is stored in CAD files and in the other forms described above. Say that an engineer is working on one component, making changes to it. This common activity presents one of the many problems that the PDM system helps address. If the engineer is working on the component and another engineer wants information about that component, what information does the second engineer get? PDM systems help manage this problem by enabling people to check *user packets* out of the system. A user packet is a body of drawings, files, and other documents that the engineer has authorization to change. The concept of checking out a packet is much richer than checking a book out of the library in that the information in the packet is tailored to the need at the moment. While this information is checked out, others can look at the old information that is still in the master file, but they cannot change it. Usually, the information in the master file also contains a note saying that it is checked out and who is working on it.

When changes are made and a packet of information is ready to be checked back in, the PDM system helps manage the approval cycle and the notification of others about the changes. This workflow management is very important because without proper approval of changes and notification of others affected by the changes there can be much wasted time and effort.

Finally, PDM systems help support the task schedule developed in Chap. 5. They generally include tools to support the development and maintenance of Gantt charts (Section 5.4), worker allocation, and task definitions.

9.7 SUMMARY

* Documentation measures progress in product development, and many types of drawings are produced during the product design phase.
* Drawings are the chief form of information communication during product design.
* Layout drawings support the development of the major components and their relationships.
* Detail drawings document the form of the individual components.

* Assembly drawings show how the components fit together.
* A bill of materials is a parts list—an index to the product.

9.8 SOURCES

American Society of Mechanical Engineers, *Dimensioning and Tolerancing,* ANSI Y14.5M-1994, ASME, New York, 1995. This volume is the standard for all dimensioning and tolerancing issues.

Engineering Drawing Practices, DOD-STD-100, U.S. Printing Office, Washington, D.C.

Foster L. W.: *Geo-Metrics III: The Application of Geometric Dimensioning and Tolerancing Techniques,* Addison-Wesley, New York, 1994. This is another good text for learning geometric dimensioning and tolerancing.

Rowbotham G. E., ed.: *Engineering and Industrial Graphics Handbook*, McGraw-Hill, New York, 1982. A large and thorough volume on all aspects of graphic representation, although weak on computer-based systems.

Ullman D. G., S. Wood, and D. Craig: "The Importance of Drawing in the Mechanical Design Process," *Computers and Graphics,* vol. 14, no. 2 (1990), pp. 263–274. A paper that itemizes the different uses of graphical representations in mechanical design.

Wilson B. A.: *Design Dimensioning and Tolerancing,* Goodheart-Wilcox, South Holland, Ill., 1992. This is a good text for learning geometric dimensioning and tolerancing.

9.9 EXERCISES

9.1. Develop a bill of materials for the following:
  (a) A stapler
  (b) A bicycle brake caliper
  (c) A hole punch

9.2. Make a layout drawing of one of the concepts developed for the original design problem (Exercise 4.1) or the redesign problem (Exercise 4.2).

9.3. Take apart a common device and produce a detail drawing of one component.

9.4. Take apart a common device and produce an assembly drawing of it.

CHAPTER 10

PRODUCT GENERATION

10.1 INTRODUCTION

The goal of this and the following chapters is to transform the concepts developed in Chaps. 7 and 8 into products that perform as designed. These concepts may be at different levels of refinement and completeness. Consider the examples in Fig. 10.1—a stick-figure representation of a mechanism and a rather complete sketch for a Splashgard concept. The Splashgard sketch shows the concept at various levels of abstraction, with fairly refined fasteners at the interface between the brushes and frame, an abstract connection between the brushes and frame, an abstract connection between the frame and the brace, and only implied information at the connection of the brace to the bike. Thus, the steps for product development must deal with concepts at many varying levels of refinement.

Refining from concept to product requires work on all the elements shown in Fig. 10.2. Central to this figure is the *function* of the product. Surrounding the function, and mutually dependent on each other, are the *form* of the product, the *materials* used to make the product, and the *production techniques* used to generate the form from the materials. Although these three were considered in conceptual design, the focus was on developing function. Now, in product design, attention turns to developing producible forms that provide the desired function.

The form of the product is roughly defined by the spatial *constraints* that provide the envelope in which the product operates. Within this envelope the product is defined as the *configuration* of *connected components*. In other words, form development is the evolution of components, how they are configured relative to each other and how they are connected to each other. This chapter covers techniques used to generate these characteristics of form.

Also shown in Fig. 10.2, decisions on production require development of how the product's components are *manufactured* from the materials and how these

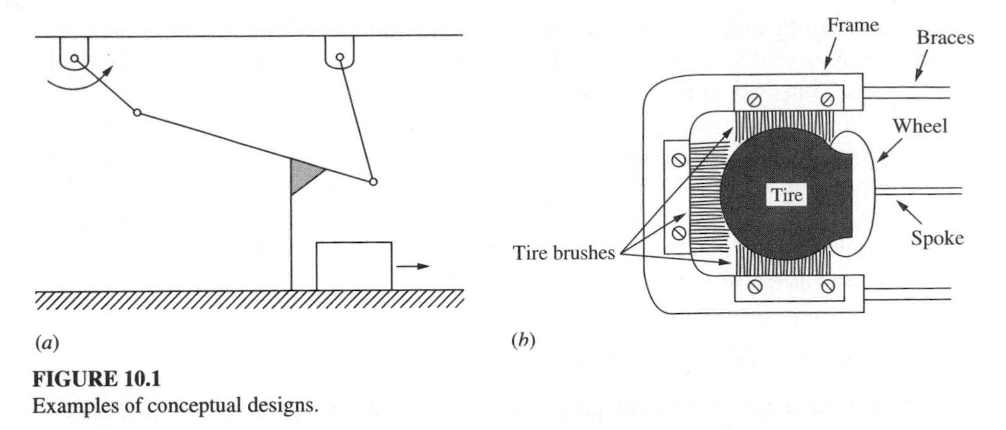

FIGURE 10.1
Examples of conceptual designs.

components are *assembled*. Simultaneous evolution of the product and the processes used to produce it is one of the key features of concurrent engineering. In this chapter the interaction of manufacturing and assembly decisions will affect the generation of the product. Production considerations will become even more important in evaluating the product (Chap. 12).

In the discussion of conceptual design, emphasis was put on developing the function of the product. It is now a reasonable question to ask what should be worked on next—the form, the materials, or the production? The answer is not easy, because even though we work from function to form, form is hopelessly interdependent on the materials selected and the production processes used. Further, the nature of the

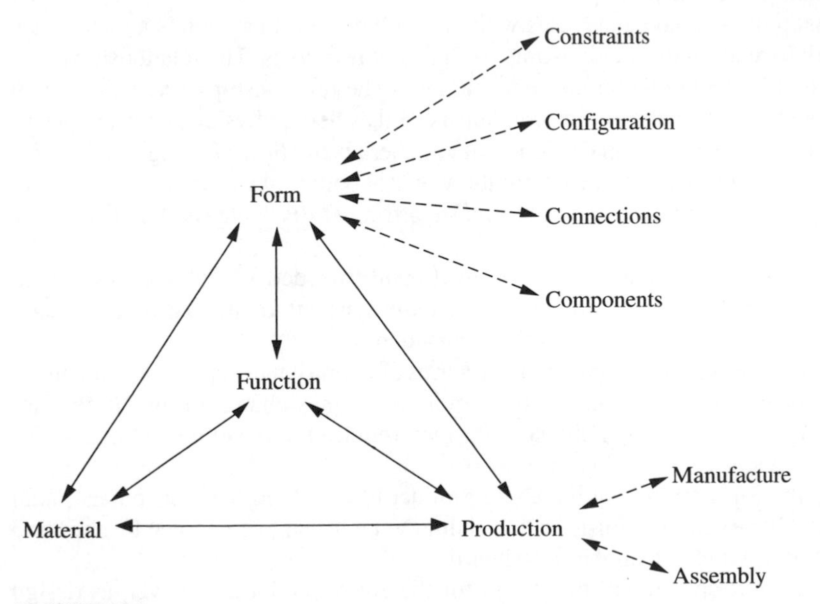

FIGURE 10.2
Basic elements of product design.

interdependency changes with factors such as the number of items to be produced, the availability of equipment, and knowledge about materials and their forming processes. Thus it is virtually impossible to give a step-by-step process for product design. Figure 10.2 shows all the major considerations in product generation. The following sections will begin with form generation and will then cover material and process selection. There is also a section on vendor development, because vendor issues affect product generation. In the following chapters product evaluation will center around the product's ability to meet the functional requirements, ease of manufacture and assembly, and cost.

10.2 FORM GENERATION

Whatever geometry was developed during conceptual design must be questioned and refined. This is usually done on the layout drawing—the drawing in which the product is refined. In this drawing the configuration of the components and the connections between the components are developed. Information generated here is further refined on the final detail drawings of the components.

The material in this section is primary for developing layout drawings of products. At the end of this chapter, the Splashgard is used as an example of the techniques presented.

10.2.1 Collect Spatial Constraints

The spatial constraints are the walls or envelope for the product. Products developed in a layout drawing must begin with knowledge about spatial constraints resulting from interfaces with other objects.

Most products have interfaces with other objects or, to state this another way, must work in relation to other existing, unchangeable objects. The relationships may define actual contact or be for needed clearance. The relationships may be based on the flow of material, energy, or information as well as being physical. For the Splashgard the interface with the bike is physical and there is the flow of energy in the form of forces. Likewise, the interface with the water is both a physical relationship and an energy-flow relationship (i.e., the Splashgard must dissipate the kinetic energy from the water).

Some spatial constraints are for functionally needed space, such as optical paths, or to clear or interfere with the flow of some material. In the case of the Splashgard there is the path of water flow to be considered.

Further, most products go through a series of operational steps as they are used. The functional relationships and spatial requirements may change during these. Taking the varying relationships into account may require the development of a series of layout drawings.

Finally, consideration of the entire product life cycle may reveal other spatial constraints. For example, consideration should to be given to clearance for the hand needed to attach and detach the Splashgard.

Initially the spatial constraints are for the entire product; however, as design decisions are made on one assembly or component, other spatial constraints are added. For large products that have independent teams working on different subassemblies, the coordination of the information can be very difficult. Interdependency can often be planned for while information-centered tasks are being developed (Section 5.4).

10.2.2 Configure Components

Configuration is the architecture, structure, or arrangement of the components and assemblies of components in the product. Developing the architecture or configuration of a product involves decisions dividing the product into individual components and developing the location and orientation of them. Even though the concept sketches probably contain representations of individual components, it is time to question the decomposition represented. A decision to decompose a product or assembly into separate components is justified by one of six reasons:

- Components must be separate if they need to move relative to each other. For example, parts that slide or rotate relative to each other have to be separate components. However, if the relative motion is small, perhaps elasticity can be built into the design to meet the need for motion. This is readily accomplished in plastic components by using elastic hinges, which are thin sections of fatigue-resistant material that act as a one-degree-of-freedom joint.
- Components must be separate if they need to be of different materials for functional purposes. For example, one area of the product is to conduct heat and another is to insulate.
- Components must be separate if they need to be moved for accessibility. For example, if the cabinet for a computer is made as one piece, this would not allow access to install and maintain the computer components.
- Components must be separate if they need to accommodate material or production limitations.
- Components must be separate if there are available standard components that can be considered for the product.
- Components must be separate if separate components would minimize costs. Sometimes it is less expensive to manufacture two simple components than it is to manufacture one complex component. This may be true in spite of the added stress concentrations and assembly costs caused by the interface between the two components.

These guidelines for defining the boundaries between components only help define part of the configuration. Equally important during configuration design are the location and orientation of the components relative to each other. *Location* is the measure of the components relative position in x, y, z space. *Orientation* refers to the angular relationship of the components. Usually components can have many different locations and orientations; layout drawings help with the search for possibilities. Configuration design was introduced in Section 2.4.2 as a problem of location and orientation.

10.2.3 Develop Connections: Create and Refine Interfaces for Functions

This is a key step when embodying a concept because *the connections or interfaces between components support their function and determine their relative positions and locations.* Here are guidelines to help develop and refine the interfaces between components:

Interfaces must always reflect force equilibrium and consistent flow of energy, material, and information. Thus they are the means through which the product will be designed to meet the functional requirements. Most design effort occurs at the connections between components, and attention to these interfaces and the flows through them is key to product development. During the redesign of an existing product, it is useful to disassemble it, note the flows of energy, information, and materials at each joint, and develop the functional model one component at a time.

After developing interfaces with external objects, consider the interfaces that carry the most critical functions. Unfortunately it is not always clear which functions are most critical. Generally, they are those functions that seem hardest to achieve (about which the knowledge is the weakest) or those described as most important in the customers' requirements.

It is important to maintain functional independence in the design of an assembly or component. This means that the variation in each critical dimension in the assembly or component should affect only one function. If changing a parameter changes multiple functions, then affecting one function without altering others may be impossible. The importance of this cannot be overemphasized.

Care must be taken in separating the product into separate components. Complexity arises since one function often occurs across many components or assemblies and since one component may play a role in many functions. For example, a bicycle handlebar (discussed in Section 2.3) enables many functions but does none of them without other components.

Creating and refining interfaces may force decompositions that result in new functions or may encourage the refinement of the functional breakdown. As the interfaces are refined, new components and assemblies come into existence. One step in the evaluation of each potential embodiment is to determine how each new component changes the functionality of the design. (This is discussed in more detail in Section 11.2.)

In order to generate the interface, it may be necessary to treat it as a new design problem and utilize the techniques developed in Chaps. 7 and 8.

When developing a connection, classify it as one or more of the following types:

- *Fixed, nonadjustable connection.* Generally one of the objects supports the other. Carefully note the force flow through the joint (see the next section). These connections are usually fastened with rivets, bolts, screws, adhesives, welds, or by some other permanent method.

- *Adjustable connection.* This type must allow for at least one degree of freedom that can be locked. This connection may be field-adjustable or intended for factory adjustment only. If it is field-adjustable, the function of the adjustment must be clear and accessibility must be given. Clearance for adjustability may add spatial constraints. Generally, adjustable connections are secured with bolts or screws.
- *Separable connection.* If the connection must be separated, the functions associated with it need to be carefully explored. See the Splashgard example in Section 10.5 for details.
- *Locator connection.* In many connections the interface determines the location or orientation of one of the components relative to another. Care must be taken in these connections to account for errors that can accumulate in joints. This will be further discussed in Chap. 11.
- *Hinged or pivoting connection.* Many connections have one or more degrees of freedom. The ability of these to transmit energy and information is usually key to the function of the device. As with the separable connections, the functionality of the joint itself must be carefully considered.

Connections are the key in the development of all products. See the discussion of the Splashgard product design generation in Section 10.5.

10.2.4 Develop Components

It has been estimated that fewer than 20 percent of the dimensions on most components in a device are critical to performance. This is because most of the material in a component is there to connect the functional interfaces and therefore is not dimensionally critical. Once the functional interfaces have been determined, designing the body of the component is often a sophisticated connect-the-dots problem.

Consider the following example of an aircraft hinge component. The spatial constraints for this component and its interface points are shown in Fig. 10.3. The major functions of this individual component are to transfer forces and clear (not interfere with) other components. The load on the component and the geometry of the interfaces are detailed in the figure. The component is a simple structural member that must transfer the load from the hinge line to the fastening area. As shown in Fig. 10.4, there are many solutions to this problem. The solutions in Fig. 10.4*a* and *b* are machined out of a solid block of material. The solution in Fig. 10.4*c* is made from welded sections of off-the-shelf extruded tubing and plate. These three solutions are good if only a few hinge plates are to be manufactured. If the number to be produced is high, then the forged component in Fig. 10.4*d* may be a good solution. Note that all four of these components have the same interfaces with adjacent components. One is fixed and may need to be removable, and the other has one degree of freedom. The only difference is in the body, the material connecting the interfaces. All of these product designs are potentially acceptable, and it may be difficult to determine exactly which one is best. A decision matrix may help in making this decision.

The material between interfaces generally serves three main purposes: (1) to carry forces or other forms of energy (heat or an electrical current, for instance) between interfaces with sufficient strength and rigidity; (2) to act as an enclosure or guide for other components (guiding air flow, for instance); or (3) to provide

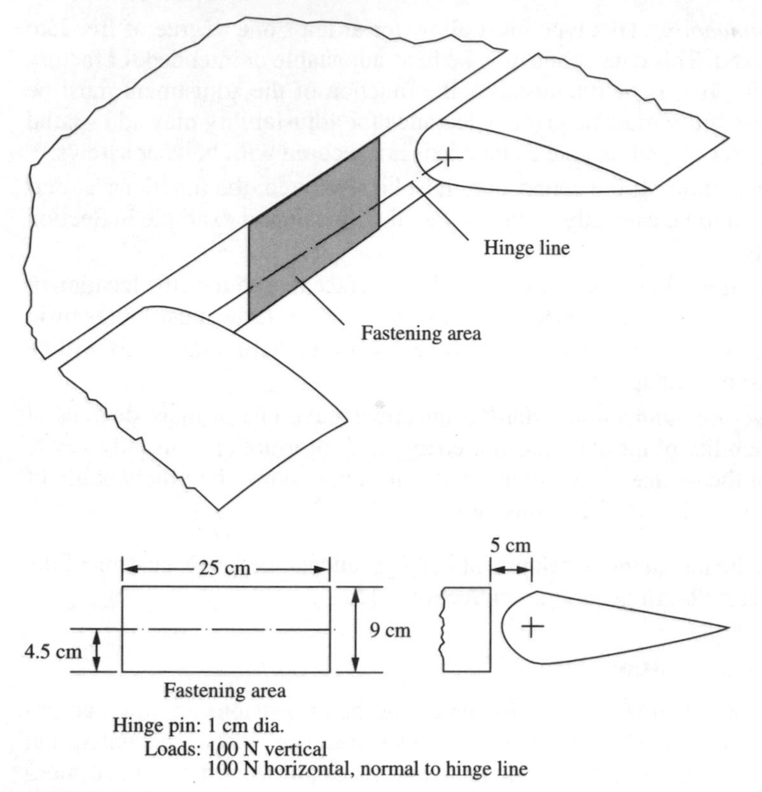

FIGURE 10.3
Requirements on an aircraft hinge plate.

appearance surfaces. We have said before that functionality occurs mainly at component interfaces; this is not always true. The exception occurs when the body of a component provides the function—for example, needed mass, stiffness, or strength—in which case shape becomes more important.

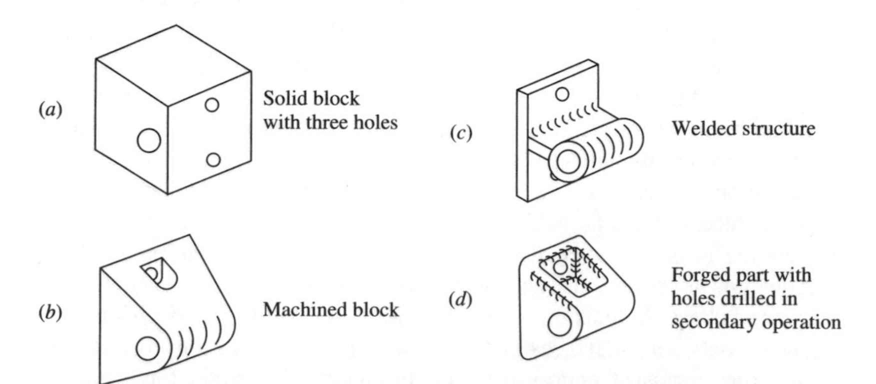

FIGURE 10.4
Potential solutions for the structure of the aircraft hinge plate.

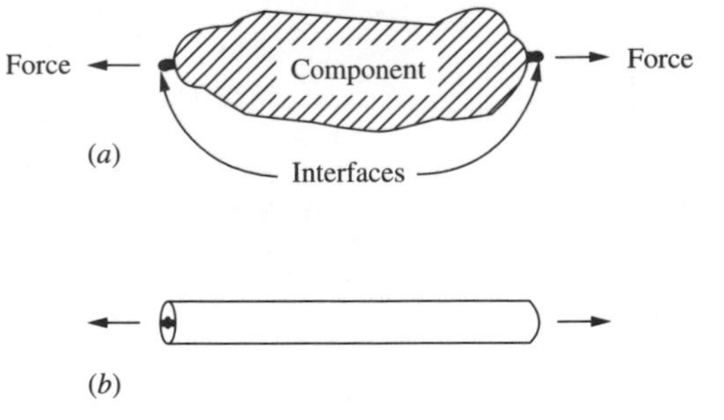

FIGURE 10.5
A bar in tension.

It is best to connect interfaces with strong structural shapes, those that carry the load from one interface to another with the least internal stress. In ideal strong shapes the force is distributed evenly throughout the entire structure. Thus, stress is evenly distributed, and failure occurs only when the stress at all locations reaches a critical value. The simplest strong shape is a rod in tension (or compression). If there are two interfaces, as shown in Fig. 10.5*a*, and a component needs to transmit a force from one to the other, the strongest shape to use is a rod in tension, Fig. 10.5*b*. Once away from the ends (interfaces), the forces are distributed as a constant stress throughout the rod. Thus, this shape provides the most efficient (in terms of amount of material used to transmit the force) shape possible. Other strong shapes are the following:

A *truss* carries all its load as tension and compression, both of which load all the material. A rule of thumb is *always triangulate the design of shapes*. This is often accomplished by providing shear webs in components to effectively act as triangulating members. The back surface in Fig. 10.6 acts as a shear web to help transmit force *A* to the bottom surface. A rib provides the same function for force *B*. An exception to this guideline is seen in the bikes shown in Fig. 2.9 and at the end of the Preface. On these, the rear wheel is mounted on a cantilever arm rather than on the traditional triangle. This allows the flexibility of the cantilever to act as a spring suspension system. The penalty for this is higher stresses where the cantilever mounts to the body of the bike.

A *hollow cylinder,* the most efficient carrier of torque, comes as close as possible to having constant stress throughout all the material. Any closed prismatic shape exhibits the same characteristic. A common example of an approximately closed prismatic shape is an automobile or van body. As the front right wheel of the van shown in Fig. 10.7 goes over a bump, a torque is put on the entire vehicle.

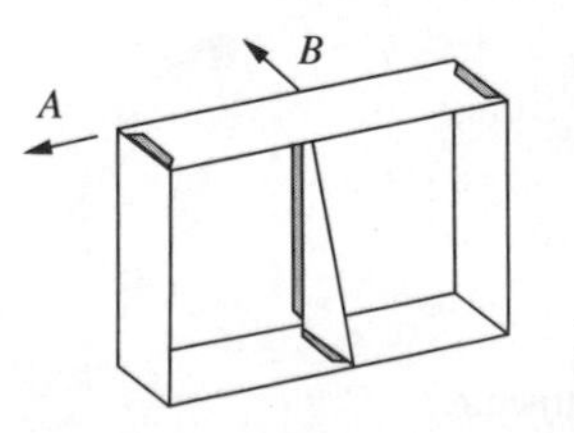

FIGURE 10.6
A triangulated component.

FIGURE 10.7
Component that carries torque.

An *I-beam* is designed to carry bending loads in the most efficient way possible, since most of the material is far away from the neutral bending axis. The principle behind the I-beam is shown in the structural shape of Fig. 10.8. Although not an I, it behaves much like one, as the majority of the material is where it will best carry the stress.

Less stress is generally developed if, when the bodies of components are being designed, direct force transmission paths are used. A good method for visualizing how forces are transmitted through components and assemblies is to use a technique called *force flow visualization*. The following rules explain the method.

1. Treat forces like a fluid that flows in and out of the interfaces and through the component.
2. The fluid takes the path of least resistance through the component.
3. Sketch multiple flow lines. The direction of each flow line will represent the maximum principal stress at the location.
4. Label the flow lines for the major type of stress occurring at the location: tension (T), compression (C), shear (S), or bending (B). Note that bending can be decomposed into tension and compression and that shear must occur between tension and compression on a flow line.
5. Remember that force is transmitted at interfaces primarily by compression. Shear only occurs in adhesive, welded, and friction interfaces.

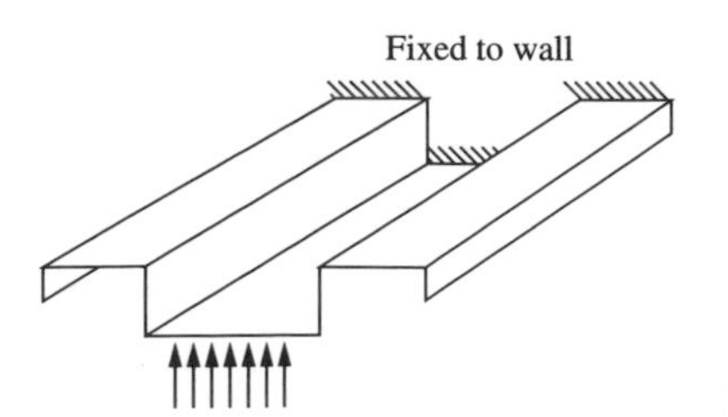

FIGURE 10.8
Example of an I-beam structure.

Two examples clearly illustrate the above rules. Consider the aft field joint of the space shuttle (Fig. 4.9) loaded in tension during launch (Fig. 10.9*a*). Assume that force in the tang is evenly distributed around the circumference at the left end of the assembly. Following the rules above, the force flow in the tang looks as shown in Fig. 10.9*b*. The flow enters (leaves) evenly at the left end and leaves (enters) at the compression interface between the tang and the pin. If the pin is replaced with a bolt and the bolt is tightened, then part of the force may be transmitted to the clevis by friction. However, it is more conservative to assume that all the load is transmitted to the pin or bolt by compression.

The flow lines can be labeled for the type of stress on the basis of our knowledge of strength of materials. The tension in the left end must change to a shear stress before becoming compressive at the interface with the pin. A tensile (compressive) force always becomes shear as it is transformed into a compressive (tensile) force.

The force flow through the clevis is similar to that in the tang. The force flow in the pin is as shown in Fig. 10.9*c*. On the basis of knowledge of strength of materials, it is compressive at each interface with shear between. Combining all three of these components, we can visualize the force flow through the entire assembly and schematically represent it by the line in Fig. 10.9*d*. The force flow is not a very

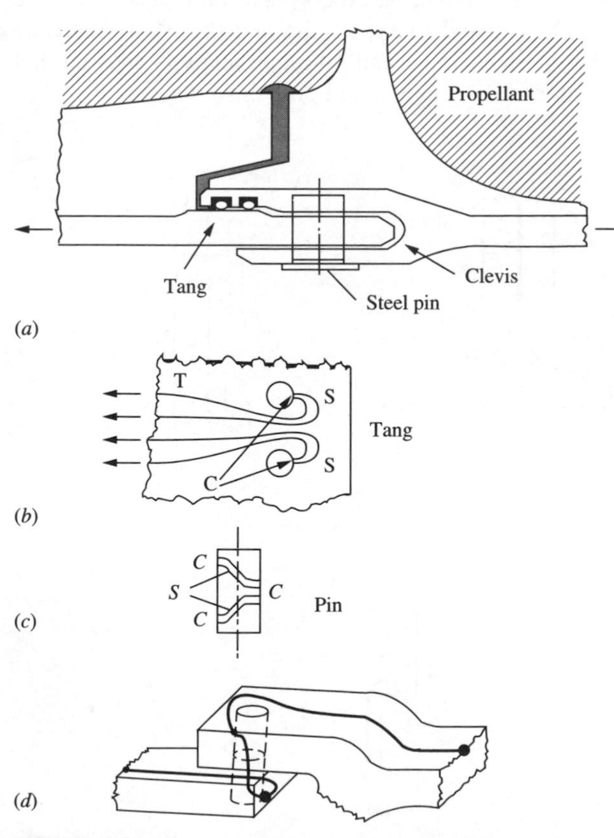

FIGURE 10.9
Force flow in the space shuttle aft field joint during launch.

direct path and leads to many points of potential failure. In fact each point labeled T, S, or C is a site of potential failure.

The tee joint in Fig. 10.10*a* represents a second example of the use of force flow visualization. Figure 10.10*b* shows two ways of representing the force flow in the flange. The left side shows the bending stress in the flange labeled B; the right side shows the bending stress decomposed into tension (T) and compression (C), which forces consideration of the shear stress. The force flow through the nut and bolt is shown in Fig. 10.10*c* and *d*. The force flow in the entire assembly is shown in Fig. 10.10*e*.

In summary, force flow helps us visualize the stresses in a component or assembly. It is best if the force paths are short and direct. The more indirect the path, the more potential failure points and stress concentrations.

In designing the bodies of components, be aware that stiffness more frequently than stress determines the adequate size. Although component design textbooks emphasize strength, the dominant consideration for many components should be their stiffness. An engineer who used standard stress-based design formulas to analyze a shaft carrying a small torque and virtually no transverse load found that it should be 1 mm in diameter. This seemed too small (a gut-feeling evaluation), so the engineer raised the diameter to 2 mm and had the system built. The first time power was put through the shaft, it flexed like a noodle and the whole machine vibrated violently. Redesign based on stiffness and vibration analysis showed that the diameter should have been at least 10 mm to avoid problems.

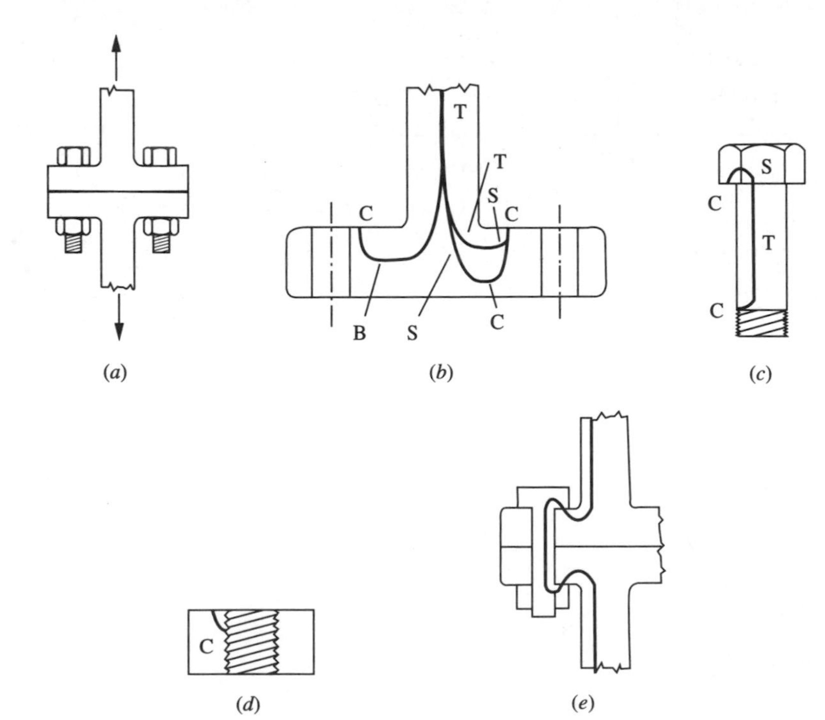

FIGURE 10.10
Force flow in a tee joint.

Finally, in designing components, use standard shapes when possible. Many companies use *group technology* to aid in keeping the number of different components in inventory to a minimum. In group technology each component is coded with a number that gives basic information about its shape and size. This coding scheme enables a designer to check whether components already exist for use in a new product.

10.2.5 Refine and Patch

Although not shown as a basic element of product design in Fig. 10.2, refining and patching are major parts of product evolution. *Refining,* as described in Section 2.8, is the activity of making an object less abstract (or more concrete). *Patching* is the activity of changing a design without changing its level of abstraction.

The importance and interrelationship of refining and patching the shape can be clearly seen in the following example. A designer was developing a small box to hold three batteries in a series. This subsystem powered the clock/calendar of a personal computer. The designer's notebook sketch of the final assembly is shown in Fig. 10.11. The assembly is composed of a bottom case, a top case, and four contacts. Figure 10.12 shows the evolution of one of the contacts (contact 1), again through the sketches and drawings made by the designer. The number beside each graphic image shows the percentage of the total design effort completed when the representation was made. The designer was simultaneously at work on other components of the product. (The circled letters in Fig. 10.12*c* were added for this discussion and were not in the original drawings.) The design of the battery contact is one of continued *refinement.* Each figure in the series moves the design closer to the final form of the component. The initial sketch (Fig. 10.12*a*) shows circles representing contact to the battery and a curved line representing current conduction. The final drawing of the contact (Fig. 10.12*e*) is a detailed design ready for prototyping. Figure 10.12*c* is of special interest, as it clearly shows the evolutionary process. The designer began by redrawing the left contact, A, from the earlier sketch (Fig. 10.12*b*). She also redrew line B, which represents an edge of the structure connecting the contact to the wire. After beginning to draw this line, she realized that, since she last worked on this component, a plastic wall, C, had been added to the product and the contact could no longer continue straight across. At this point, she *patched* the design by tilting the connecting structure B up to position D. The sketch was then completed, with the wire connection still represented by an arc (E). Moments later, the designer further patched the component by combining the wire and the connecting structure, making the structure between contacts all one component (F), and then immediately redrew it, as in Fig. 10.12*d*. The elimination of the wire simplified the component; there was no reason for a separate wire in the first place. The battery contact was patched by combining two components. The component was then refined to a fully dimensioned form (Fig. 10.12*e*).

From this example and others we can identify many different types of patching:

- *Combining* is making one component serve multiple functions or replace multiple components. Combining will be strongly encouraged when the product is evaluated for its ease of assembly (Section 12.5).

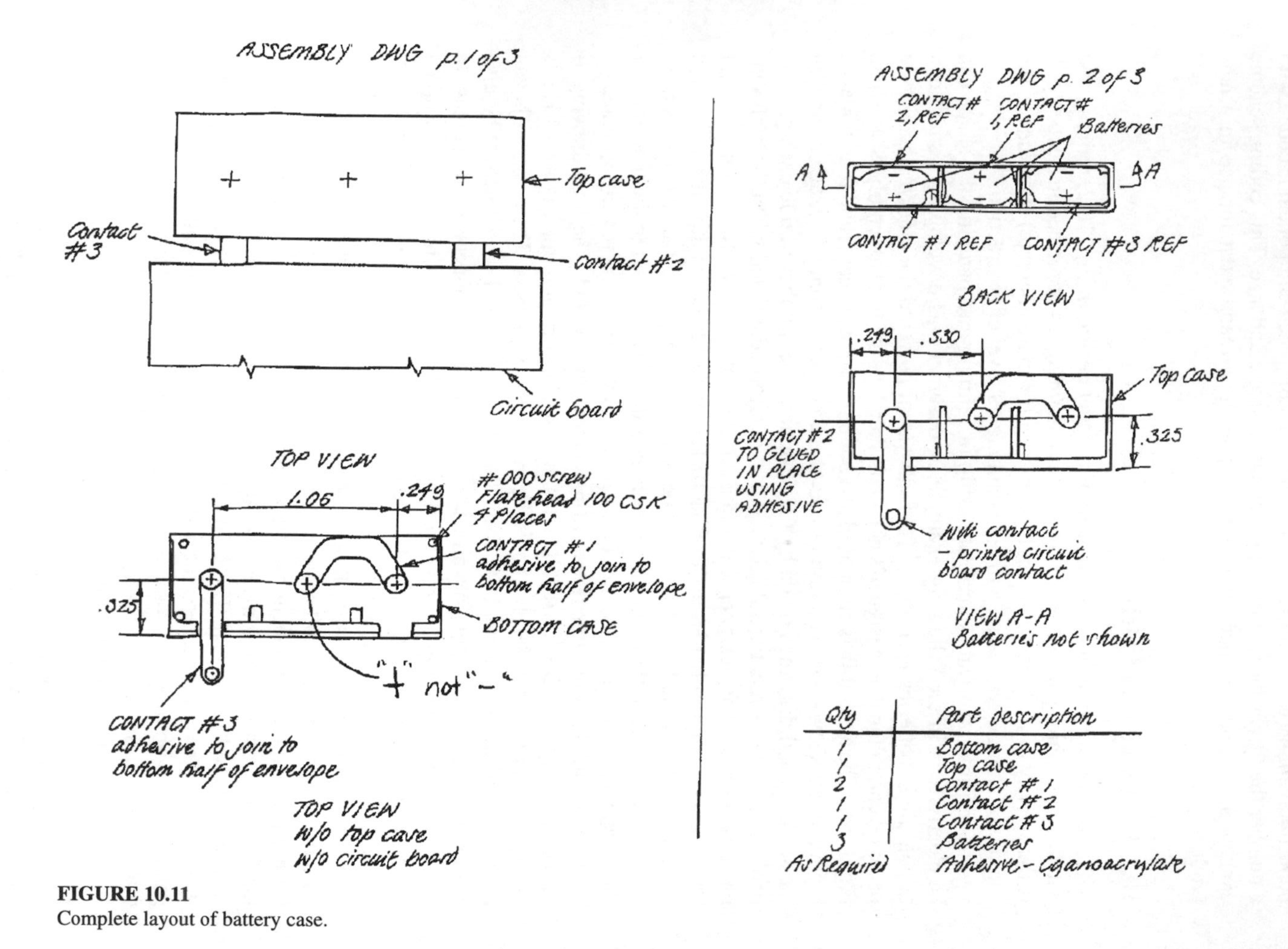

Qty	Part description
1	Bottom case
1	Top case
2	Contact #1
1	Contact #2
1	Contact #3
3	Batteries
As Required	Adhesive - Cyanoacrylate

FIGURE 10.11
Complete layout of battery case.

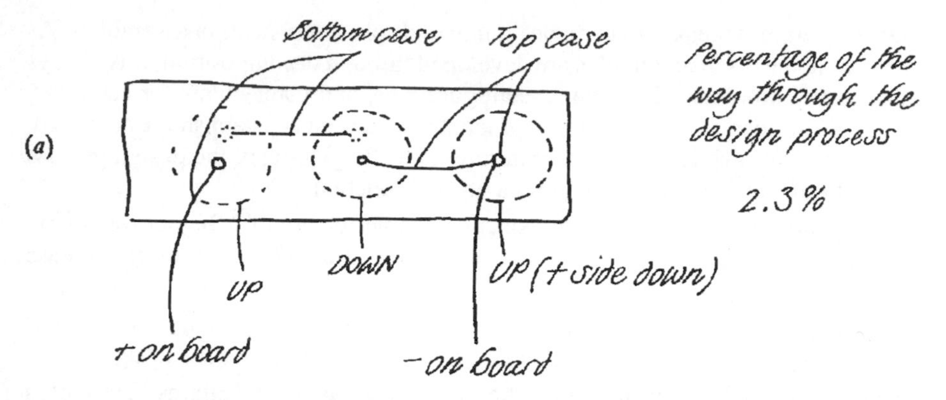

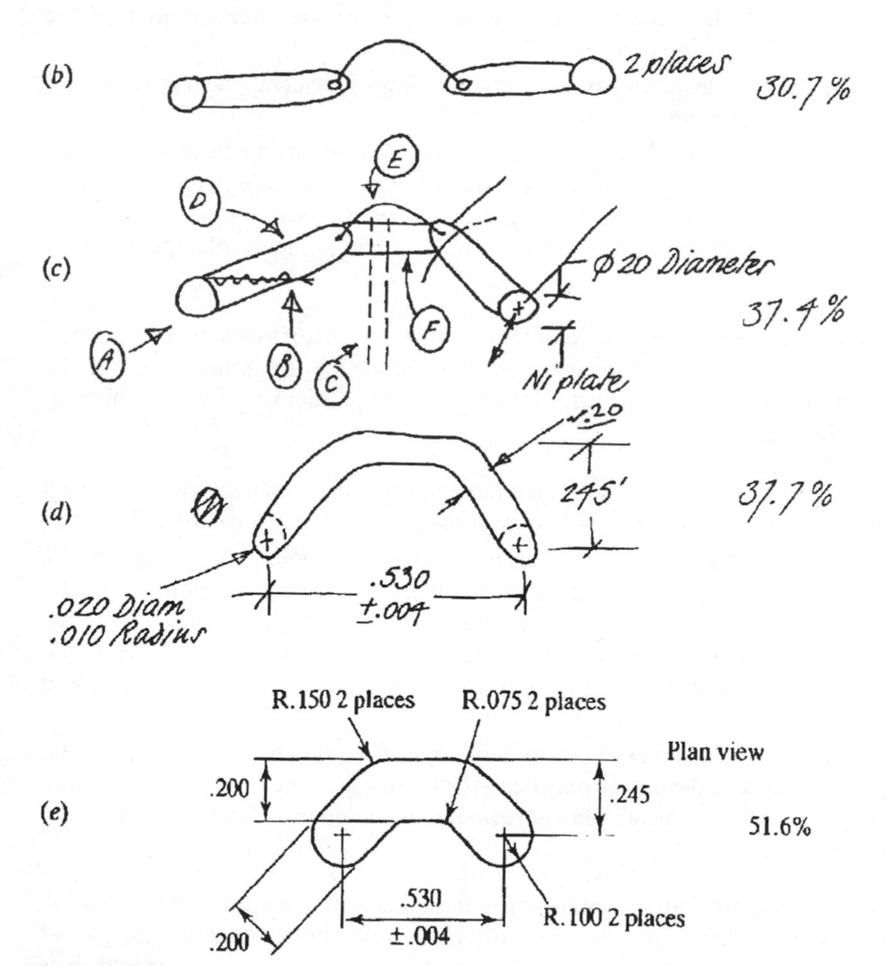

FIGURE 10.12
Evolution of a battery contact.

- *Decomposing* is breaking a component into multiple components or assemblies. As new components or assemblies are developed through decomposition, it is always worthwhile to review constraints, configurations, and connections for each one. Because the identification of a new component or assembly establishes a new need, it is even worthwhile to consider returning to the beginning of the design process with it and considering new requirements and functions.
- *Magnifying* means making a component or some feature of it bigger relative to adjacent items. Exaggerating the size or number of a feature will often increase one's understanding of it.
- *Minifying* means making a component or some feature of it smaller. Sometimes eliminating, streamlining, or condensing a feature will improve the design.
- *Rearranging* means reconfiguring the components or their features. This often leads to new ideas, because the reconfigured shapes force rethinking of how the component fulfills the functions. It may be helpful to rearrange the order of the functions in the functional flow.
- *Reversing* means transposing or changing the view of the component or feature; it is a subset of rearranging.
- *Substituting* means identifying other concepts, components, or features that will work in place of the current idea. Care must be taken because new ideas sometimes carry with them new functions. Sometimes the best approach here is to revert to conceptual design techniques in order to aid in the development of new ideas.

Excessive patching implies trouble. If design progress is stuck on one function or component and patching does not seem to be resolving the difficulty, it may be a waste of time to continue the effort. To relieve the problem, apply the following three suggestions.

- Return to the techniques in conceptual design; try to develop new concepts based on the functional breakdown and the resources for ideas given in Chap. 7.
- Consider that certain design decisions have altered or added unknowingly to the functions of the component. As products evolve, many design decisions are made; it is easy to unintentionally change the function of a component in the process. It is always worthwhile, when stuck on finding a quality solution, to investigate what functions the component is fulfilling. (This topic is further addressed in Section 11.2.)
- If investigating the changes in functionality does not aid in resolving the problem, the requirements on the design may be too tight. It is possible that the targets based on engineering requirements were unrealistic; the rationale behind them should be reviewed.

The results of efforts to refine or patch any aspect of the product can lead in either of two directions. First, and most often, the refinement or patching is part of the generate/evaluate loop in product design. After each patch or refinement, it is good practice to revisit the decisions that have been made in developing the product to this point before reevaluating. As the product becomes more refined, evaluation usually requires more time and resources; therefore, double-checking can lead to savings. Second, if no satisfactory solution can be found, the result of the refining or patching effort requires a return to an earlier phase of the design process.

10.3 MATERIALS AND PROCESS SELECTION

At the same time form is being developed, it is important to identify materials and production techniques and to be aware of their specific engineering requirements. An experienced designer has a short list of materials and processes in mind even with the earliest concepts.

In developing an understanding of the product, we may have set requirements on materials, manufacturing, and assembly. At a minimum we did competitive benchmarking on similar devices, studying them for conceptual ideas and for what they were made of and how they were made. All this information influences the embodiment of the product in several ways:

First, the *quantity of the product to be manufactured* greatly influences the selection of the manufacturing processes to be used. For a product that will only be built once, it is difficult to justify the use of a process that requires high tooling costs. Such is the case with injection molding, in which the mold cost almost exclusively determines the component cost for low-volume production (see Section 12.2.4). In general, injection-molded plastic components are only cost-effective if the production run is at least 15,000.

A second major influence on the selection of a material and a manufacturing process is *prior-use knowledge* for similar applications. This knowledge can be both a blessing and a curse. It can direct selection to reliable choices, yet it may also obscure new and better choices. In general it is best to be conservative, and

When in doubt,
Make it stout,
Out of things
You know about.

When studying existing mechanical devices, get into the habit of determining what kind of materials were used for what types of functions. With practice, the identity of many different types of plastics and, to some degree, of the type of steel or aluminum can be determined simply by sight or feel.

Appendix A provides an excellent reference for material selection. It includes two types of information: a compendium of the properties of the 25 materials most often used in mechanical devices and a list of the materials used in common mechanical devices. The 25 most commonly used materials include eight steels and irons, five aluminums, two other metals, five plastics, two ceramics, one wood, and two other composite materials. The properties listed include the standard mechanical properties, along with cost per unit volume and weight. This list is intended to serve as a starting place for material selection. Detailed information on the many thousands of different materials available can be found in the list of references given at the end of Appendix A. Additionally, the appendix contains a list of materials used in common products. Since many different materials can be used in the manufacture of most products, this list gives only those most commonly used.

Knowledge and experience are the third influence on the choice of materials and manufacturing processes. Limited knowledge and experience limit choices. If only available resources can be utilized, then the materials and the processes are limited by these capabilities. However, knowledge can be extended by including on the design team vendors or consultants who have more knowledge of materials and manufacturing processes, so the number of choices can be increased.

Probably the most compelling point in the selection of a material is its *availability*. A product that has a very small production run would probably use off-the-shelf materials. If the design requires structural shapes (I-beams, channels, or L shapes) that must be light in weight, then extruded aluminum shapes could be used. This decision, however, limits the material choices. Aluminum extrusions are readily available in only a few alloy/temper combinations (6061-T6, 6063-T6, and 6063-T52). Other alloys are available on special order. There is a setup charge to obtain these, and a minimum order of a few hundred pounds—a complete run of material—would also apply. If the available alloy/tempers have properties needed by the product, they can be used. If they do not, the product shape may need to be changed.

In concurrent engineering the material and production processes selected must evolve as the shape of the product evolves. As a product matures, its layout, details, materials, and production techniques are refined (become less abstract). At the same time a product is refined, changes are sometimes patched with no accompanying refinement. Suppose the material initially chosen for a component was identified only as "aluminum"; this selection must now be refined and may be patched. For example, the refining/patching history of the selection of material for one component is

$$\text{“Aluminum”} \rightarrow 2024 \rightarrow 6061 \rightarrow 6061\text{-T6}.$$

That is, the selection of "aluminum" was refined to a specific alloy 2024, which was changed (patched) to a different alloy, 6061, which was then refined by identifying its specific heat treatment, T6. This evolution is typical of what occurs as a product is refined toward a final configuration.

Sometimes during the design of a new product, the requirements cannot be met with existing materials or production techniques, no matter how much patching and shape modification occurs. This situation gives rise to the development of new materials and manufacturing processes. Until recently, the thought of designing the materials and processes to meet the product design needs meant postponing the design project so that material or production technology could reach maturity (Section 8.4). However, recent advancements in the knowledge of metal and plastic materials have, to a certain extent, allowed for material and process design on demand.

10.4 VENDOR DEVELOPMENT

Systems, assemblies, and components fall into two classifications: use of what is available from vendors, or design of new hardware. Mechanical designers seldom design basic mechanical components for each new product, since these components are readily available from vendors. For example, few engineers outside of fastener manufacturing companies design new fasteners. Similarly, few designers outside of gear companies design gears. When such basic components are needed in a product, they are usually specified by the designer and purchased from a vendor who specializes in manufacturing them. In general, finding an already existing product that meets the needs in the design is less expensive than designing and manufacturing it, since the companies that specialize in making a specific component have many advantages over an in-house design-and-build effort:

- They have a history of designing and manufacturing the product, so they already have the expertise and machinery to produce a quality product.
- They already know what can go wrong during design and production. A new design effort would require extensive time and experience before reaching the same level of expertise.
- They specialize in the design and manufacture of the component, so they can make it in volumes high enough to keep the cost below what can be achieved through an in-house effort.

Additionally, even if the exact product is not available, most vendors can help develop products or components that are similar to what they already manufacture.

In past times it was common for a company to send detail drawings of components to a number of vendors and select the vendor that quoted the lowest cost. This philosophy has given way to concurrent engineering, which involves a small number of vendors in the design process from the beginning and includes them in the decisions that affect what they will be supplying. In fact, large companies have reduced the number of vendors by an order of magnitude since the mid 1980s. Some companies financially invest in their vendors, and vice versa, to further improve the bond. These tight relationships lead to improved product quality, as will be discussed in Section 13.7.

10.5 GENERATING PRODUCT DESIGN FOR THE SPLASHGARD

We now look at results of the design efforts in the Splashgard project as an example of concurrent design. Only small portions of the design team's work will be presented; the entire history of the design process is too lengthy to present here.

During the conceptual design effort, the design team decided to develop a Splashgard between the wheel and rider that would fasten onto the seat post using a two-part clip. To begin developing the layout drawing the designer first determined the spatial constraints. For the Splashgard, the important external spatial constraints identified were the wheel envelope, seat tube dimensions, seat configuration, and brake mechanism area. These were used to initialize the layout drawing shown in Fig. 10.13. Another important spatial constraint, based on function, was defined by the need to keep the wheel spray off the rider. This constraint determined how far the Splashgard should project back behind the seat. From this layout, the design team concluded that the Splashgard had to project back at least 42.24 cm (16.63 in) from the seat post.

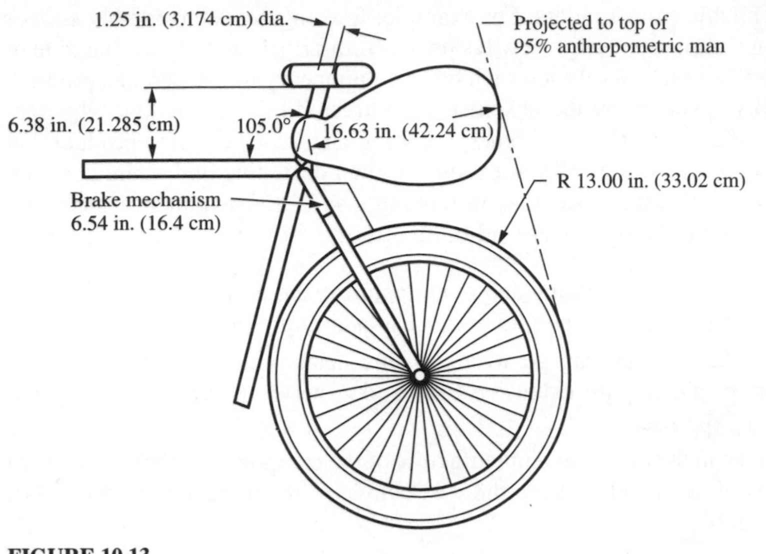

FIGURE 10.13
Initial layout drawing for the Splashgard.

The design team then developed the configuration for the components in the product. On the basis of work to date, the team identified three separate components: the keeper, the clip, and Splashgard body. The architecture for these was very simple. The keeper had to attach to the seat post or seat and thus had to be separate from the Splashgard. The clip was to attach to the keeper. The clip could be part of the body (the part that would actually deflect the water) or could be a separate component. This decision was highly dependent on the manufacturing processes chosen. The design team initially specified that the clip would be made by injection molding and that the body would be made of sheet material. This decision necessitated a connection between these two components.

Based on this decomposition, they identified six different connections, or interfaces, to be designed (Fig. 10.14). Interface 1 is between the keeper and the seat or seat post. Interface 2 is between the installer of the keeper and the keeper itself. Even though a customer survey showed the installation of the keeper to be less important than the mounting of the Splashgard itself, this interface needs at least some attention. During its design, the functional steps identified in the lower half of Fig. 7.5 apply; namely, the keeper must be grasped, positioned, attached, and

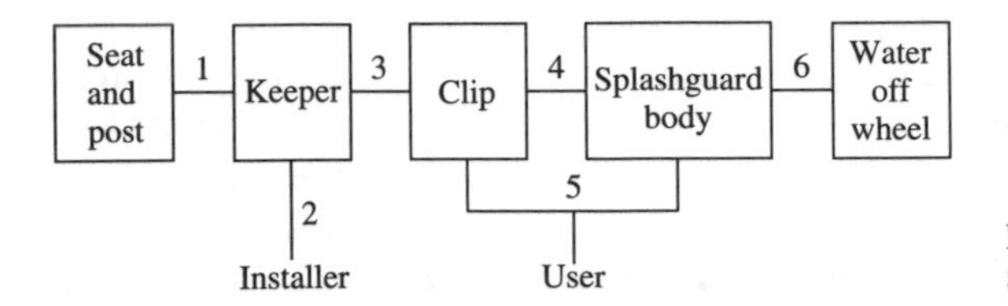

FIGURE 10.14
Splashgard connections.

verified. Interface 3 is between the keeper and the clip, which would be similar to a Fastex clip noted in Fig. 7.7 and would require extensive design effort. A decision has already been made (Table 8.4) to use this type of connection. Interface 4, between the clip and the body, can possibly be eliminated if they are one component. Interface 5 is between the user and the assembly that is to be attached and detached. Finally, interface 6, the main purpose of the Splashgard, is between the water and the Splashgard body. Note that interfaces 1, 2, 5, and 6 are between the Splashgard and the external environment and that interfaces 3 and 4 are internal.

Each of these interfaces defines a subsystem that the team designed using the same process as in the design of the total system. Specifically, for each interface the engineering specifications and multiple concepts were developed. Often this was accomplished with less formality than for the entire system; however, in each case care was taken to ensure that they understood *what* the interface had to accomplish (i.e., specifications and functions) before they spent time designing *how* the hardware would look.

The design team refined the interfaces as much as possible, beginning with external interfaces 1 and 6. The engineering target for the force to cause release was 5 N (1.1 1b) (Fig. 6.3). In further thinking about this requirement, the team realized that there might also be forces to the side and downward. The Splashgard had to be either stiff enough to withstand these forces without breaking or flexible enough to deflect in any direction without damage. Thus, the actual design of the component itself forced modification of the requirements to read, "Lateral or vertical force at tip of Splashgard without damage." The team decided that 5 N was too low and raised it to 15 N. They also decided that it would be easier to design a flexible system; they noted that this decision was based on weak knowledge and might need to be changed later.

The free-body diagram of the vertical plane forces on interfaces 1 and 6 is shown in Fig. 10.15. The distance *d*, the height of the keeper interface with the seat post, is assumed to be 5 cm, and thus the forces for equilibrium are as shown in the figure. The design team established the function of this joint prior to developing any concepts for it. The functions they found, shown in Fig. 10.16, are not the same as those developed during conceptual design, Figs. 7.5 and 7.6, since the decision to add the keeper to the product came *after* the initial functional decomposition and imposed new functions for interface 1. The design team identified two fairly standard concepts to fill these functions (Fig. 10.17*a* and *c*). Here (*b*) is the side view for both top views (*a*) and (*c*). (Utilizing the techniques of Chap. 7, the team might have

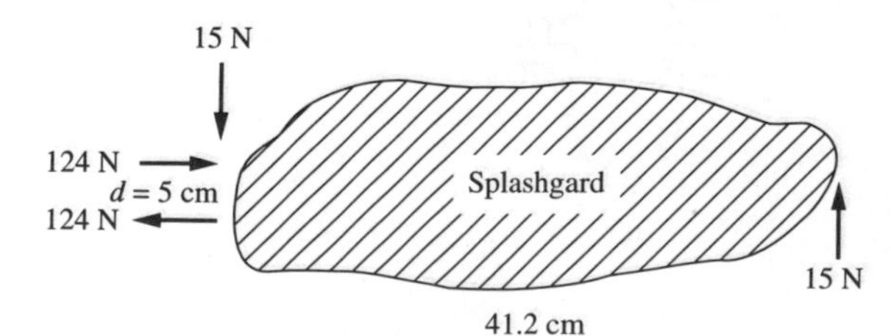

FIGURE 10.15
Free-body diagram of the Splashgard.

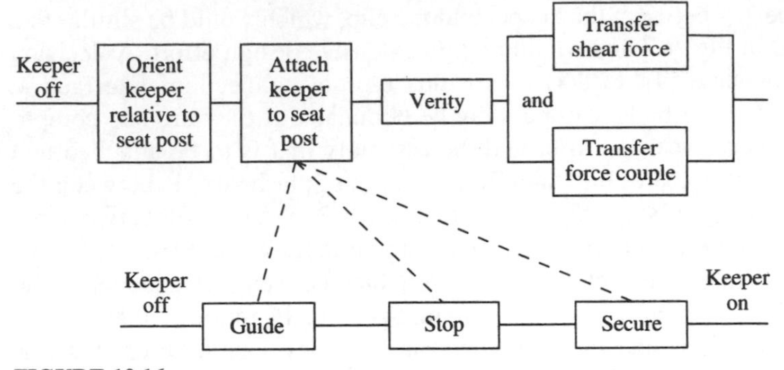

FIGURE 10.16
Functional decomposition of interface 1.

developed other, more unique concepts.) These two interface designs were evaluated later in the design process and one chosen for the final product.

Since connection 3 was critical to the perceived quality of the product, the team took care to develop engineering specifications for this interface. In Fig. 6.3, the house of quality for the entire system, many of the engineering specifications relate directly to this connection, such as steps to attach, time to attach, steps to detach, and time to detach. Additionally, now that attention was focused on this joint, other customers' requirements and engineering specifications were developed. A complete QFD for this joint is shown in Fig. 10.18. Note that the items marked by a ~ are from Fig. 6.7. Most of the other entries were not possible when the original house was developed and have resulted from decisions made in the interim.

In order to refine connection 3, the team investigated available products. As noted on the original sketch for the fastening idea (Fig. 7.7), two-part clips used on backpacks and bike helmets are generically called "Fastex" clips. The name of the manufacturer of Fastex clips, ITW Nexus, was found on the back of a sample, and the company was located in the *Thomas Register's* company index. A phone call to its sales department resulted in securing a catalog (one page from which is shown in Fig. 10.19); it revealed that the company made only the clips that connected to webbing, as shown in the figure. Thus, if the Fastex-type fastener was to connect the Splashgard to the seat post, the team would have to design a custom fastener. The

(*a*)

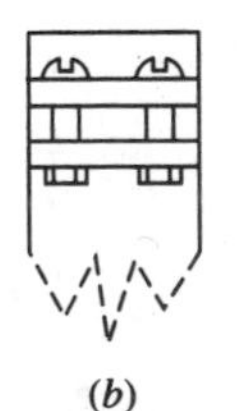

(*b*)

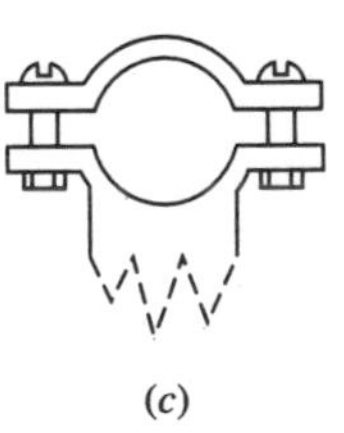

(*c*)

FIGURE 10.17
Concepts for interface 1.

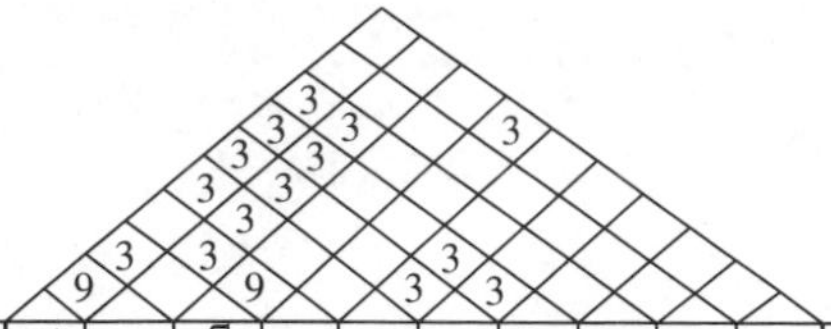

	Joe	Jane	Fred	Number of steps to attach	Time to attach (sec)	Number of steps to detach	Time to detach (sec)	Vertical alignment tolerance (mm)	Horizontal alignment tolerance (mm)	Angular alignment tolerance (deg)	Force to connect (N)	Clear verification signal (y = 1, n = 0)	Thumb/finger squeeze force to detach (N)	Whale tail	Norco	Fastex clip
Functional performance																
Fast to attach	5	4	8	3	9			9	9	9	9	1		1	4	4
Fast to detach	9	5	10			3	9						9	2	4	4
Can attach when dirty	7	13	12	3	3			9	9	9	9	3		3	3	4
Can detach when dirty	11	12	13			3	9						9	3	3	4
Human factors																
Can attach with one hand																
Easy to align	10	11	8					9	9	9				1	1	3
Easy to connect	8	8	8								9			1	2	4
Connection is verified	5	10	10									9		2	2	4
Can easily detach with one hand	7	6	4										9	1	4	3
Whale tail				5	25	2	5	4	1	1	10	0	5			
Norco				3	5	1	3	2	3	30	10	0	5			
Fastex dip				1	1	2	2	5	5	20	10	1	3			
Target				1	2	2	3	5	5	20	10	1	3			

FIGURE 10.18
QFD for connection 3.

team also found two patent numbers (4,150,464 and 4,171,555) stamped on the back of each Fastex clip. They obtained these patents as a source of ideas and to ensure that the resulting design did not infringe on the property rights of ITW Nexus. As it turned out, only patent number 4,150,464 pertained to the clip itself; the other patent related to the connection to the webbing.

The design team studied the patent drawings for patent number 4,150,464, partially shown in Fig. 10.20. As seen in drawing figure 2, features 58, 60, 42, and 44 orient the clip; in drawing figure 4, features 54 and 56 guide it in slot 48; in drawing figure 1, features 18 and 24 stop it; and in drawing figure 2, features 30, 34, and 38 secure it. The snap noise of the spring fingers 34 and 36 slapping against sides 24 and 26 verify that the clip is latched. The other features in the patent drawings helped with other added functions, such as connecting the webbing and deflecting and stopping the spring fingers.

To generate conceptual ideas for interface 3, the design team used the morphology approach (Table 10.1). The main difference between this figure and Table

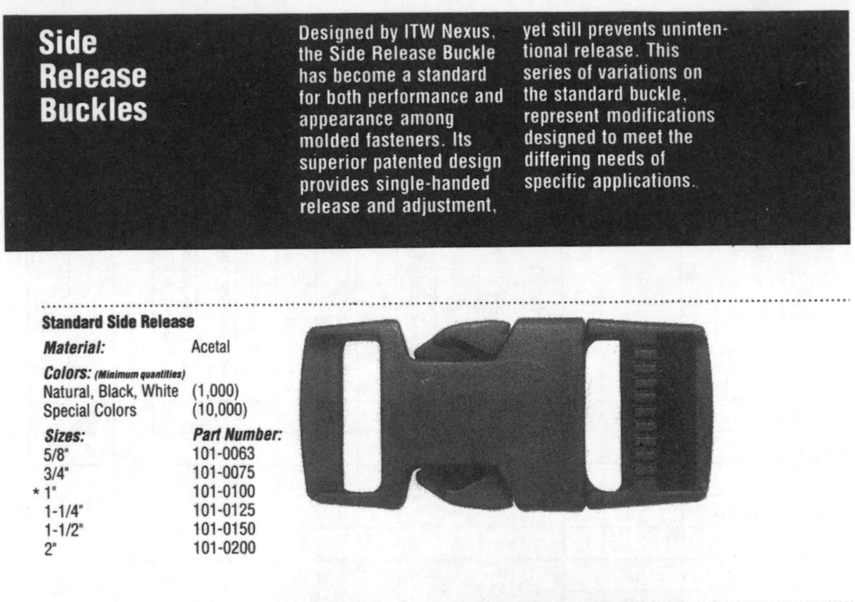

Side Release Buckles
Designed by ITW Nexus, the Side Release Buckle has become a standard for both performance and appearance among molded fasteners. Its superior patented design provides single-handed release and adjustment, yet still prevents unintentional release. This series of variations on the standard buckle, represent modifications designed to meet the differing needs of specific applications.
Standard Side Release
Material: Acetal
Colors: (Minimum quantities)
Natural, Black, White (1,000)
Special Colors (10,000)
Sizes: Part Number:
5/8" 101-0063
3/4" 101-0075
* 1" 101-0100
1-1/4" 101-0125
1-1/2" 101-0150
2" 101-0200

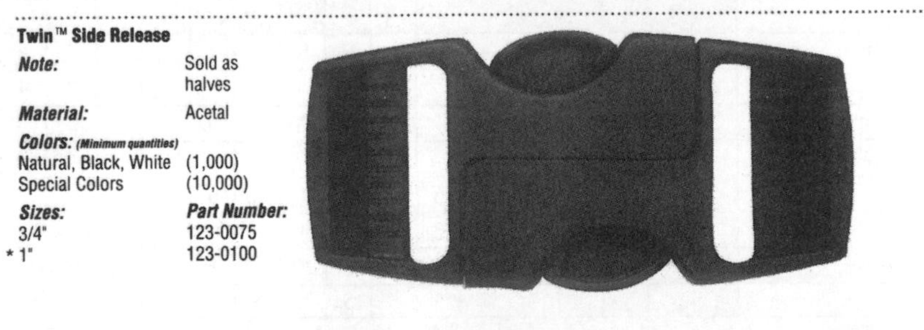

Twin™ Side Release
Note: Sold as halves
Material: Acetal
Colors: (Minimum quantities)
Natural, Black, White (1,000)
Special Colors (10,000)
Sizes: Part Number:
3/4" 123-0075
* 1" 123-0100

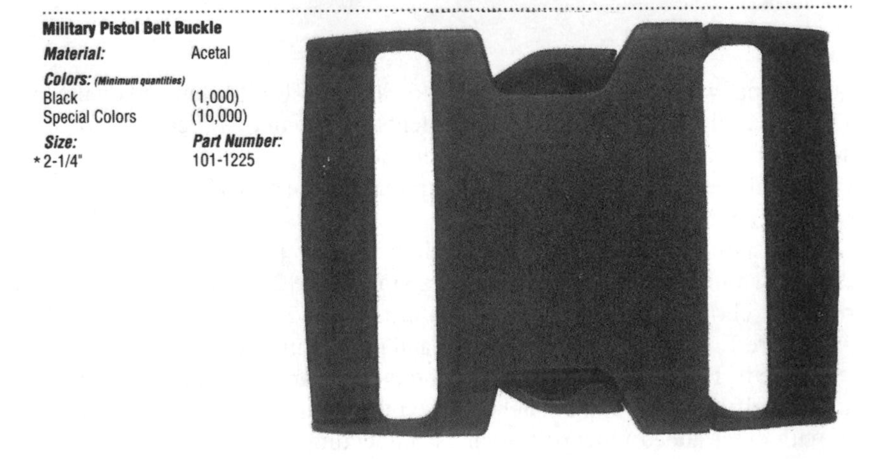

Military Pistol Belt Buckle
Material: Acetal
Colors: (Minimum quantities)
Black (1,000)
Special Colors (10,000)
Size: Part Number:
* 2-1/4" 101-1225

* Pictured

FIGURE 10.19
A page from the ITW Nexus catalog.

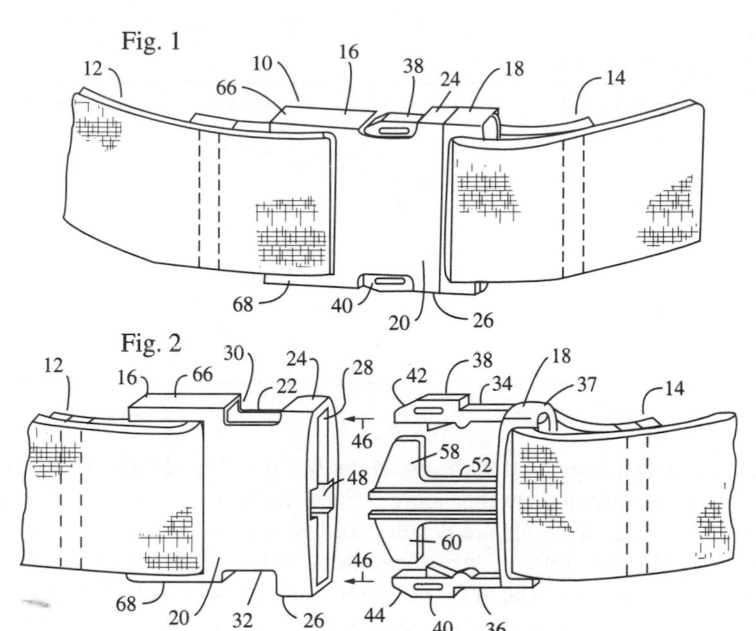

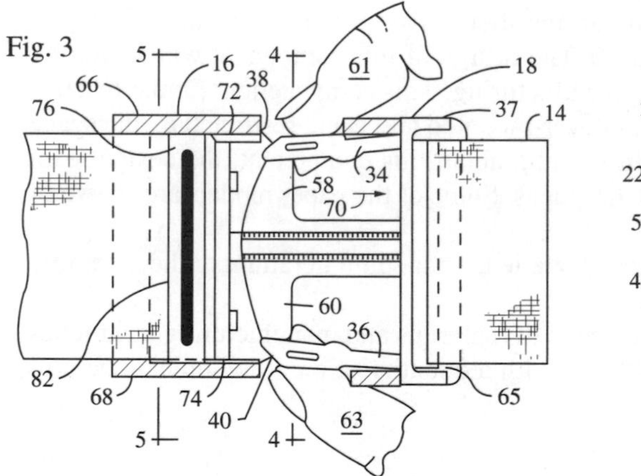

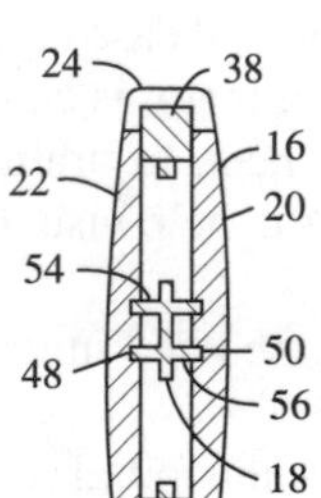

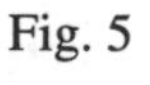

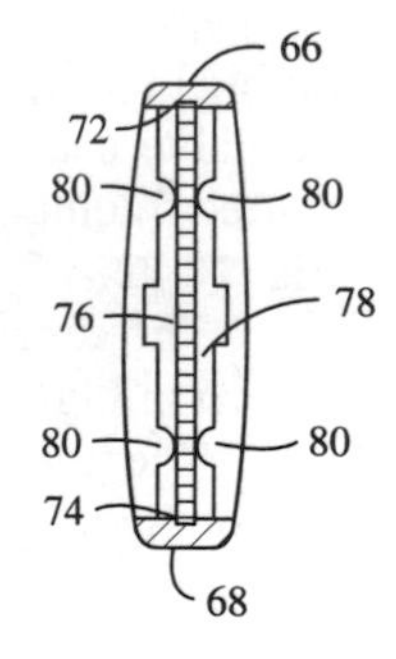

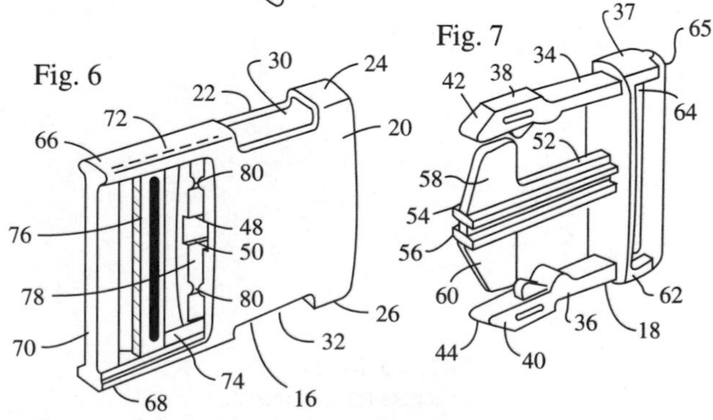

FIGURE 10.20
Drawings for the Fastex clip, patent 4,150,464.

TABLE 10.1
Morphology for interface 3

Orient (1.0)	Straight-sided wall	Tapered wall	Notch	Hook	Tongue and groove
Guide (2.1)	Straight central rail	Pivot	Curved rail		
Secure (2.3)	Snap on top and bottom	Snap on side	Hook on bottom and snap on top	Snap on bottom and hook on top	

7.2, the initial morphology, is that the concepts listed in Fig. 10.21 are focused on the two-part spring clips and are thus more refined. In Table 10.1 only the orienting, guiding, and securing functions are developed, because the other functions copied the Fastex design. This listing was developed iteratively with the development of the Splashgard body. Thus the shape of the body and interface 4 evolved at the same time as this morphology. Concepts for the interface resulting from this study are shown in Fig. 10.21.

Work on design of the components themselves centered on the Splashgard body, between interfaces 4, 5, and 6. The design of this component was driven by stiffness, appearance, and ease of manufacturing. This component is similar to a traditional fender except that it is cantilever-mounted and is not restricted by the space available near the wheel. In brainstorming about this component, the design team started folding paper to help visualize ideas. Some of the paper models are shown in Fig. 10.22.

The concept embodied in the above led, after some iteration, to the assembly shown in Fig. 10.23.

The team made the initial selection of materials and manufacturing techniques for the Splashgard. They reasoned that with a quantity of a million units to be pro-

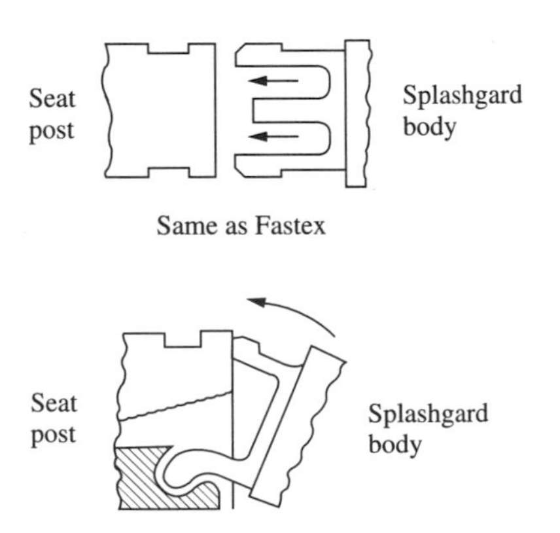

FIGURE 10.21
Concepts for interface 3.

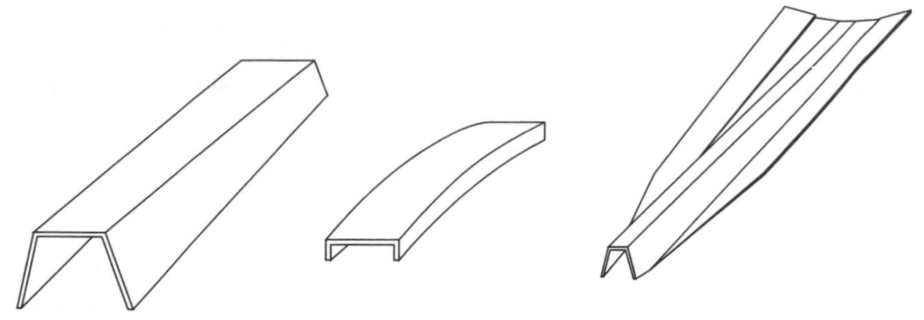

FIGURE 10.22
Models for the Splashgard.

duced over the life of the product (200,000 per year for five years), almost any material/production combination suitable for high volume could be specified. They found that many bike parts are made of ABS (see App. A). Fastex clips are standardly produced by injection-molding Acetal. The manufacturer assured the design team that they could be made in many other materials. The manufacturing techniques most associated with bicycle parts are injection-molding and sheet-forming processes.

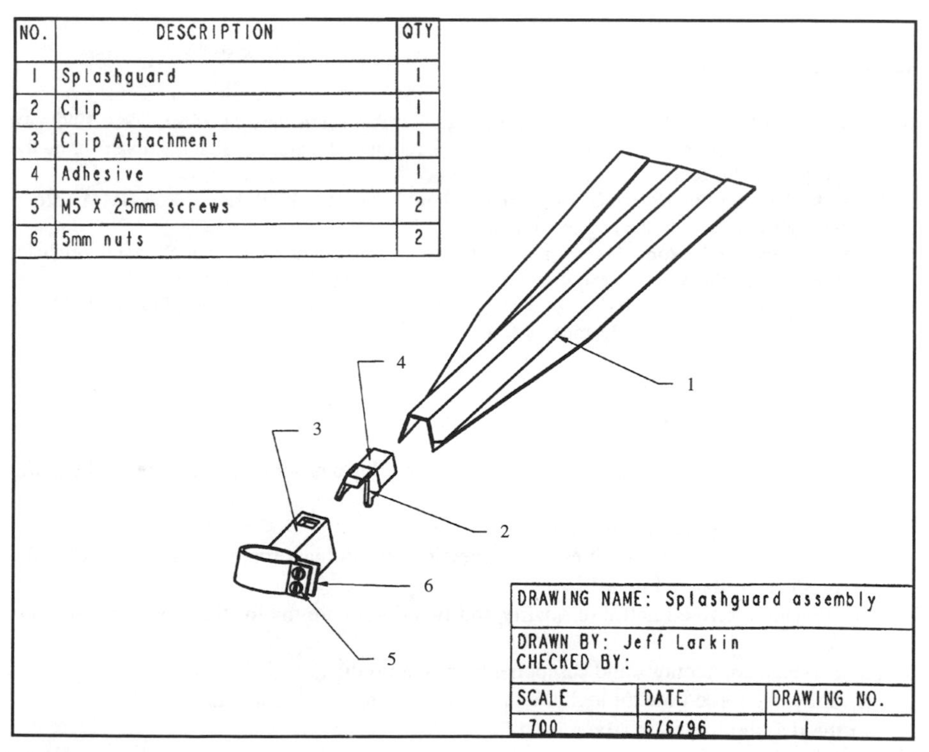

NO.	DESCRIPTION	QTY
1	Splashguard	1
2	Clip	1
3	Clip Attachment	1
4	Adhesive	1
5	M5 X 25mm screws	2
6	5mm nuts	2

FIGURE 10.23
Assembly drawing of the Splashgard.

Thus the team concluded that ABS would be their first material choice and that components would be sheet-formed or injection-molded and bonded at connection 3 with an adhesive specifically formulated for ABS. These choices would be refined as the process continued.

10.6 SUMMARY

* Products must be developed from concepts through concurrent development of *form, material,* and *production* methods. This process is driven by the functional decomposition discussed in Chap. 7.
* Form is bound by the geometric *constraints* and defined by the *configuration* of *connected components.*
* The development of most components and assemblies starts at their interfaces, or connections, since for the most part function occurs at the interfaces between components.
* Product development is an iterative loop that requires the development of new concepts, the decomposition of the product into subassemblies and components, the refinement of the product toward a final configuration, and the patching of features to help find a good product design.

10.7 SOURCES

M. F. Ashby, *Materials Selection in Mechanical Design,* Pergamon Press, Oxford, U.K., 1992. An excellent text on materials selection. There is a computer program available implementing the approach in this text.

K. Budinski, *Engineering Materials: Properties and Selection,* Reston, Reston, Va., 1979. A text on materials written with the engineer in mind.

C. S. Snead, *Group Technology: Foundations for Competitive Manufacturing,* Van Nostrand Reinhold, New York, 1989. An overview of group technology for classifying components.

E. Tjalve, *A Short Course in Industrial Design,* Newnes-Butterworths, London-Boston, 1979. An excellent book on the development of form.

10.8 EXERCISES

1. For the original design problem (Exercise 4.1), develop a product layout drawing by doing the following:
   (a) Develop the spatial constraints.
   (b) Draw a figure that identifies the connections for at least three configurations for the product similar to Fig. 10.14.
   (c) Develop a refined house of quality and function diagrams for the most critical interface.
   (d) Develop connections and components for the product.
   (e) Show the force flow through the product for its most critical loading.
2. For the redesign problem (Exercise 4.2):
   (a) Identify the spatial constraints for all important operating sequences.
   (b) At critical interfaces, identify the energy, information, and material flows.

   (c) Develop a refined house of quality and function diagrams for the most critical interface.
   (d) Develop new connections and components for the product.
   (e) Show the force flow through the product for its most critical loading.
3. Determine the force flow in the following:
   (a) A bicycle chain
   (b) A car door being opened
   (c) A paper hole punch
   (d) Your body while holding a 5-kg weight straight out in front of you with your left hand

CHAPTER 11

PRODUCT EVALUATION FOR PERFORMANCE

11.1 INTRODUCTION

The primary goal in this chapter is to compare the performance of the product to the engineering specifications developed earlier in the design project. Performance is the measure of behavior, and the behavior of the product results from the design effort to meet the intended function. Thus, part of the goal is to track and ensure understanding of the functional development of the product. If the functional development is not understood, the product may exhibit unintended behaviors. Finally, another subgoal is to design in quality. Although this chapter is about "evaluation for performance," it gives another opportunity to be sure that a quality product is developed—that it will always work as it was designed to. These goals are met through the first three best practices identified in Table 11.1 and covered in this chapter. The remainder of the best practices listed in the table are aimed at other, nonperformance product evaluation techniques and are covered in the next chapter. Although all of these best practices are discussed as techniques for product evaluation, they all contribute to the generation of the product as part of the iterative generate/evaluate cycle.

11.2 THE IMPORTANCE OF FUNCTIONAL EVALUATION

Although the main goal of evaluation is comparing product performance with engineering targets, it is equally important to track changes made in the function of the product. Conceptual designs were developed first by functionally modeling the problem and then, on the basis of that model, developing potential concepts to

TABLE 11.1
The process of product evaluation

Monitoring functional change (Sec. 11.2)
Evaluating product performance (Sec. 11.3–11.5)
Analytical model development
Physical model development
Graphical model development
Robust design (Sec. 11.6–11.10)
Design for cost (DFC) (Sec. 12.2)
Value engineering (Sec. 12.3)
Design for manufacture (DFM) (Sec. 12.4)
Design for assembly (DFA) (Sec. 12.5)
Design for reliability (DFR) (Sec. 12.6)
Design for test and maintenance (Sec. 12.7)
Design for the environment (Sec. 12.8)

fulfill these functions. This transformation from function to concept does not end the usefulness of the functional modeling tool. As the form is refined from concept to product, new functions are added. As demonstrated in the design of the Splashgard interfaces in the previous chapter, the evolution of the functionality was integrated into the development of concepts for components and assemblies.

An obvious question about this process arises: What benefit is there in refining the function model as the form is evolving? The answer is that by updating the functional breakdown, the functions that the product must accomplish can be kept very clear. Nearly every decision about the form of an object adds something, either desirable or undesirable, to the function of the object. It is important not to add functions that are counter to those desired. For example, in the design of the Splashgard, the decision to use the flexing spring clip necessitated a certain interface between the user and the clip. The exact steps a user must go through to use this clip were made clear by refining the function occurring at the interfaces between the user and the Splashgard. Besides tracking the functional evolution of the product, the refinement of the functional decomposition also aids in the evaluation of potential failure modes (Section 12.6).

Finally, tracking the evolution of function means continuously updating the flow models of energy, information, and materials. It is these flows that determine the performance of the product. As the product matures, the intended function and actual behavior merge and so what was, in conceptual design, concern for "the desired" now turns to measuring "the reality."

11.3 THE GOALS OF PERFORMANCE EVALUATION

In Chap. 6 we developed a set of engineering requirements based on the needs of the customer. For each of these requirements, a specific target was set. The goal now is to evaluate the product design relative to these targets. Since the targets are represented as numerical values, the evaluation can only occur after the product is refined to the point that numerical engineering measures can be made. In Chap. 8

the concepts developed were not yet refined enough to compare with the targets and were thus compared with more abstract measures by using the first four techniques shown in Fig. 8.1. Evaluation can now be based on comparison with the engineering requirements, the last technique listed in Fig. 8.1. Beyond comparison to the requirements, effective evaluation procedures should clearly show what should be altered (patched) in order to make deficient products meet the requirements, and they should demonstrate the product's insensitivity to variation in the manufacturing processes, aging, and operating environment. Restated, the evaluation of product performance must support the following:

1. Evaluation must result in numerical measures of the product for comparison with the engineering requirement targets developed during problem understanding. These measurements must be of sufficient accuracy and precision for the comparison to be valid.
2. Evaluation should give some indication of which features of the product design to modify, and by how much, in order to bring the performance on target.
3. Evaluation procedures should include the influence of manufacturing variations, aging effects, and environmental changes. Insensitivity to these "noises" while meeting the engineering requirement targets results in a quality product.
4. Evaluation should use the best method available: investigation through analytical models, testing of physical models, and representation with graphical models.

To formalize these four points and add structure to this chapter, consider Fig. 11.1, the P-diagram (*P* stands for product or process). This figure shows a typical block diagram with the input signal on the left and the output response on the right. The center of the box may be the entire product or some system, subsystem, or process within it. The input signal is the energy, material, or information that causes a response in the product. The output response must be of some attribute that is measurable and comparable to the engineering specifications. The output is sometimes called the *quality characteristic* because it is the measure of product or system quality.

In traditional systems modeling, the block has only an input and an output, but here we also have control parameters and noise affecting the product. Control parameters are the design factors that the engineer can specify on drawings or other specification documents. These are the factors that can be changed during the design process. *Noise* refers to the uncontrollable factors that cause variation in the performance of the product. Noise is a major topic in the next section.

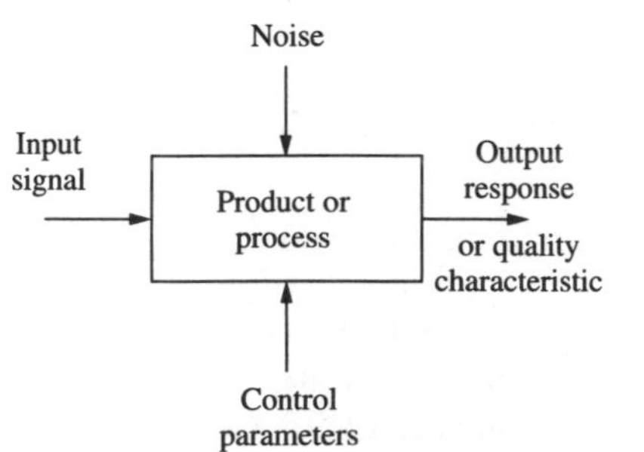

FIGURE 11.1
The P-diagram.

To meet the product performance evaluation goals listed above requires more than throwing together a prototype and seeing if it will work, completing a set of drawings, or running a computer simulation. Meeting the goals requires an understanding of concepts such as *sensitivity analysis, robust design, parameter design, tolerance design,* and *design of experiments.* The remainder of this chapter is focused on these techniques. This phase of the design process is the last chance to design quality into the product.

11.4 ACCURACY, VARIATION, AND NOISE IN MODELING PRODUCTS

In Section 5.3 we discussed modeling using physical prototypes, analysis, and graphical representation. Regardless of the type of model, the goal of modeling is to find the easiest method by which to evaluate the product for comparison with the engineering targets using available resources. To compare the product under development with the engineering targets means that numerical values must be produced; even a rough value is better than no value at all.

In any model (rough or very precise) two kinds of errors may occur: *errors due to inaccuracy* and *errors due to variation.* Accuracy is the correctness or truth of the model's estimate. If there is a distribution of results (each time we measure the performance, we get a different number), then the estimate is the mean value of the distribution. With an accurate model, the best estimate (mean) will be a good predictor of product performance; with an inaccurate model it will be a poor one. The variation in the results obtained from the model refers to the statistical variation of the results about the mean value. The terms *precision, resolution, range,* and *deviation* are also used to refer to the distribution of the evaluation. Where accuracy tells "how much," distribution tells "how sure." In Fig. 11.2, the inaccurate estimate is shown

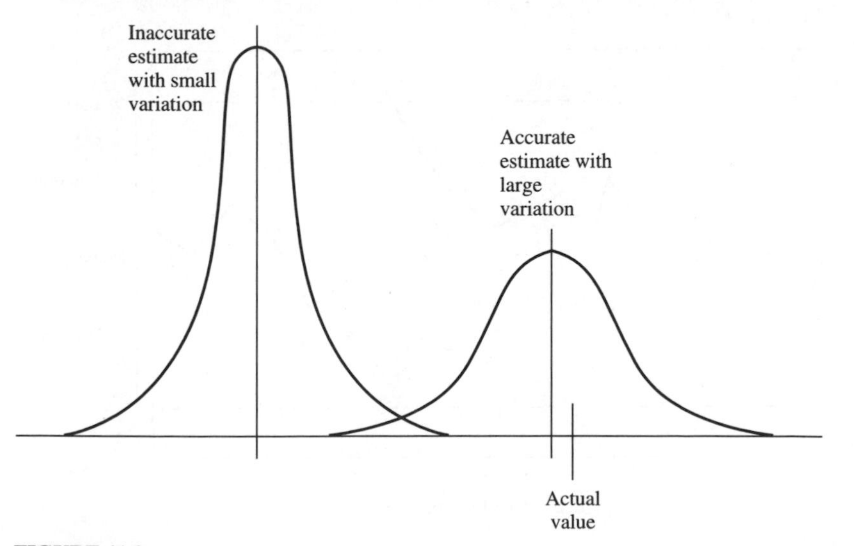

FIGURE 11.2
Relation of accuracy and resolution of error in modeling.

with a small variation and the accurate estimate with a large one. The obvious goal in modeling is to develop an accurate model with a small variation. The next best model is accurate with a large variation. *An inaccurate model is inaccurate no matter how small the variation.*

Why so much concern about variation? Attributes of prototypes and of final production samples may vary greatly from the desired mean. During production, not all samples of the product are exactly the same size, are made of exactly the same material, or behave in exactly the same way. For example, the actual dimensions of components that were specified on a drawing to be 38.1 ± 0.06 mm are shown in Fig. 11.3. The target for this dimension was 38.1 mm. However, during manufacturing, the values ranged from 0.03 mm below the target to 0.07 mm above. It was only during inspection that the tool cutting the component was found to be worn, thus causing the 0.07-mm deviation from the mean. The tool was replaced.

Another example of variation's importance during design is shown in the data in Fig. 11.4. These data represent the tensile strength for 913 samples of 1035 hot-rolled steel. The data have been grouped to the closest 1 kpsi. The tensile strength varies by as much as 10 kpsi from the mean. These data are replotted in Fig. 11.5 on normal-probability paper. Since a straight line fits the data in this figure, the tensile strength of the sample material is normally distributed. From Fig. 11.5, the mean strength is 86.2 kpsi (the 50 percent point) and the standard deviation is 3.9 kpsi. (Details on normal distributions are in App. B.)

A number of points can be made about these examples that are important to the understanding of the designer's goals in modeling the product:

The importance of normal distributions in modeling. The data for the steel in Figs. 11.4 and 11.5 were fit by a normal distribution. The data in Fig. 11.3 for the dimension of a component, although skewed when the tool wore, were also close

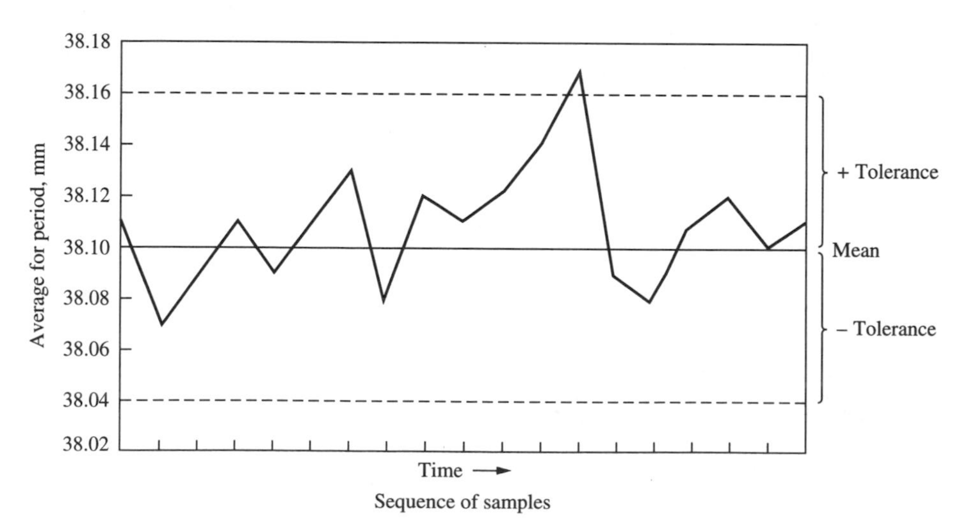

FIGURE 11.3
Manufactured component distribution relative to design specification.

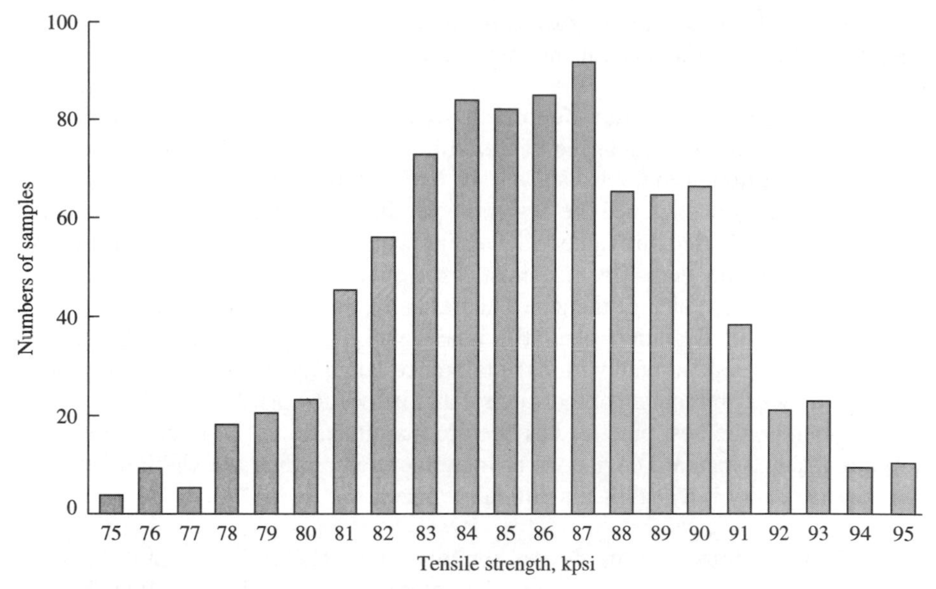

FIGURE 11.4
Distribution of tensile strength of 1035 steel.

to being normally distributed. For most design parameters, variations in value are considered as normal distributions fully characterized by the mean and variance or standard deviation.

However, most analytical models are *deterministic*—that is, each variable is represented by a single value. Since *all* parameters are really distributions, this single value is generally assumed to be the mean. Calculations performed with only mean value information may or may not give accurate estimates. Regardless of accuracy, these models give no information on the variation of the estimated value. There are,

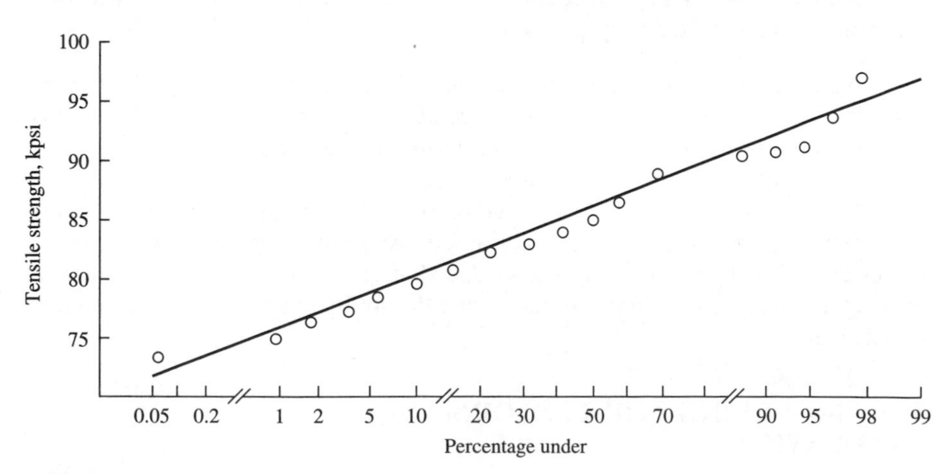

FIGURE 11.5
Steel data plotted on normal-distribution paper.

however, *nondeterministic*, or *stochastic*, analytical methods that account for both the mean and the variation by using methods from probability and statistics.

The difference between manufacturing variations and tolerance. The data in Fig. 11.3 show the manufacturing variation in components that are all supposed to have the same dimension. Also shown are lines ±0.06 mm from the mean value. These represent the tolerance the designer specified for the dimension. The manufacturing engineer used this value to determine which manufacturing machine to use in making the component. The machinist or the quality control inspectors used it for determining when to change the tool. Theoretically, the tolerance is assumed to represent ±3 standard deviations about the mean value. This implies that 99.68 percent of all the samples should fall within the tolerance range. In actuality, the variation is controlled by a combination of tool control and inspection, as shown in Fig. 11.3.

Recently, the best practice has been to manufacture to "6-sigma." This term implies that six standard deviations of manufactured product are within tolerance. Six-sigma quality is a goal that is difficult to reach.

The effect of "noise" on the design performance. The variations in the dimensions and in the material properties are examples of types of noise that affect the performance of a product. Noise affecting the design parameters is classified as follows:

- *Manufacturing, or unit-to-unit, variations,* including dimensional variations, variations in material and other properties, and process variations such as those in manufacturing and assembly.
- *Aging, or deterioration, effects,* including etching, corrosion, wear, and other surface effects, along with material property or shape (creep) changes over time.
- *Environmental, or external, conditions,* including all effects of the operating environment on the product. Some environmental conditions, such as temperature or humidity variations, affect the material properties; others, such as the amount of paper in the tray of a paper feeder or the amount of load on a walkway, affect the operating stresses, strains, or positions.

All of these types of noises are inherent in the final product. They all affect the variation in the product's performance. *A quality product is one that is insensitive to noise and thus has a small variation in performance. A product model must also represent the effects of these noise factors.*

Noises that affect strength are often accounted for by using a *factor of safety.* Two methods for calculating the factor of safety are given in App. C. In both of these, noises caused by uncertainties in knowledge about the material properties, the load causing the stress, unit-to-unit variations, and the ability to analyze failure are taken into account.

11.5 MODELING FOR PERFORMANCE EVALUATION

The concerns for accuracy and variation are reflected in the concerns for choosing the best modeling technique for each situation.

Although each modeling challenge is different, the steps discussed below give order to the considerations taken into account during evaluation. The discussion is centered on analytical and physical modeling. Graphical representations are often used in support of these models. Drawings aid in understanding the results of analysis and represent hardware to be built for physical models.

STEP 1: IDENTIFY THE OUTPUT RESPONSES (I.E., THE CRITICAL OR QUALITY PARAMETERS) THAT NEED TO BE MEASURED. Often the goal in evaluation is to see if a new idea is feasible. Even with this ill-defined goal, the important critical parameters, those that demonstrate the performance, must be clearly identified. In developing engineering requirements and targets during the specification development phase of the design process, many of the parameters of interest are identified. As the product is refined, other important requirements and targets arise. Thus, throughout the development of the product, the parameters that demonstrate the performance of the product are identified and measured during product evaluation.

STEP 2: NOTE HOW ACCURATE THE OUTPUT NEEDS TO BE. Early in the product refinement, it may be sufficient to find only the order of magnitude of some parameters. Back-of-the-envelope calculations may be sufficient indicators of performance for relative comparisons. As the product is refined, the accuracy of the evaluation modeling must be increased to enable comparison with the target values. It is important to realize the degree of accuracy needed before beginning the evaluation. Much effort spent on a finite element model would be wasted if a rough calculation using classical strength-of-materials techniques or a simple laboratory test of a piece of actual material would be sufficient.

STEP 3: IDENTIFY THE INPUT SIGNAL, THE CONTROL PARAMETERS AND THEIR LIMITS, AND NOISES. It is important, before beginning to model a system, that a P diagram be drawn and the factors affecting the output be at least initially identified and classified. Input signals are the energy, information, and materials modified by the product or process. Usually these signals are important; however, they may be secondary to the control parameters and ignored in many design situations. For example, if we are trying to evaluate the ease of attaching the Splashgard, the design of the interfaces over which the designer has control is more important than the energy necessary to do the attaching.

Many noises can be controlled if desired; however, controlling them when a design can be made robust is needlessly expensive. Thus, list as noises all unit-to-unit, aging, and environmental variations that can be identified. Then decide which may have an impact on the output (this may be dependent on the outcome of evaluation).

Control parameters are sometimes difficult to identify, and it is not until a model (either analytical or physical) is built and tested that some dependencies are discovered. One may build a model only to find that the variables thought to be important are not and other, more important variables have been left out of consideration.

It is important to list the control parameters and their upper and lower limits. Considering these limits helps in understanding the design and aids in the development of the layout drawing. The physical limits on these parameters give the limits on patching the design during iteration. Knowledge about limits is one measure of technology readiness discussed in Section 8.4.

STEP 4: UNDERSTAND ANALYTICAL MODELING CAPABILITIES. Generally, analytical methods are less expensive and faster to implement than physical modeling methods. However, the applicability of analytical methods depends on the level of accuracy needed and on the availability of sufficient methods. For example, a rough estimate of the stiffness of a diving board can be made using methods from strength of materials. In this analysis the board is assumed to be a cantilever beam, made of one piece of material, of constant prismatic cross section, and with known moment of inertia. Further, the load of a diver bouncing on the end of the board is estimated to be a constant point load. With this analysis, the important dependent variables—the energy storage properties of the board, its deflection, and the maximum stress—can be estimated.

Using more sophisticated modeling techniques from advanced strength of materials, the accuracy of the model is improved. For example, the taper of the diving board, the distributed nature of the diver in both time and space, and the structure of the board can be modeled. The dependent variables remain unchanged. More independent parameters can now be utilized in a more laborious and more accurate evaluation.

Finally, using finite element methods, even more accuracy can be achieved, though at a higher cost in terms of time, expertise, and equipment. If the diving board is made of a composite material, it may even be that no finite element methods are yet available to allow for sufficiently accurate evaluation.

Each of these three analytical evaluation methods (basic strength of materials, advanced strength of materials, and finite element analysis) provides a single answer for the stiffness of the diving board. Although the accuracies of the models vary, none gives an indication of the variance in the stiffness; all three are deterministic and provide single answers insensitive to the variations in the dependent variables. Methods are presented later in this chapter for using deterministic analytical models that give stochastic information and thus data on the variance of the dependent parameters.

In the above discussion on analytical modeling, a number of issues were raised:

- What level of accuracy is needed? Analytical models can be used instead of physical models only when there is a high degree of confidence in their accuracy.
- Are analytical models available that can give the needed accuracy? If not, then physical models are required. Often it is valuable to do both to confirm one's understanding of the product.
- Are deterministic solutions sufficient? They probably are in the early evaluation efforts. However, as the product is finalized, they are not sufficient, as knowledge of the effect of noises on the dependent parameters is essential in developing a quality product.
- If no analytical techniques are available, can new techniques be developed? In developing a new technology, part of the effort is often devoted to generating

analytical techniques to model performance. During a design effort there is usually no time to develop very sophisticated analytical capabilities.

- Can the analysis be performed within the resource limitations of time, money, knowledge, and equipment? As discussed in Chap. 1, time and money are two measures of the design process. They are usually in limited supply and greatly influence the choice of the modeling technique used. Limitations in time and money can often overwhelm the availability of knowledge and equipment.

STEP 5: UNDERSTAND THE PHYSICAL MODELING CAPABILITIES. Physical models, or prototypes, are hardware representations of all or part of the final product. Most design engineers would like to see and touch physical realizations of their concepts all the way through the design process. However, time, money, equipment, and knowledge—the same resource limitations that affect analytical modeling—control the ability to develop physical models. Generally, the fact that physical models are expensive and take time to produce controls their use.

The ability to develop physical prototypes of complex components has improved greatly since the mid 1980s. During this period *rapid prototyping* methods were developed. These systems use CAD models of components to deposit materials or laser-harden polymers to rapidly make a physical model. The components made by some of the methods are actually usable in tests; others are only visual and usable to test fit and interference.

STEP 6: SELECT THE MOST APPROPRIATE MODELING METHOD. There is nothing as satisfying in engineering as modeling a system both analytically and physically and having the results agree! However, resources rarely allow both modeling methods to be pursued. Thus, the method that yields the needed accuracy with the fewest resources must be selected.

STEP 7: PERFORM THE ANALYSIS OR EXPERIMENTS. Most of the rest of this chapter focuses on this step.

STEP 8: VERIFY THE RESULTS. Document that the targets have been met or that the model has given a clear indication of what parameters to alter, which direction to alter them in, and how much to alter them. In evaluating models, not only are the results as important as in scientific experimentation, but since the results of the modeling are used to patch or refine the product, the model must also give an indication of what to change and by how much. In analytical modeling this is possible through sensitivity analysis, as will be discussed in the next section. This is more difficult with physical models. Unless the model itself is designed to allow easily changed parameters, it may be difficult to learn what to do next.

11.6 ROBUST DESIGN

In the previous chapters of this text we have developed methods to generate, evaluate, and embody concepts, taking care with materials, cost, assembly, and other considerations. Even after all these steps have been taken, there is still no guarantee that a quality product has been developed. Now we address two interdependent design issues that are a key part of product design: (1) determining final values for dimensions, material properties, and other control parameters and (2) establishing the best manufacturing tolerances to specify. These two issues, parameter design and tolerance design, greatly affect the final product quality.

The goal of *parameter design* is best explained through a simple design problem. Consider the design of a tank to hold liquid. Conceptual design of the tank has resulted in a cylindrical shape with an internal radius r and an internal length l. Thus the volume of the tank V can be written as

$$V = \pi r^2 l$$

Additionally, a customer's requirement is to design the "best" tank to hold "exactly" 4 m^3 of liquid. This seems a simple enough problem with r and l as the control parameters, V as the quality measure, and its target response as

$$V = 4m^3$$

Then

$$r^2 l = 1.27m^3$$

As can be seen from Fig. 11.6, there are an infinite number of solutions to the problem. The tank at point A, for example, is short and fat, and the one at point B is long and thin. It is not clear which point on the curve might be "best" in terms of holding "exactly" 4 m^3 of liquid. Obviously some more thought on what is meant by the term "exactly" is necessary.

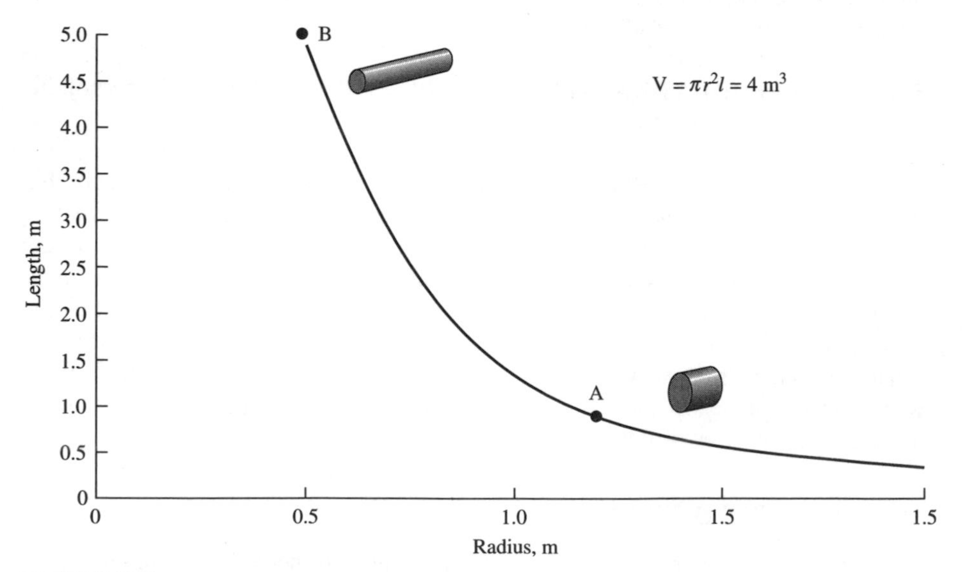

FIGURE 11.6
Potential solutions for the tank problem.

Because the length and radius of the tank must be manufactured, they are not exact values, as shown in Fig. 11.6, but have manufacturing variations. Thus the control parameters, r and l, are actually distributions about some nominal value. If r and l are distributions, then the dependent parameter, the volume V, must also be a distribution and thus cannot be "exact." The problem is now reduced to determining the dependence of the distribution of V on r and l and finding the values of r and l that make V as exact as is possible.

Making this problem even more difficult, the liquid that will be stored in the tank is corrosive and over time will etch the inside of the tank, increasing the values of r and l. Additionally, the tank will operate at a wide range of temperatures, so r and l will vary. Even with the effects of manufacturing variance, the aging effects due to etching, and the environmental effects of the temperature variation, it is still our goal to keep the volume as close to 4 m^3 as possible. Thus, we want to find the values for l and r that make V the least sensitive to manufacturing, aging, and environmental noises or variations.

In general, there are three ways to deal with these noises. The first is to keep them small by tightening manufacturing variations (generally expensive). The second is shielding the product from aging and environmental effects (sometimes difficult and maybe impossible). The third is to make the product insensitive to the noises. A product that is insensitive to manufacturing, aging, and environmental noises is considered *robust* and will be perceived as a high-quality product. If robustness is accomplished, the product will assemble as designed and will be reliable once in operation. Thus the key philosophy of robust design is to

> Determine values for the parameters based on easy-to-manufacture tolerances and default protection from aging and environmental effects so that the best performance is achieved. The term "best performance" implies that the engineering requirement targets are met and the product is insensitive to noise. If noise insensitivity cannot be met by adjusting the parameters, then tolerances must be tightened or the product shielded from the effects of aging and environment.

With such a philosophy, quality can be designed into a product. For example, recall that Xerox's 1981 line fallout was 30 components per thousand (Fig. 1.3). This means that 1 out of every 33 components did not fit in the product during assembly. This failure to fit was discovered either during inspection or by the inability of the assembly personnel or machine to mate the components to the product. By 1995, using the robust design philosophy, Xerox had reduced the line fallout to about 30 components per million.

Concern about tolerances on dimensions and other variables that affect the product is the focus of *tolerance design.* If the nominal tolerances do not give sufficient performance of the quality measures, then tolerances need to be changed to meet the targets. As seen in the example above, the values of the parameters are dependent on knowledge about tolerances. A drawing of a component or an assembly to be manufactured is incomplete without tolerances on all the dimensions. These tolerances act as bounds on the manufacturing variations such as shown in Fig. 11.3. However, studies have shown that only a fraction of the tolerances on a typical component actually affect its function. The remainder of the dimensions on a typical product could be outside the range set by their tolerances and it would still operate satisfactorily. Thus, when specifying tolerances for noncritical dimensions, always use those that are nominal for the manufacturing process specified to make the component. For example, as shown in Fig. 11.7, machining operations have *nominal tolerances*. These values for steel reflect the expected variation if standard practices are followed. If tighter values are held, the cost will increase (as will be demonstrated in the next chapter). Most companies have specifications for tolerances that are within the nominal variations for each process.

Once the designer has specified a tolerance on a drawing, what does it mean downstream in the product life cycle? First, it communicates information to manufacturing that is essential in helping to determine the manufacturing processes that will be used. Second, tolerance information is used to establish *quality-control* guidelines. In the 1920s, when mass production was instituted on a broad scale, quality control by inspection was also begun. This type of quality assurance is often called "on-line," as it occurs on the production line. Most production facilities have

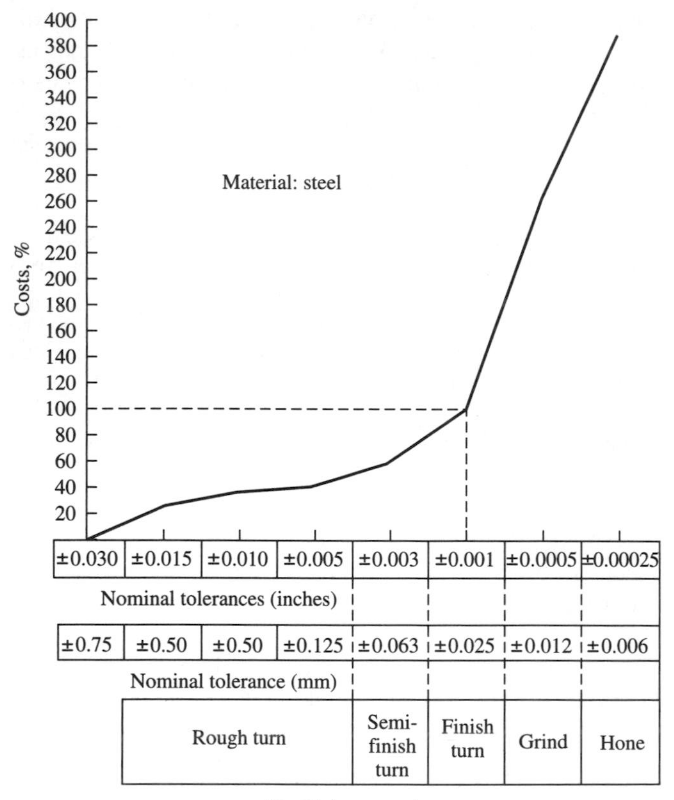

FIGURE 11.7
Tolerance versus manufacturing process.

quality-control inspectors whose job it is to verify that produced products are within specified tolerances. This effort to increase product quality through inspection is not very robust, because poor manufacturing process control and poor design can make quality inspection very difficult.

In the 1940s much effort was expended on designing the production facilities to turn out more uniform components. This moved some of the responsibility of quality control from on-line inspection to off-line design of production processes. To keep manufactured components within their specified tolerances, many statistical methods were developed for manufacturing process control. However, even if a production process can keep a manufactured component within the specified tolerances, there is no assurance of a robust, quality product. Thus, it was realized in the 1980s that quality control is really a design issue. If robustness is designed in, the burden of quality control is taken off production and inspection.

Robust design is often called *Taguchi's method* after Dr. Genichi Taguchi, who popularized the robust design philosophy in the United States and Japan. It must be noted that this philosophy is different from that traditionally used by designers. Traditionally, parameter values are determined without regard for tolerances or other noises and the tolerances are added on afterward. These tacked-on tolerances are usually based on company standards. This philosophy does not lead to a robust design and may require tighter tolerances to achieve quality performance.

The implementation of robust design techniques is fairly complex. To ease our explanation of the techniques here, we will make two simplifying assumptions. First, we will consider only noise due to manufacturing variations; second, the only parameters that are considered are dimensions. To further aid our understanding of robust design, we consider dimensional tolerances and sensitivity analysis before we proceed to robust design itself. Additionally, we first develop these techniques analytically so that the philosophy is better understood. The actual methods Taguchi developed are based on experimental methods and require a background in statistical data reduction beyond the scope of this text. Thus, these experimental methods are only briefly introduced in Section 11.9. The tank design problem, started above, is completed later in the chapter.

11.7 SENSITIVITY ANALYSIS: A WINDOW ON QUALITY

Sensitivity analysis is a technique for evaluating the statistical relationship of control parameters (e.g., dimensions) and their tolerances in a design problem. In this section we explore the use of sensitivity analysis for a simple dimensional problem and then apply the method to the problem of the tank volume.

Consider the joint shown in Fig. 11.8, one concept for attaching the cover on a pressure vessel. Here a bolt is used to hold a gasketed cover to a flange on the vessel body. The gasket must be held within the tolerances shown or it will either be crushed or leak. Note that other noises on this joint besides the dimensional tolerances are changes in temperature and pressure, which would cause the components to change length. These effects are not considered in this example.

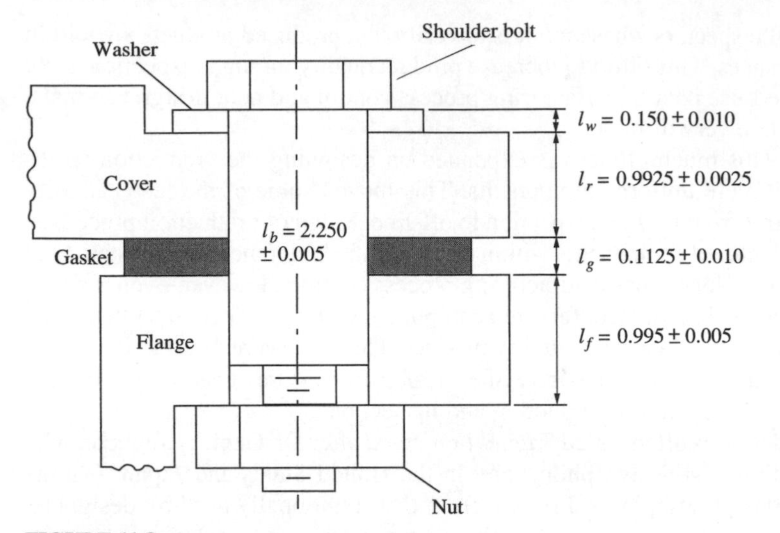

FIGURE 11.8
Pressure-vessel joint showing tolerance stacking.

A shoulder bolt has been specified to ensure that the gasket will not be crushed. (The nut's travel is limited by the step in diameter of the bolt—the shoulder.) The figure shows the dimensions and tolerances for each component in the assembly. The question is, How do these tolerances for the washer, cover, gasket, and vessel flange add together, or *stack up*, to affect the performance of the joint? Analysis of tolerance stack-up is the most common form of *tolerance analysis*. For this analysis, the following notation is used:

l = actual axial length of component
$\bar{l}$ = mean length
t = tolerance on dimension
s = standard deviation of dimension (assume $s = t/3$)

Subscripts refer to

b = bolt shank length
w = washer
c = cover
g = gasket
f = flange
gg = gasket gap (space for gasket to fit into)

When the joint is assembled, the length of the bolt must equal the total length of the other components:

$$l_b = l_w + l_c + l_{gg} + l_f \qquad (11.1)$$

Since our primary concern is the length (thickness) of the gasket gap, rewrite the equation with l_{gg} as the dependent parameter:

$$l_{gg} = l_b - (l_c + l_w + l_f) \tag{11.2}$$

If l_{gg} is within the range of thicknesses specified for l_g, then the gasket will seal the joint and not be crushed.

Assume, as an example, that a randomly selected bolt is within the tolerances and at the maximum length ($l_b = 2.255$ in.) and that the other components selected just happen to be at their minimum length ($l_w = 0.140$, $l_c = 0.990$, and $l_f = 0.990$ in.). This situation could happen if the components are selected at random, yet the odds of selecting the longest possible bolt and thinnest joint components are low. Suppose it did occur; then, from Eq. (11.2), $l_{gg} = 0.135$ in. With the bolt tightened snugly against the shoulder, the gap for the gasket is 0.0125 in. larger than the thickest gasket $[0.135 - (0.1125 + 0.010) = 0.0125]$. Thus no gasket is able to seal the joint. Conversely, if the shortest bolt and the thickest washer, cover, gasket, and flange are chosen, then the shoulder on the bolt stops the nut when the gap is 0.090 in. This is 0.0125 in. below the minimum thickness allowable on the gasket, and any gasket chosen will be crushed.

The method of adding the maximum and minimum dimensions to estimate the stack-up is called *worst-case analysis.* This technique assumes that the shortest and longest components are as likely to be chosen as some intermediate value. The odds are that the components will be nearer to the mean than to either of their extreme values. In other words, even though the gasket might leak or be crushed as calculated above, the probability of these two assemblies occurring from the random selection of components is very small.

A more accurate estimate of the total thickness of the joint can be found statistically, in a form of stochastic analysis. Consider a stack-up problem composed of n components, each with mean length $\bar{l}_i$, tolerance t_i (assumed symmetric about the mean), and $i = 1, \ldots, n$. If one dimension is identified as the dependent parameter (in the joint example, the gap for the gasket), then its mean dimension can be found by adding and subtracting the other dimensions, as in Eq. (11.2):

$$\bar{l} = \bar{l}_1 \pm \bar{l}_2 \pm \bar{l}_3 \pm \cdots \pm \bar{l}_n \tag{11.3}$$

The sign on each term depends on the structure of the device. Similarly, the standard deviation is

$$s = ({s_1}^2 + {s_2}^2 + \cdots + {s_n}^2)^{1/2} \tag{11.4}$$

where the signs are always positive. (This basic statistical relation is discussed in App. B.) Generally, "tolerance" is assumed to imply three standard deviations about the mean value. Thus a tolerance of ± 0.009 in. means that $s = 0.003$ and that 99.68 percent of all samples should be within tolerance (i.e., within $\pm 3\sigma$). Since $s = t/3$, Eq. (11.4) can be rewritten as

$$t = (t_1^2 + t_2^2 + \cdots + t_n^2)^{1/2} \tag{11.5}$$

For the example,

$$\bar{l}_{gg} = \bar{l}_b - \bar{l}_c - \bar{l}_w - \bar{l}_f \tag{11.6}$$

and

$$t_{gg} = (t_b{}^2 + t_c{}^2 + t_w{}^2 + t_f{}^2)^{1/2} \tag{11.7}$$

Substituting in values,

$$\bar{l}_{gg} = 2.250 - 0.150 - 0.9925 - 0.995 = 0.1125 \text{ in.} \tag{11.8}$$

and

$$t_{gg} = (0.005^2 + 0.010^2 + 0.0025^2 + 0.005^2)^{1/2} = 0.0125 \text{ in.} \tag{11.9}$$

These results imply that the gasket, as designed, has a mean equal to the average gasket gap, but that only a percentage of the joints will operate as designed because the tolerance on the gasket gap, 0.0125 in., is larger than the 0.010-in. tolerance on the gasket itself. The distribution of the gap dimension is shown in Fig. 11.9, along with the distribution of the gasket itself.

The standard deviation on the gap ($\frac{1}{3}$ the tolerance calculated) is 0.0125/3 = 0.00417 in. And so the 0.010-in. tolerance on the gasket is 2.4 standard deviations of the gap tolerance (0.010/0.00417). Thus, the probability of the gasket filling the gap is 99.34 percent. Increased quality could be achieved by inspecting each joint and reworking those that do not meet the specification or swapping components between joints to meet them. Another way to increase the quality is to use the results of the analysis to redesign the joint. This is accomplished through sensitivity analysis.

Sensitivity analysis allows the contribution of each parameter to the variation to be easily found. Rewriting Eq. (11.4) in terms of $P_i = s_i{}^2/s^2$,

$$1 = P_1 + P_2 + \cdots + P_n \tag{11.10}$$

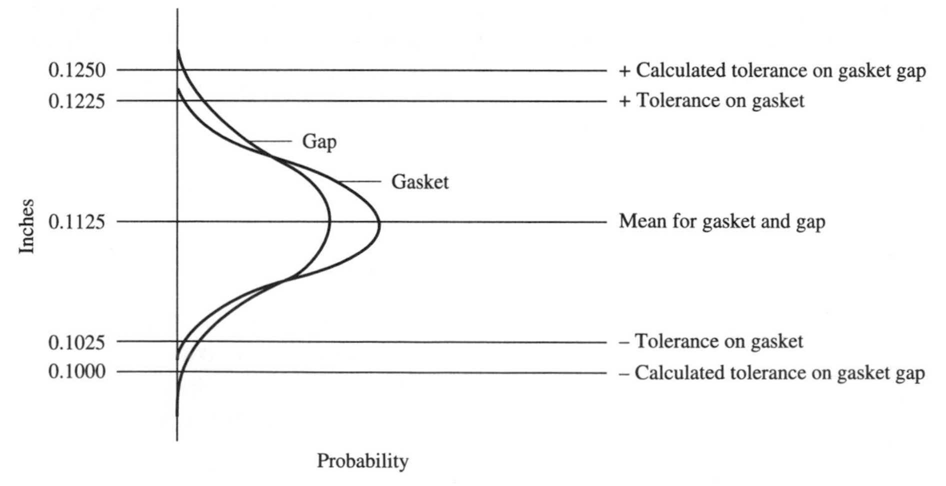

FIGURE 11.9
Comparison of gap to gasket.

where P_i is the percentage contribution of the ith term to the tolerance (or variance) of the dependent variable. For the current example, these are

$$P_b = \frac{0.005^2}{0.0125^2} = 0.16 = 16 \text{ percent}$$

$$P_w = \frac{0.010^2}{0.0125^2} = 0.64 = 64 \text{ percent}$$

$$P_c = \frac{0.0025^2}{0.0125^2} = 0.04 = 4 \text{ percent}$$

$$P_f = \frac{0.005^2}{0.0125^2} = 0.16 = 16 \text{ percent}$$

$$\text{Total} = 1.00 = 100 \text{ percent}$$

This result clearly shows that the tolerance on the washer thickness has the greatest effect on the gap for the gasket. For one-dimensional tolerance stack-up problems such as this, the results of the sensitivity analysis can be used for *tolerance design.* Since the washer causes 64 percent of the noise in the joint and is a simple component, it is the most likely candidate for patching. Using Eqs. (11.7) and (11.9),

$$t_{gg}^{\ 2} = 0.005^2 + t_w^{\ 2} + 0.0025^2 + 0.005^2 \tag{11.11}$$

With $t_{gg} = t_g = 0.010$,

$$t_w = 0.0066 \text{ in.}$$

If the tolerance on the washer is lowered from ± 0.010 in. to ± 0.0066 in., the gasket will seal the joint over 99.7 percent (± 3 standard deviations) of the time. Thus we have used sensitivity analysis to accomplish the design of the tolerance on the washer. This technique will work on all one-dimensional problems in which all the parameters are dimensions on the product. To summarize:

Step 1. Develop a relationship between the dependent dimension and those it is dependent on, as in Eq. (11.3) or (11.6). Using each independent dimension's mean value, calculate the mean value of the dependent dimension.

Step 2. Calculate the tolerance on the dependent variable using Eq. (11.5) or work in terms of the standard deviations (Eq. (11.4)).

Step 3. If the tolerance found is not satisfactory, identify which independent dimension has the greatest effect, using Eq. (11.10), and modify it if possible. Depending on the ease (and expense), it may be necessary to choose a different dimension to modify.

Problems of two or three dimensions are similarly solved. The equations relating the variables become complex for all but the simplest multidimensional systems.

If the variables are not related in a linear fashion, the equations given above are modified. This is best shown through the tank-volume problem introduced earlier in this chapter. The major difference is that the parameters r (the radius) and l (the length) are not linearly related to the dependent variable, V (the volume), as can be seen in Fig. 11.6. The method shown below is a generalization of the method for the linear problem. It is good for investigating any functional relationship, whether the parameters are dimensions or not.

Consider a general function

$$F = f(x_1, x_2, x_3, \ldots, x_n) \tag{11.12}$$

where F is a dependent parameter (dimension, volume, stress, or energy) and the x_i's are the control parameters (usually dimensions and material properties). Each parameter has a mean $\overline{x}_i$ and a standard deviation s_i. In this more general problem, the mean of the dependent variable is still based on the mean of the independent variables, as in Eq. (11.3). Thus,

$$\overline{F} = f(\overline{x}_1, \overline{x}_2, \overline{x}_3, \ldots, \overline{x}_n) \tag{11.13}$$

Here, however, the standard deviation is more complex:

$$s = \left[\left(\frac{\partial F}{\partial x_1}\right)^2 {s_1}^2 + \cdots + \left(\frac{\partial F}{\partial x_n}\right)^2 {s_n}^2\right]^{1/2} \tag{11.14}$$

Note that if $\partial F/\partial x_i = 1$, as it must in a linear equation, then Eq. (11.14) reduces to Eq. (11.4). Equation (11.14) is only an estimate based on the first terms of a Taylor series approximation of the standard deviation. It is generally sufficient for most design problems.

For the tank problem, the independent parameters are r and l. The mean value of the dependent variable V is thus given by

$$\overline{V} = 3.1416\overline{r}^2\overline{l} \tag{11.15}$$

To evaluate this, we must consider specific values of r and l. There is an infinite number of these pairs that meet the requirement that the mean volume be 4 m^3. For example, consider point A in Fig. 11.10 (which is Fig. 11.6 with added information), with $\overline{r} = 1.21$ m and $\overline{l} = 0.87$ m. From Eq. (11.15), $\overline{V} = 4$ m^3.

The tolerances on these parameters can be based on what is easy to achieve with nominal manufacturing processes. For example, take $t_r = 0.03$ m ($s_r = 0.01$) and $t_l = 0.15$ m ($s_l = 0.05$). These values are shown in the figure as an ellipse around point A. Using formula (11.14), the standard deviation on this volume is

$$s_v = \left[\left(\frac{\partial V}{\partial l}\right)^2 {s_l}^2 + \left(\frac{\partial V}{\partial r}\right)^2 {s_r}^2\right]^{1/2} \tag{11.16}$$

where

$$\frac{\partial V}{\partial r} = 6.2830rl$$

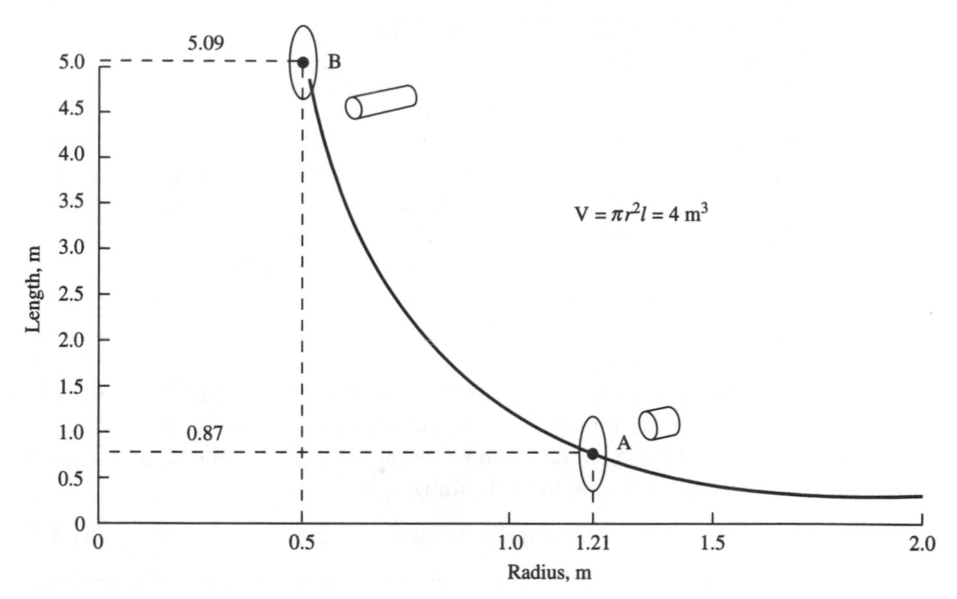

FIGURE 11.10
Effect of noise on the potential solutions for the tank problem.

and

$$\frac{\partial V}{\partial l} = 3.1416r^2$$

For the values in this example, $\partial V/\partial r = 6.61$ and $\partial V/\partial l = 4.60$, so $s_v = 0.239 \text{ m}^3$. Thus 99.68 percent (three standard deviations) of all the vessels built will have volumes within $0.717 \text{ m}^3 (3\times 0.239)$ of the target 4 m^3. Also, the percentage contribution of each parameter can be found as in Eq. (11.10). Here the length contributes 92.3 percent of the variance in volume. However, noting that the tolerance on the length is much larger than that on the radius and considering the shape of the curve in Fig. 11.10, it is evident that a longer vessel with a smaller radius might yield a smaller variance in volume. If the control parameters are taken at $r = 0.50$ m and $l = 5.09$ m (point B in Fig. 11.10), the mean volume is still 4 m^3. Now $\partial V/\partial r = 16.00$ and $\partial V/\partial l = 0.78$, so $s_v = 0.166 \text{ m}^3$, which is 31 percent smaller than at point A. Also, now the tolerance on r contributes 94 percent to the variance in the volume. Note that we achieved *the reduction in variance not by changing the tolerances on the parameters, but by changing only their nominal values.* The second design has higher quality because the volume is always closer to 4 m^3. If we can find the values of the parameters r and l that give the smallest variance on the volume, then we are employing the philosophy of robust design. To find the best parameter values requires some knowledge of optimization.

11.8 PARAMETER AND TOLERANCE DESIGN FOR QUALITY

We have seen that by merely changing the shape of the tank we could improve the quality of the design. The tank with the greater length had less sensitivity to the large tolerance on the length, so the tank's volume varies less. Our goal now is to combine the techniques of sensitivity analysis and optimization to give a method for determining the most robust values for the parameters. Following this we will consider tightening the tolerances to make the best tank possible.

Consider the initial problem: the goal was to have $V = 4\ \text{m}^3$, exactly. This is impossible, as V is dependent on r and l and they are random variables, not exact values. Thus, the "best" we can do is to keep the absolute difference between V and $4\ \text{m}^3$ as small as possible, in other words, minimize the standard deviation of V. We must accomplish this minimization while keeping the mean volume at $4\ \text{m}^3$. Defining the difference between the mean value ($3.1416\bar{r}^2\bar{l}$) and the target T ($4\ \text{m}^3$) as the bias, the objective function to be minimized is

$$C = \text{standard deviation} + \lambda \times \text{bias} \qquad (11.17)$$

where λ is a Lagrange multiplier.[1] This can also be written in terms of the variance, the standard deviation squared.

In general, using Eq. (11.14) and optimizing the variance instead of the standard deviation, this looks like

$$C = \left[\left(\frac{\partial F}{\partial x_1}\right)^2 {s_1}^2 + \cdots + \left(\frac{\partial F}{\partial x_n}\right)^2 s_n^2\right] + \lambda(F - T) \qquad (11.18)$$

For the tank,

$$C = (2\pi rl)^2 {s_r}^2 + (\pi r^2)^2 {s_1}^2 + \lambda(\pi r^2 l - T)$$

The minimum value of the objective function can now be solved. With known standard deviations on the parameters s_r and s_l (or tolerances t_r and t_l) and a known target T, values for the parameters r and l can be found from the derivatives of the objective function with respect to the parameters and the Lagrange multiplier:

$$\frac{\partial C}{\partial r} = 0 = 2r(2\pi l)^2 {s_r}^2 + 4r^3\pi^2 {s_l}^2 + \lambda 2\pi rl$$

$$\frac{\partial C}{\partial l} = 0 = 2l(2\pi r)^2 {s_r}^2 + \lambda\pi r^2$$

$$\frac{\partial C}{\partial \lambda} = 0 = \pi r^2 l - 4$$

Solving simultaneously results in

$$r = 1.414l\left(\frac{s_r}{s_l}\right) \qquad (11.19)$$

[1] Many different optimization methods could be used. Lagrange's method is well suited to this simple problem.

and

$$l = \left[\frac{2}{\pi} \left(\frac{s_l}{s_r} \right)^2 \right]^{1/3} \tag{11.20}$$

Thus, for any ratio of the standard deviations or the tolerances, the parameters are uniquely determined for the best (most robust) design. For the values of $s_r = 0.01$ ($t_r = 0.03$ m) and $s_l = 0.05$ ($t_l = 0.15$ m), the above equations result in $r = 0.71$ m and $l = 2.52$ m. Substituting these values into Eq. (11.16), the standard deviation on the volume is $s_v = 0.138$ m^3. Comparing this to the results obtained in the sensitivity analysis, 0.239 and 0.165 m^3, the improvement in the design quality is evident.

If the radius were harder to manufacture than the length, say $s_r = 0.05$ and $s_l = 0.01$, then, using Eqs. (11.19) and (11.20), the best values for the parameters would be $r = 2.06$ m and $l = 0.29$ m. The resulting standard deviation on the volume would be 0.233 m^3.

In summary, the tolerance or standard deviation information on the dependent variables has been used to find the values of the parameters that minimize the variation of the dependent variable. In other words, the resulting configuration is as insensitive to noise as possible and is thus a robust, quality design.

If the standard deviation on the volume is not small enough, then the next step is to tighten the tolerances. Sensitivity analysis can be used to determine which control variable will give the greatest change to the volume.

Robust design can be summarized as a three-step method:

Step 1. Establish the relationship between quality characteristics and the control parameters [for example, Eq. (11.12)]. Also define a target for the quality characteristic, making it as large as possible, as small as possible, or a specific value.

Step 2. Based on known tolerances (standard deviations) on the control variables, generate the equation for the standard deviation of the quality characteristic [for example, Eq. (11.14) or (11.16)].

Step 3. Solve the equation for the minimum standard deviation of the quality characteristic subject to this variable being kept on target. For the example given, Lagrange's technique was used; other techniques are available, and some are even included in most spreadsheet programs. There are usually other constraints on this optimization problem that limit the values of the parameters to feasible levels. For the example given, there could have been limits on the maximum and minimum values of r and l.

There are some limitations on the method developed here. First, it is only good for design problems that can be represented by an equation. In systems in which the relation between the variables cannot be represented by equations, experimental methods must be used (Section 11.9). Second, Eq. (11.17) does not allow for the inclusion of constraints in the problem. If the radius, for example, had to be less than 1.0 m because of space limitations, Eq. (11.17) would need additional terms to include this constraint.

11.9 ROBUST DESIGN THROUGH TESTING

It is often impossible to analytically evaluate a proposed design because no mathematical models of the system exists. In many cases, even when analysis is possible, the analytical model of the system may not allow determination of the effect of the noise on a proposed design. In either case it is necessary to design and build a physical model for experimental testing. In Chap. 8 physical models were used to verify the concept; now they are needed to refine the product. The material in this section is an introduction to Taguchi's method for the robust design of products. Like many other topics, this subject has entire books devoted to it (see Section 11.12, Sources); however, this material is sufficient to make us appreciate the strength of the method and its complexity and enable us to apply it to the design of the tank as a simple example. We will assume that we do not know the formula $V = \pi r^2 l$ and only know $V = f(r, l)$. To experimentally find dimensions for radius and length, we could begin by building a tank with some best-guess dimensions and then measuring the volume. Then, if the volume was too high, we could build new models, one with a smaller radius and another with a shorter length, and then measure the volumes. Based on these new measurements, we could try to estimate the dependence of the volume on each of the dimensions and iterate (i.e., patch) our way to the target volume. This is the way most experiments are run. This "random walk" toward a solution may require many models, so it is not very efficient. Additionally, the solution found could be anywhere on the curve shown in Figs. 11.6 and 11.10; there is no guarantee that the final design will be the most robust. The steps below can overcome these drawbacks.

Step 1. Identify signals, noise, control, and quality factors (i.e., independent parameters). Referring back to the P-diagram in Fig. 11.1, it is necessary to list all the dependent and independent parameters related by the product or system. Then it is necessary to decide which of these are critical to the evaluation of the product. Sometimes this is not easy, and critical parameters or noises may be overlooked. This may not become evident until data are taken and the results are found to have wide distribution, implying that the model is not complete or the experiments have been poorly done. It is essential to take care here to understand the system.

The P-diagram for the tank (Fig 11.11) shows that the designer has control over the length and radius and that there are many noises that affect the volume of liquid held. The function of the tank is to "hold liquid," and its performance is measured by

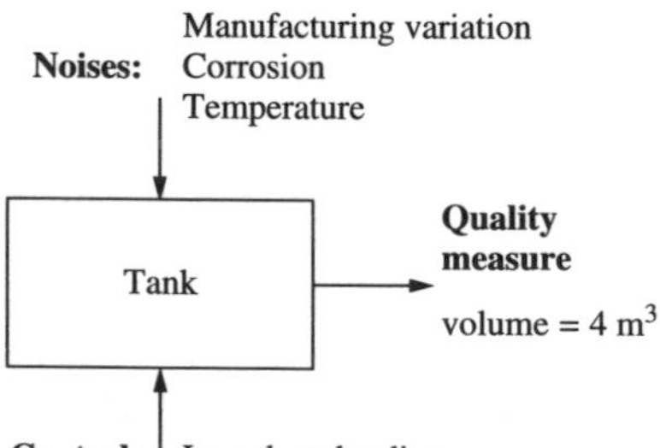

FIGURE 11.11
P-diagram for tank problem.

how accurately the tank can be held to the target value of 4 m^3. The noises include the manufacturing variations on the radius and length, and the aging and environmental effects not considered here.

Step 2. For each quality measure (i.e., output response) to be evaluated, recall or determine its target value and the nature of the quality loss function (see Table 11.2). During the development of the QFD, target values were determined and the shape of the loss function was identified. If this information has not been previously generated for the parameter being measured, do this before the experiment is developed.

Loss is proportional to the mean square deviation, MSD, the average amount the output response is off the target. This amount is also often referred to as the signal-to-noise ratio, or S/N ratio. Generally the S/N ratio is $-10 \log$ (MSD). The minus is included so that the maximum S/N ratio is the minimum quality loss, the 10 is used to get the units to decibels, and the logarithm is used to compress the values.

The MSD and S/N for the three most common types of targets identified in Section 6.9 are shown in Table 11.2. For the smaller-is-better target, the larger the value of the output, y, the larger the MSD and the smaller the signal-to-noise ratio. In other words, larger values of y are noise, so the signal is weaker relative to that noise. For the larger-is-better case smaller values of y are seen as noise.

The nominal-is-best target is more complex; there are many ways to calculate the S/N ratio. The most common is shown here. As shown in Table 11.2, the mean square deviation is simply the sum of the variation about the mean and the accuracy about the target. This distribution is the same as the "sensitive" function loss curve in Fig. 9.5. Generally, only the sum of the variation is used in calculating the S/N ratio, as shown in the table.

For the tank problem, 4 m^3 is a nominal-is-best target.

Parameter design is based on maximizing the S/N ratio and then tuning the parameters to bring the design on target. In other words, the goal is to find the conditions that make the product insensitive to noise and then use parameters that do not affect the S/N ratio to bring the quality functions to the desired value. The use of this philosophy will become clear in the example problem.

TABLE 11.2
Formulas for means and S/N ratios

Quality loss function	Mean square deviation (MSD)	S/N ratio
Smaller-is-better	$\frac{1}{n}\sum_{i=1}^{n} y_i^2$	$-10 \log\left(\frac{1}{n}\sum_{i=1}^{n} y_i^2\right)$
Larger-is-better	$\frac{1}{n}\sum_{i=1}^{n}\left(\frac{1}{y_i^2}\right)$	$-10 \log\left(\frac{1}{n}\sum_{i=1}^{n}\frac{1}{y_i^2}\right)$
Nominal-is-best	$\frac{1}{n}\sum_{i=1}^{n}(y_i-\overline{y})^2+(\overline{y}-m)^2$ m = target value	$-10 \log \frac{1}{n}\sum_{i=1}^{n}(y_i-\overline{y})^2$

Step 3. Design the experiment. The goal is to design an experiment that forces what ever can happen, to happen. It is not sufficient to design a simple experiment in which the model is patched and patched until it works once. This does not lead to a robust design. Instead, the experiment should be designed so that the results give a clear understanding of the effects on the output response of changing control parameters and an understanding of the effects of noise. An ideal experiment will show how to adjust the control parameter to meet the target and show which one to choose so that the resulting system is insensitive to noise.

The physical model of the product or system must be designed so that the following can be achieved:

- Control factors can be changed to represent the options available. This may mean designing a number of different physical devices or designing one with changeable parts or configuration. This model may not be very representative of the final product because its main goal is to support the collection of data.
- Noises can be controlled over the expected range. This may require precision components made to match the upper and lower bounds of tolerances. It may require the use of an environmental chamber capable of temperature, humidity, or other noise control. It may require the components to be artificially aged, corroded, or worn. The noises must be forced to expected extremes so that the effect on output responses can be measured.
- The output responses can be measured accurately. Note that in measuring the output, additional noise is added by the instrumentation. Ensure that this noise is of a lower order of magnitude than the effect of the noise and control variables.

Suppose there are n control factors and data are taken for each at two different settings, there are m noise variables also to be tested at two levels, and, for accuracy, there are k repetitions to be run for each condition. Then there are $k \cdot 2^n \cdot 2^m$ experiments to perform. For example, if there are two control factors, two noises, and three repetitions for each condition, then there are 48 output responses to be recorded. To keep the number of experiments to a reasonable level, on large problems there are statistically based techniques for choosing a subset of experiments to run. These experiments allow the missing data to be inferred (see the books by Barker, by Phadke, and by Fowlkes and Creveling in Sources, Section 11.12).

Table 11.3 shows a layout for an experiment with two control factors each tested at two levels with two noises also each at two levels. The results for the output response, F, are shown for the 16 experiments. If, for example, there were three repetitions of experiment $F21_{12}$ (control factor 1 at level 2, control factor 2 at level 1, noise 1 at level 1, and noise 2 at level 2), then there would be three $F21_{12}$ values. If all the experiments were run three times, there would be 48 experiments. The mean value and S/N ratio for each control-factor combination are calculated in the last two columns.

For experiments with more than two control factors, with control factors run at more than two levels, or for more than two noises, Table 11.3 is easily extended. Again, for a large number of control factors or noises there are methods of reducing the number of experiments.

TABLE 11.3
Layout for a two-control-factor experiment

	Noise 1:	Level 1	Level 1	Level 2	Level 2		
	Noise 2:	Level 1	Level 2	Level 1	Level 2		
Control factor 1	**Control factor 2**					**Mean**	**S/N**
Level 1	Level 1	$F11_{11}$	$F11_{12}$	$F11_{21}$	$F11_{22}$	$\overline{F11}$	S/N11
Level 1	Level 2	$F12_{11}$	$F12_{12}$	$F12_{21}$	$F12_{22}$	$\overline{F12}$	S/N12
Level 2	Level 1	$F21_{11}$	$F21_{12}$	$F21_{21}$	$F21_{22}$	$\overline{F21}$	S/N21
Level 2	Level 2	$F22_{11}$	$F22_{12}$	$F22_{21}$	$F22_{22}$	$\overline{F22}$	S/N22

For the tank problem, experimental models are built to allow accurate setting of the length and radius. This may require one model for each experiment, or a model may be designed that allows these values to be changed with sufficient accuracy. In Table 11.4 values of $r = 0.5$ and $r = 1.5$ are chosen as the two levels for the radius. These were chosen as the extreme values of Fig 11.10 and are only a starting place. Likewise, $l = 0.5$ and 5.5. The noises are set at the tolerance levels representing the length as harder to manufacture than the radius: $l = \pm 0.15$ and $r = \pm 0.03$. These values are entered into Table 11.4. For the output response for cell $F21_{12}$, the experiment needs a tank made as precisely as possible with $r = 1.53$ m and $l = 0.35$ m.

Step 4. Take and reduce data. Mean values and S/N ratios are calculated for repetitions of each set of control and noise conditions.

The measured volumes of the tank are shown in Table 11.4 along with the calculated values of the mean and nominal-is-best S/N ratio. Two of the mean values are fairly close to the target of 4 m^3. This was the result of luck in choosing the starting values for r and l. In fact, this result raises the question of which one is best, because they have vastly different values for radius and length.

Step 5. Analyze the results, and select new test conditions if needed. The first set of experiments may not yield satisfactory results. The goal is to maximize the S/N ratio

TABLE 11.4
Tank experiment results

	∂r (m):	0.03	0.03	−0.03	−0.03		
	∂l (m):	0.15	−0.15	0.15	−0.15		
r (m)	l (m)					Mean (m^3)	S/N, db
0.5	0.5	0.57	0.31	0.45	0.244	0.396	3.74
0.5	5.5	5.00	4.76	3.91	3.69	4.34	11.87
1.5	0.5	4.81	2.59	4.39	2.40	3.55	4.40
1.5	5.5	41.89	39.53	38.46	36.13	39.00	19.48

and then bring the mean value on target. For analytical problems we can find the true maximum (Section 11.8); here we can only estimate when we reach that point.

For the tank problem, the experiment with the radius $r = 0.5$ m and the length $l = 5.5$ m gives results near the target and with the highest S/N (11.87 db). The experiments could be stopped here if a mean value of 4.34 m^3 is close enough. The information in the table could also be used to adjust the control parameters to bring the product closer to target. Since experiments with $l = 5.5$ resulted in better S/N values, r can be estimated to bring the output to 4 m^3. However, how much to change r may not be evident from the data. A better idea is to perform experiments by setting new values for r and l around the values found above and taking new readings. This iteration would eventually lead to a volume $V = 4$ m^3 and an S/N ratio of 13.69 at $r = 0.71$ m and $l = 2.52$ m, the same values found analytically. Note that the S/N value for this final result is only 1.78 db higher than the first experimental value found. This implies only a mean square deviation change of 50 percent (working the S/N equation in Table 11.2 backward, $(V_i - \overline{V})^2 = 10^{(1.78/10)}$).

11.10 EVALUATION OF THE SPLASHGARD

Even though the Splashgard is a relatively simple product, there are many features that need evaluation. In reviewing Fig. 10.14, which shows the connections between the components of the Splashgard, the design team concluded the following:

Connection 3. The interface between the keeper and the clip (interface 3) was critical and would require evaluation to ensure that the Splashgard had a long life.

Connection 3/5. The interface between the keeper and the clip (3) combined with the interface between the clip body and the user (5) would require evaluation to ensure that the Splashgard could meet the engineering requirements for attachment and removal.

Splashgard body and connection 6. The shape of the Splashgard body would need verification to ensure its strength, its longevity, and its ability to deflect water and mud.

Connection 4. The joint between the Splashgard body and the clip (4) would need to be designed.

The team's approach to each of these is discussed below.

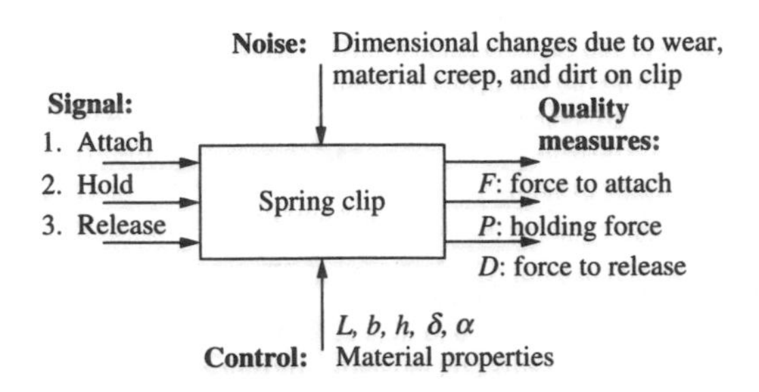

FIGURE 11.12
P-diagram for interface 3 spring clip.

The spring clip is the key feature of interface 3. Its performance is critical to the quality of this product. Thus the team spent considerable time evaluating it.

- (Step 1) The output responses most critical to the user of the Splashgard are the force to attach, the force to release, and the amount of load the clip will withstand without releasing accidentally.
- (Step 2) Evaluation within ±20 percent would be sufficient for this part because it does not have to meet any exact standards or other constraints.
- (Step 3) The P-diagram in Fig. 11.12 shows the input signals, control parameters, and noises. The signals are the three operating conditions developed during functional modeling: attachment, holding, and release. The parameters that the designer has control over are the dimensions of the clip and, to some degree, the material properties. The control variables are defined in Fig. 11.13. The noises are the uncertainties in dimensions, material properties, and the ability to model the situation. Additionally, wear, the effect of dirt and water, and creep could have been included.
- (Steps 4, 5, and 6) This situation is easy to model analytically using basic strength-of-materials techniques. However, it would take molding or machining to build a physical model. Thus, for this case, analysis is the clear choice.
- (Step 7) Analysis for the spring clip was based on modeling it as a cantilever spring with the force to release it at the end. Deterministic methods were used to estimate dimensions, and the methods in Section 11.8 were used to refine these for the most robust design.

As shown in Fig. 11.13 the clip is a cantilever spring. Force F is the force exerted to engage the clip in its keeper. Force D is the force needed to release the clip. And force P is the force exerted on the clip to hold it in place when there is a downward load on the Splashgard. In order for the spring clip to engage or disengage, it

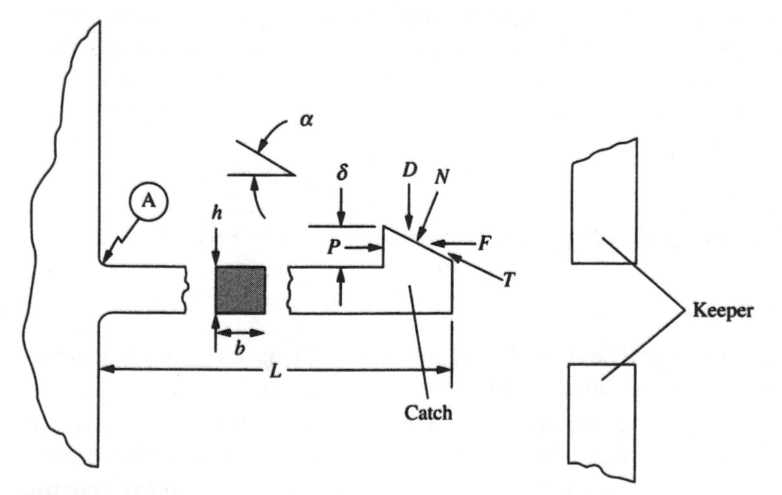

FIGURE 11.13
A cantilever spring clip.

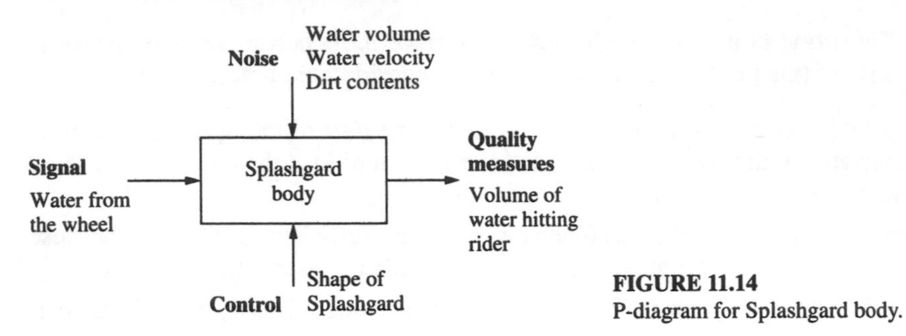

FIGURE 11.14
P-diagram for Splashgard body.

has to deflect δ mm. There are an infinite number of possible values for the variables that will meet the requirements. The team explored the design space using analytical methods presented earlier in this chapter.

The other three critical evaluations require experimental methods because there are no good analytical methods to evaluate them. Whereas the discussion above concerned the structural design of the clip, the team also performed an experimental study to verify that the product could meet the requirements for attachment and removal. A starting point for this evaluation was to review the human factors guidelines like those given in App. D. Then they developed an experimental study to verify that a wide variety of users found that the design was easy to use and met the other customers' requirements.

The design of the body shape and the water interface was also evaluated experimentally. In fact, the evaluation was combined with refinement of the shape of the fender through the use of designed experiments. Following the steps in Section 11.9, the team first constructed the P-diagram shown in Fig. 11.14. Here, the designers can control the shape, which must be strong and rigid enough to deflect the water off the wheel and withstand variations (i.e., noises) in the dirt contents, speed, and volume of the water. In the original QFD (Fig. 6.3) the target for a "water hitting rider" was targeted at 0 percent. In terms of S/N ratios, this translates to a "smaller-is-better," and therefore, as seen in Table 11.2, the S/N ratio is merely the sum of squares of the volume of water hitting the rider. The team then designed an experimental apparatus that enabled them to easily try different Splashgard shapes, to vary the volume, velocity, and dirt contents of the water, and to measure the volume of the water getting past the Splashgard.

The design of the joint between the body and the clip (interface 4) also needed experimental work. Although vendors recommended various adhesives to bond the material together, a program of experiments was established to verify the choice.

11.11 SUMMARY

* Product evaluation should be focused on comparison with the engineering requirements and also on the evolution of the function of the product.
* Products should be refined to the degree that their performance can be represented as numerical values in order to be compared with the engineering requirements.
* P-diagrams are useful for identifying and representing the input signals, control parameters, noises, and output response.

* Physical and analytical models allow for comparison with the engineering requirements.
* Concern must be shown for both the accuracy and the variation of the model.
* Parameters are stochastic, not deterministic. They are subject to three types of noises: the effects of aging, of environment change, and of manufacturing variation.
* Robust design takes noise into account during the determination of the parameters that represent the product. Robust design implies minimizing the variation of the critical parameters.
* Tolerance stacking can be evaluated both by the worst-case method and by statistical means.
* Both analytical and experiment methods exist for finding the most robust design.

11.12 SOURCES

Barker, T. B.: *Quality by Experimental Design,* Marcel Dekker, New York, 1985. A basic text on experimental design methods.

Fowlkes, W. Y., and C. M. Creveling: *Engineering Methods for Robust Product Design: Using Taguchi Methods in Technology and Product Design,* Addison-Wesley, New York, 1995. An outstanding, readable book on robust design oriented toward the statistically unsophisticated.

Haugen, E. B.: *Probabilistic Mechanical Design,* Wiley Interscience, New York, 1980. A text on stochastic methods used in the design of mechanical components.

Mischke, C. R.: *Mathematical Model Building,* Iowa State University Press, Ames, 1980. An introductory text on the basics of building analytical models.

Papalambros P., and D. Wilde: *Principles of Optimal Design: Modeling and Computation,* Cambridge University Press, New York, 1988. An upper-level text on the use of optimization in design.

Phadke, M. S.: *Quality Engineering Using Robust Design,* Prentice Hall, Englewood Cliffs, N.J., 1989. A basic book on robust design.

Rubenstein, M. F.: *Patterns of Problem Solving,* Prentice Hall, Englewood Cliffs, N.J., 1975. An introductory book on analytical modeling.

Siddal, J. N.: *Probabilistic Engineering Design,* Marcel Dekker, New York, 1983. A text on stochastic methods used in the design of mechanical components.

Tjalve, E., M. M. Andreasen, and F. F. Schmidt: *Engineering Graphic Modeling,* Newnes-Butterworths, London, 1979. An excellent tutorial on the different types of drawings and drawing techniques that support modeling requirements in mechanical design.

11.13 EXERCISES

11.1. For the original design problem (Exercise 4.1):
- (a) Identify the critical parameters and interfaces for evaluation.
- (b) Develop a P-diagram for each.
- (c) Choose whether to build physical models for testing or run an analytical experiment for each.
- (d) Perform the experiments or analysis and develop the most robust product.

11.2. For the redesign problem (Exercise 4.2), repeat the steps in Exercise 1.

11.3. You have just designed a tennis-ball serving machine. You take it out to the court, turn it on, and quickly run to the other side of the net to wait for the first serve. The first serve is right down the middle, and you return it with brilliance. The second serve is out to the left, the third is long, and the fourth hits the net.

(a) Does your machine have an accuracy or a variation problem?
(b) Itemize some of the potential causes of each type of error. Consider the types of "noise" discussed in Section 11.4.

11.4. Convince yourself about the applicability of normal distribution by doing the following:
(a) Measure some feature of at least twenty people and plot the data on normal-distribution paper. Easy measurements to make are weight, height, length of forearm, shoe size, or head circumference.
(b) Take a sample of fifty identical washers, bolts, or other small objects and weigh each on a precision scale. Plot the weights on normal-distribution paper and calculate the mean and standard deviation.

11.5. For the following design problems discuss the trade-offs between using analytical models and using experimental models.
(a) A new, spring-powered can opener.
(b) A diving board for your new swimming pool.
(c) An art nouveau shelf bracket.
(d) A pogo-stick spring.

CHAPTER 12

PRODUCT EVALUATION FOR COST, MANUFACTURE, ASSEMBLY, AND OTHER MEASURES

12.1 INTRODUCTION

In the previous chapter we considered the best practices for evaluating the product design relative to changes in function and for measuring performance and robustness. Also of importance are the evaluations introduced in this chapter: evaluation for cost, ease of assembly, reliability, testability and maintainability, and environmental friendliness.

12.2 COST ESTIMATING IN DESIGN

One of the most difficult and yet important tasks for a design engineer in developing a new product is estimating its production cost. It is important to generate a cost estimate as early in the design process as possible and to compare with the original cost requirements. In the conceptual phase or at the beginning of the embodiment phase, a rough estimate of the cost is first generated, and then as the product is refined, the cost estimate is refined as well. For redesign problems, where changes are not extreme, early cost estimates may be fairly accurate, because the current costs are known.

As the design matures, cost estimations converge on the final cost. This often requires price quotes from vendors and the aid of a cost estimation specialist. Most manufacturing companies, even small ones, have a purchasing or cost-estimating department whose responsibility it is to generate estimates for the cost of manufactured and purchased components. However, the designer shares the responsibility, especially when there are many concepts or variations to consider and when the potential components are too abstract for others to cost-estimate. Before we describe cost-estimating methods for use by designers, it is important to understand what control the design engineer has over the manufacturing cost and selling price of the product.

12.2.1 Determining the Cost of a Product

The total cost of a product to the customer (i.e., the list price) and its constituent parts are shown in Fig. 12.1. All costs can be lumped into two broad categories, direct costs and indirect costs. *Direct costs* are those that can be traced directly to a specific component, assembly, or product. All other costs are called *indirect costs*. The terminology generally used to describe the costs that contribute to the direct and indirect costs is defined below. Each company has its own method of bookkeeping, so the definitions given here may not match every accounting scheme. However, every company needs to account for all the costs discussed.

A major part of the direct cost is the *material costs*. These include the expenses of all the materials that are purchased for a product, including the expense of the waste caused by scrap and spoilage. Scrap is often an important consideration. For most materials the scrap can be reclaimed, and the return from the reclamation can be deducted from the material costs. Spoilage includes parts and materials that may not be usable because of manufacturing defects, deterioration, or other damage. Part fallout (see the Xerox part fallout history in Fig. 1.3), those components that cannot be assembled because of poor fit, also contributes to spoilage.

Components that are purchased from vendors and not fabricated in-house are also considered direct costs. At a minimum, this *purchased-parts cost* includes fasteners and the packaging materials used to ship the product. At a maximum, all

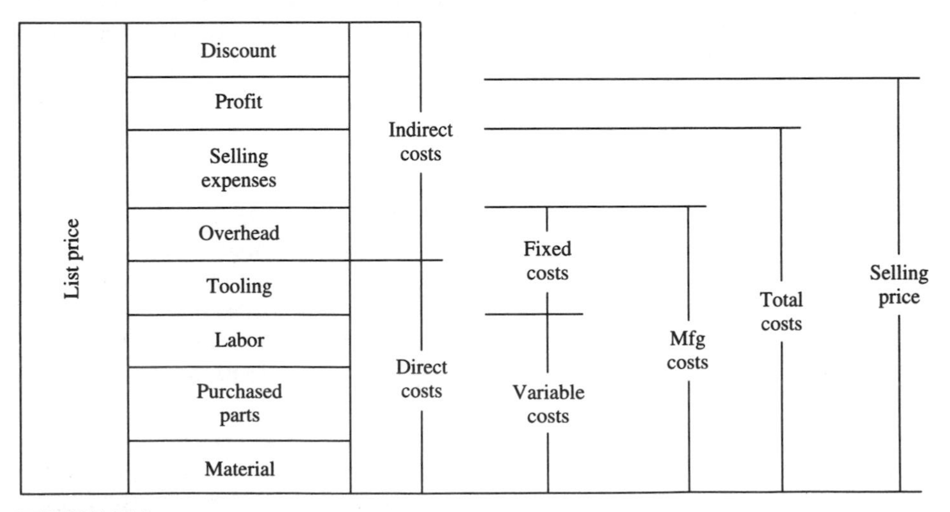

FIGURE 12.1
Product cost breakdown.

components may be made outside the company with only the assembly performed in-house. In this case there are no material costs.

Labor cost is the cost of wages and benefits to the workforce needed to manufacture and assemble the products. This includes the employees' salaries as well as all fringe benefits, including medical insurance, retirement funds, and vacation times. Additionally, some companies include overhead (to be defined below) in figuring the direct labor cost. With fringe benefits and overhead included, the labor cost of one worker will be two to three times his or her salary.

The last element of direct costs is the *tooling cost.* This cost includes all jigs, fixtures, molds, and other parts specifically manufactured or purchased for the production of the product. For some products these costs are minimal; very few items are being made, the components are simple, or the assembly is easy. On the other hand, for products, such as the bicycle Splashgard, that have injection-molded components, the high cost of manufacturing the mold will be a major portion of the part cost. However, a large volume of Splashgards is projected to be built (200,000 per year for five years), and the cost of these molds can be distributed over the one million units. If the total volume were to be only 20,000 units, the tooling cost alone might prohibit the use of injection-molded parts. (A cost estimate for the Splashgard is further developed in the next section.)

Figure 12.1 shows that the sum of the material, labor, purchased parts, and tooling used is the *direct cost.* The *manufacturing cost* is the direct cost plus the *overhead,* which includes all cost for administration, engineering, secretarial work, cleaning, utilities, leases of buildings, utilities, and other costs that occur day to day, even if no product rolls out the door. Some companies subdivide the overhead into engineering overhead and administrative overhead, the engineering portion including all expenses associated with research, development, and the design of the product. Many companies subdivide overhead into fixed and variable portions, items such as shop supplies, depreciation on equipment, equipment lease costs, and human resource costs being variable.

The manufacturing cost can be broken down in another important way. The material, labor, and purchased-parts costs are *variable costs,* as they vary directly with the number of units produced. For most high-volume processes, this variation is nearly linear: it costs about twice as much to produce twice as many units. However, at lower volumes the costs may change drastically with volume. This is reflected in the price quote made by a vendor for a small electric motor shown in Fig. 12.2.

Other manufacturing costs such as tooling and overhead are *fixed costs,* because they remain the same regardless of the number of units made. Even if production fell to zero, funds spent on tooling and the expenses associated with the facilities and nonproduction labor would still remain the same.

In general, the cost of a component, C, is given by

$$C = C_m + \frac{C_c}{n} + \frac{C_l}{\dot{n}}$$

where C_m is the cost of materials needed for the component (raw materials minus salvage price for scrap), C_c is the capital cost of tooling and a fraction of the cost of the machines and facilities needed, n is the number of components to be made, C_l

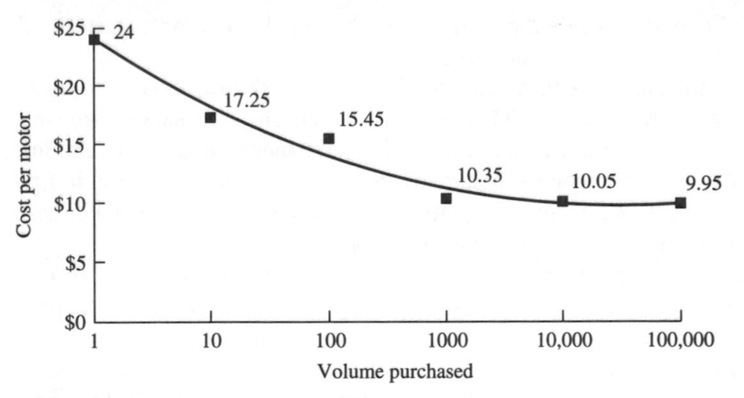

FIGURE 12.2
Sample of cost per volume purchased for a component.

is the cost of labor per unit time, and $\dot{n}$ is the number of components per unit time. Additionally, if the firm is buying from a vendor, the paperwork and other overhead of selling a small quantity of an item may also appear in C_c. The curve that results from this equation generally looks like that in Fig. 12.2, where the second term is very large at low volume and becomes small relative to the others at high volume.

The *total cost* of the product is the manufacturing cost plus the selling expenses. It accounts for all the expenses needed to get the product to the point of sale. The actual *selling price* is the total cost plus the *profit.* Finally, if the product has been sold to a distributor or a retail store (anything other than direct sales to the customer), then the actual price to the consumer, the list price, is the selling price plus the *discount.* Thus the discount is the part of the list price that covers the costs and profits of retail sales. If the design effort is on a manufacturing machine to be used in-house, then costs such as discount and selling expenses do not exist. Depending on the bookkeeping practices of the particular company, there may still be profit included in the cost.

The salaries for the designers, drafters, and engineers and the costs for their equipment and facilities are all part of the overhead. Designers have little control over these fixed expenses, beyond using their time and equipment efficiently. The designer's big impact is on the direct costs: tooling, labor, material, and purchased-parts costs. Reconsider Fig. 1.1, reprinted here as Fig. 12.3. These data from Ford show the manufacturing cost, emphasizing the low cost of design activities. If it is

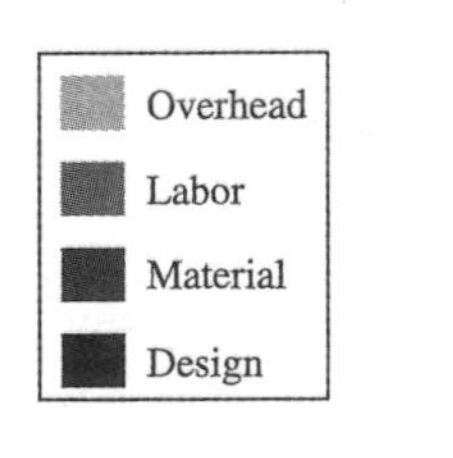

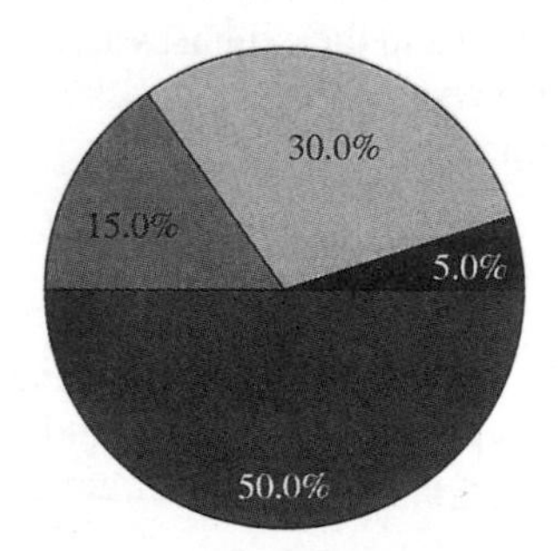

FIGURE 12.3
The cost of design.

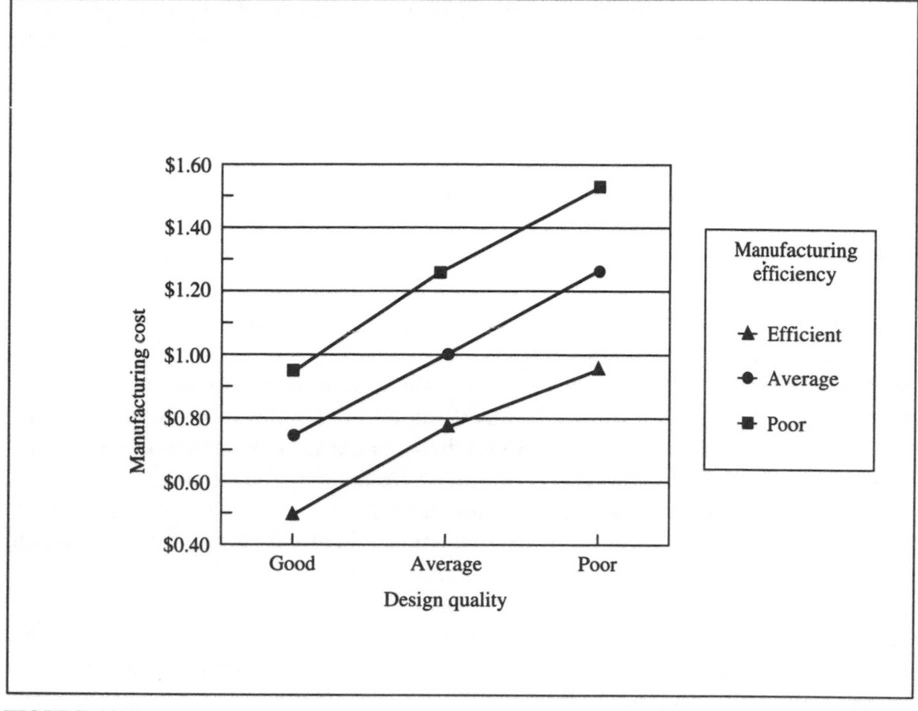

FIGURE 12.4
The effect of design quality on manufacturing cost.

assumed that the costs of purchased parts and tooling are included in the material costs, then these account for about 50 percent of the manufacturing costs. The labor is about 15 percent, and the overhead, including design expenses, is 35 percent. As a rule of thumb, for companies whose products are manufactured mainly in-house and in high volume, the manufacturing cost is approximately three times the cost of the materials. Also, the selling price is approximately nine times the material cost, or three times the manufacturing cost. This is sometimes called the material-manufacturing-selling 1-3-9 rule. This ratio varies greatly from product to product. The Ford data in Fig. 12.3 show a 1:2 ratio between materials cost and manufacturing cost, less than the rule would predict.

Figure 1.1, reprinted here as Fig. 12.4, shows the influence of design quality on manufacturing cost. As mentioned above, the designer can influence all the direct costs in a product, including the types of materials used, the purchased parts specified, the production methods, and thus the labor hours and the cost of tooling. Management, on the other hand, has much less influence on the manufacturing costs. They can negotiate for lower prices on a material specified by the designer, negotiate lower wages for the workers, or try to trim overhead. With these considerations, it is not surprising that data in Fig. 12.4 show that 50 percent of the influence on the manufacturing cost is controlled by design.

One final term that should be understood by engineers is *margin.* This is calculated by taking the ratio of profit to selling price. Typically, for product-generating companies, a margin of 40–50 percent will generate a good profit. However, for high-volume production this may drop to 10 percent, and for custom production it may be as high as 60–70 percent.

12.2.2 Making a Cost Estimate

It is the responsibility of the engineer to know the manufacturing cost of components designed. The ability to make these estimations comes with experience and from help from experienced team members and vendors. In many companies cost-estimating is accomplished by a professional who specializes in determining the cost of a component whether it is made in-house or purchased from a vendor. This person must be as accurate as possible in his or her estimates, as major decisions about the product are based on these costs. Cost estimators need fairly detailed information to perform their job. It is unrealistic for the designer to give the cost estimator 20 conceptual designs in the form of rough sketches and expect any cooperation in return. In most small companies, all cost estimations are done by the engineer.

The first estimations should be made early in the product design phase and be precise enough to be of use in making decisions about which designs to eliminate from consideration and which designs to continue refining. At this stage of the process, cost estimates within 30 percent of the final direct cost are possible. The goal is to have the accuracy of this estimate improve as the design is refined toward the final product. The more experience one has in estimating similar products, the more accurate the early estimates will be.

The cost-estimating procedure depends on the source of the components in the product. There are three possible options for obtaining the components: purchase finished components from a vendor, have a vendor produce components designed in-house, or manufacture components in-house.

As discussed in Chap. 10, there are strong incentives to buy existing components from vendors. If the quantity to be purchased is large enough, most vendors will work with the product designer and modify existing components to meet the needs of the new product.

If existing components or modified components are not available off the shelf, then they must be produced, in which case a decision must be made as to whether they should be produced by a vendor or made in-house. This is the classic "make or buy" decision, a complex decision that is based on the cost of the component involved as well as the capitalization of equipment, the investment in manufacturing personnel, and plans by the company to use similar manufacturing equipment in the future.

Regardless of whether the component is to be made or bought, cost estimates are vital. We look now at cost-estimating for two primary manufacturing processes: machining and injection molding.

12.2.3 The Cost of Machined Components

Machined components are manufactured by removing portions of the material not wanted. Thus the costs for machining are primarily dependent on the cost and shape of the stock material, the amount and shape of the material that needs to be removed, and how accurately it must be removed. These three areas can be further decomposed into seven significant control factors that determine the cost of a machined component:

1. *From what material is the component to be machined?* The material affects the cost in three ways: the cost of the raw material, the value of the scrap produced, and the ease with which the material can be machined. The first two are direct material costs, and the last affects the amount of labor, the amount of time, and the choice of machines that are used manufacturing the component.
2. *What type of machine is used to manufacture the component?* The type of machine—lathe, horizontal mill, vertical mill, etc.—used in manufacture affects the cost of the component. For each type there is not only the cost of the machine time itself but also the cost of the tools and fixtures needed.
3. *What are the major dimensions of the component?* This factor helps determine what size of machines of each type will be required to manufacture the component. Each machine in a manufacturing facility has a different cost for use, depending on the initial cost of the machine and its age.
4. *How many machined surfaces are there, and how much material is to be removed?* Just knowing the number of surfaces and the material removal ratio (the ratio of the final component volume to the initial volume) can aid in giving a good estimate for the amount of time required to machine the part. More accurate estimates require knowing exactly what machining operations will be used to make each cut.
5. *How many components are made?* The number of components in a batch has a great effect on the cost. For one piece, fixturing is minimal, though long setup and alignment times are required. For a few pieces, simple fixtures are made. For a high volume, the manufacturing process is automated, with extensive fixturing and numerically controlled machining.
6. *What tolerance and surface finishes are required?* The tighter the tolerance and surface finish requirements, the more time and equipment are needed in manufacture.
7. *What is the labor rate for machinists?*

As an example of how these seven factors affect the cost of machined components, consider the component in Fig. 12.5. For this component the seven significant factors affecting cost are as follows:

1. The material is 1020 low-carbon steel.
2. The major manufacturing machine is a lathe. Two additional machines need to be used to mill the flat surfaces and drill the hole.
3. The major dimensions are a 57.15-mm diameter and a 100-mm length. The initial raw material must be larger than these dimensions.
4. There are three turned surfaces and seven other surfaces to be made. The final component is approximately 32 percent the volume of the original.
5. The number of components to be made is discussed in the next paragraph.

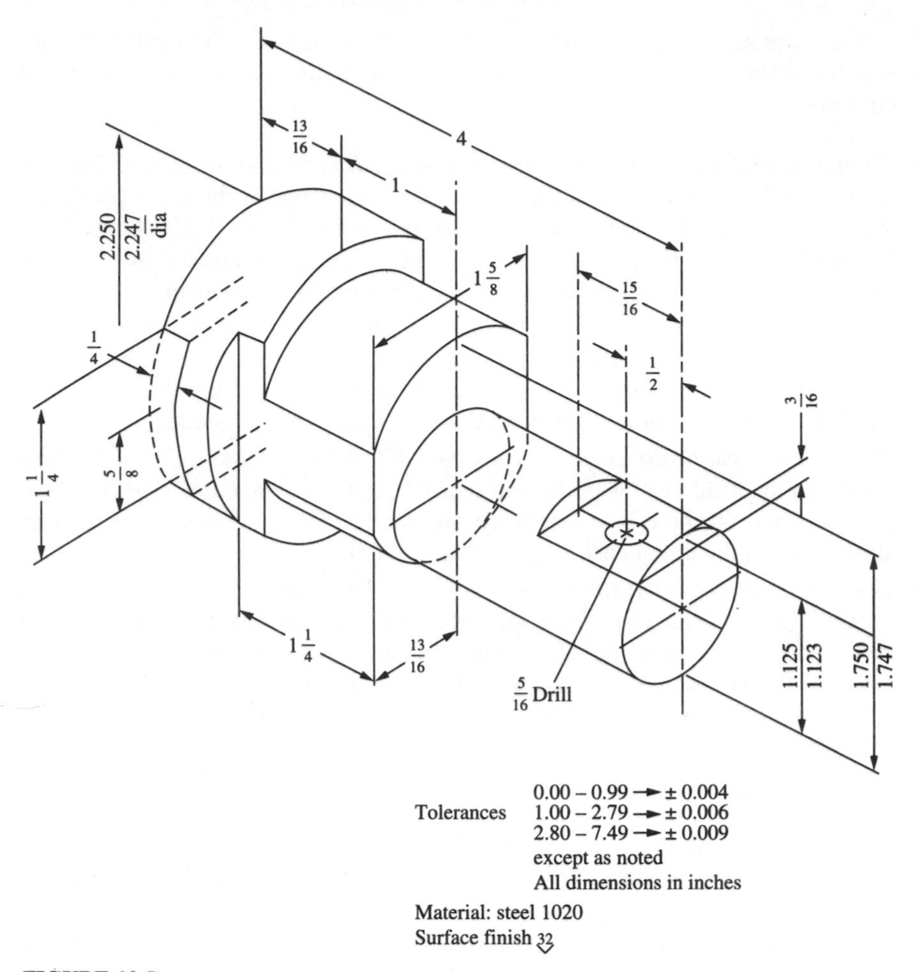

FIGURE 12.5
Sample component for evaluating machining cost.

6. The tolerance varies over the different surfaces of the component. On most surfaces it is nominal, but on the diameters it is a fit tolerance. The surface finish, 800 μmm (32 μin.), is considered intermediate.
7. The labor rate used is $35 per hour; this includes overhead and fringe benefits.

Figure 12.6 shows the cost of this component for various manufacturing volumes. The values are the total manufacturing cost per component. The cost of materials per component remains fairly constant at $1.48, but the labor hours and thus the cost of labor drop with volume. For machined components the cost dependence on volume is small in quantities above 10 because of the use of computer-aided manufacturing, CAM.

The dependence of the manufacturing cost on other variables is shown in Table 12.1, in which the tolerance, finish, and material are varied. The first three lines show

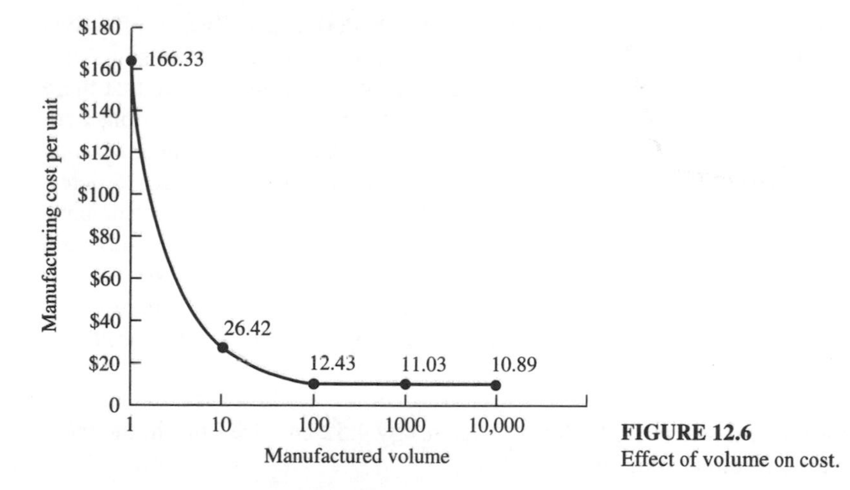

FIGURE 12.6
Effect of volume on cost.

the change with tolerance. A fine tolerance was used for the data in Fig. 12.6 and is shown in line 1. As the tolerance was relaxed to nominal (2) and then to rough (3), the cost dropped.

The fourth and fifth lines show the effect of surface finish on the manufacturing cost. The data in Fig. 12.6 were based on an intermediate surface finish, as specified in the drawing. As this was improved (4), the manufacturing cost rose, and as it was reduced to "as turned" (5), the cost dropped dramatically. Also shown in Table 12.1 is the effect of changing the material from low-carbon steel to high-carbon steel (6), which doubles the cost when compared to line 1 in part because of an increase in material cost (+ $4.00) and an increase in the machining time.

12.2.4 The Cost of Injection-Molded Components

Probably the most popular manufacturing method for high-volume products is plastic injection molding. This method allows for great flexibility in the shape of the

TABLE 12.1
Effect of tolerance, finish, and material on cost

Control parameters		
Tolerance	**Surface finish**	**Manufacturing cost**
1. Fine	Intermediate	$11.03
2. Nominal	Intermediate	$8.83
3. Rough	Intermediate	$7.36
4. Fine	Polished	$14.85
5. Fine	As turned	$8.17
6. High-carbon steel		$22.45

Note: For 1000 units.

components and, for manufacturing volumes over 100,000, is usually cost-effective. On a coarse level, all the factors that affect the cost of machined components also affect the cost of injection-molded components. The only differences are that there is only one type of machine, an injection-molding machine, and the questions concerning geometry are modified. Beside the major dimensions of the component, it is important to know the wall thickness and component complexity in order to determine the size of the molding machine needed; the time it will take the components to cool sufficiently for ejection from the machine; the number of cavities in the mold (the number of components molded at one time); and the cost of the mold.

To demonstrate the effect of the factors, we show the cost for a component developed for the Splashgard. The component is shown in Fig. 12.7. The significant factors affecting cost are as follows:

1. The overall dimensions are 9.46 cm (3.72 in.) by 4.52 cm (1.77 in.) in the mold plane and 4.13 cm (1.6 in.) deep.
2. The wall thickness is 3.2 mm (0.125 in.).
3. The number of components to be manufactured is 1 million.

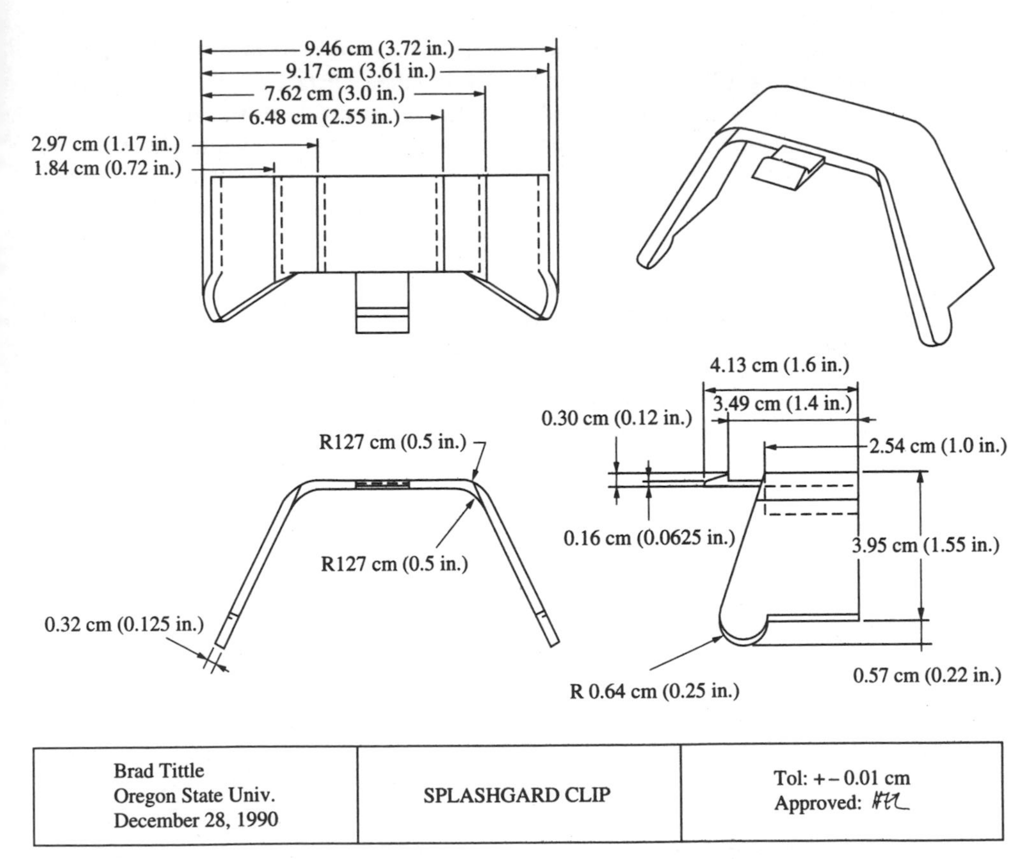

FIGURE 12.7
Splashgard component for cost estimation.

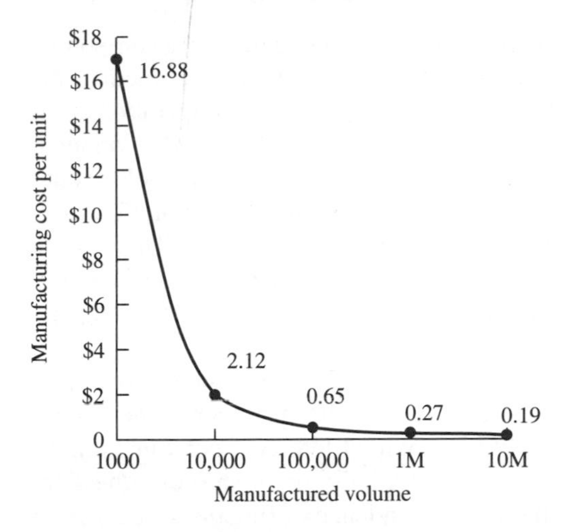

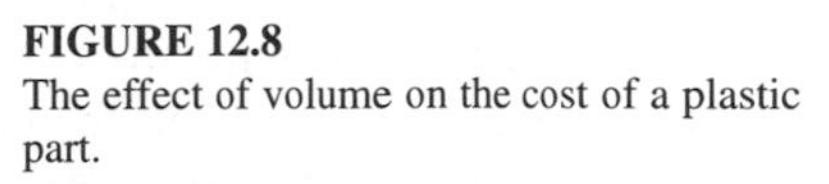
FIGURE 12.8
The effect of volume on the cost of a plastic part.

4. The labor hourly rate is $35.
5. The tolerance level is intermediate.
6. The surface finish is not critical.

The cost of manufacturing the component in Fig. 12.7 is shown in Fig. 12.8 for varying production volumes. The capital cost of making a mold is high enough to dominate the cost of the component at low volumes. This is why making just 1000 injection-molded plastic parts would be very expensive.

The manufacturing cost can be affected by the wall thickness. In the drawing the thickness is 3.2 mm. If this is lowered to 2.5 mm, the part cost will drop about 18 percent. This is primarily because the time needed in the mold for cooling drops from 18 sec to 13 sec, saving cycle time.

12.3 VALUE ENGINEERING

The concept of value engineering (also called value analysis) was developed by General Electric in the 1940s and evolved into the 1980s. Value engineering is a customer-oriented approach to the entire design process. It is a predecessor to many of the techniques in this book. One key point of value engineering is that it is not sufficient to only find cost—it is necessary to find the value of each feature, component, and assembly to be manufactured. Value is defined as

$$\text{Value} = \frac{\text{Worth of a feature, component, or assembly}}{\text{Cost of it}}$$

The worth of a feature of a component, for example, is determined by the functionality it provides to the customer. Thus, a refined definition for value is *function provided per dollar of cost.*

The value formula is used as a theme through the value engineering steps suggested below. These steps are focused on features of components. The method can also be applied to components and assemblies.

STEP 1. To ensure that all the functions are known, for each feature of a component ask the question, What does it do? If a feature provides more than one function, the fact must be noted. Features that result from a specific manufacturing operation are at the finest level of granularity that should be considered. For the machined component in Fig. 12.5, each turned diameter and face, each milled surface, and the hole should be considered. For the Splashgard clip, the injection-molded plastic part in Fig. 12.7, the 6.4-mm-radius round feature at the bottom is a good feature to query. This feature provides a number of functions. Looking back at the morphology for interface 3 in Table 10.1, it contributes to orienting, guiding, and securing the clip in the keeper.

STEP 2. Identify the life-cycle cost of this feature. This cost should include the manufacturing cost as well as any other downstream costs to the customer. If the feature provides multiple functions, the cost should be divided into cost per function. To do this, consider an equivalent feature that provides only the function in question. Although it is not accurate because of the interdependence of functions, it gives an estimate.

The cost of the round feature (R = 0.64 cm) in Fig. 12.7 is not evident. Consultation with tooling and manufacturing engineers revealed that, for a volume of 100,000 components ($0.65 component cost in Fig. 12.8) $0.02 was due to this feature. Their logic was that the feature does not contribute to labor cost because the cycle time would not change if the feature were removed. They estimated that, since the feature was hard to machine in the mold, it contributed about 5 percent to the mold cost. Amortized over the production volume, this gives $0.017. Finally, the material used for this feature is worth $0.003. So the feature costs $0.02 total. It could be argued that the structure of body of the component should be included because it contributes to the function of the round feature. A decision has to be made as to where to allocate all the costs in the component, one of the challenges of value engineering.

STEP 3. Identify the worth of the function to the customer. In an ideal world we would be able to ask customers how much each function was worth to them. However, this is not realistic. To obtain at least a qualitative indication of function worth, the information developed in the QFD is used. If no formal method was used to develop the customers' requirements and measures of importance, then the best that can be done is to ask, How important is this feature to the customer?

The Splashgard feature contributes to a number of functions that are very important to the customer (as developed in the QFD diagram in Fig. 10.18). To complicate matters, each of these functions involves other features. The best that can be done is to say that the functions contributed to by the round feature are worth a great deal to the customer. A customer will not pay as much for a product that is hard to attach, so the engineers estimated the worth at $2.00. Since this method compares relative values, it is only consistency in estimating the customer values, and not the values themselves, that is important.

STEP 4. Compare worth to cost to identify features that have low relative value. If one feature costs more than the others and is worth more—provides important function to the product—then its value may be as high as or higher than the others. On the other hand, if its costs outweigh its worth, then it has low value and should be redesigned.

The round feature contributes to a number of important functions for very low cost and thus is considered to be of high value.

The concept of value is further discussed in Section 12.5, on design for assembly. In that section features are added to ease assembly. Even though these features cut assembly time and thus cost, they often raise the manufacturing cost. Whether to use these features is best judged by considering their value.

12.4 DESIGN FOR MANUFACTURE

The term *design for manufacture,* or DFM, is widely used but poorly defined. Manufacturing engineers often use this term to include all or some of the best practices discussed in this book. Here we will define DFM as *establishing the shape of components to allow for efficient, high-quality manufacture.* Notice that the subject of the definition is *component.* In fact, DFM could be called DFCM, design for component manufacture, to differentiate it from design for assembly, DFA, the assembly of components covered in the next section.

The key concern of DFM is in specifying the best manufacturing process for the component and ensuring that the component form supports the manufacturing process selected. For any component there are many manufacturing processes that could be used in its manufacture. For each manufacturing process there are design guidelines that, if followed, would result in consistent components and little waste. For example, the best process to manufacture the clip on the bicycle Splashgard is injection molding. Thus, the form of the clip will need to follow design guidelines for plastic injection molding if the product is to be free from sink marks, surface finish blemishes, and other problems causing low-quality results.

A part of matching the component to the manufacturing process includes concern for tooling and fixturing. Components must be held for machining, released from molds, and moved between processes. The design of the component can affect all these manufacturing issues. Further, the design of the tooling and fixturing should be treated concurrently with the development of the component. The design of tooling and fixturing follows the same process as the design of the component: establish requirements, develop concepts, and then the final product.

In the days of over-the-wall product design processes, design engineers would sometimes release drawings to manufacturing for components that might be difficult or impossible to make. The concurrent engineering philosophy, with manufacturing engineers as members of the design team, helps avoid these problems. With thousands of manufacturing methods it is impossible for a designer to have sufficient knowledge to perform DFM without the assistance of manufacturing experts.

12.5 DESIGN-FOR-ASSEMBLY EVALUATION

Part of the labor cost in manufacturing a product is the cost of assembling it; for some products it is a significant part of the labor expense.

Design for assembly, DFA, is the best practice used to measure the ease with which a product can be assembled. Since virtually all products are assembled out of many components and assembly takes time (that is, costs money), there is a strong incentive to make products as easy to assemble as possible.

Throughout the 1980s many methods evolved to measure the assembly efficiency of a design. All of these methods require that the design be a fairly refined product before they can be applied. The technique presented in this section is based on these methods. It is organized around 13 design-for-assembly guidelines, which form the basis for a worksheet (Fig. 12.9). Before we discuss these 13 guidelines, we mention a number of important points about DFA.

Assembling a product means that a person or a machine must (1) *retrieve* components from storage, (2) *handle* the components to orient them relative to each other, and (3) *mate* them. Thus the ease of assembly is directly proportional to the number of components that must be retrieved, handled, and mated, and the ease with which they can be moved from their storage to their final, assembled position. Each act of retrieving, handling, and mating a component or repositioning an assembly is called an *assembly operation.*

Retrieval usually starts at some type of component feeder; this can range from a simple bin of loose bulk components to an automatic machine that feeds one component at a time in the proper orientation for a robot to handle.

Component handling is a major consideration in the measure of assembly quality. Handling encompasses maneuvering the retrieved component into position so that it is oriented for assembly. For a bolt to be threaded into a tapped hole, it must first be positioned with its axis aligned with the hole's axis and its threaded end pointed toward the hole. A number of motions may be required in handling the component as it is moved from storage and oriented for mating. If component handling is accomplished by a robot or other machine, each motion must be designed or programmed into the device. If component handling is accomplished by a human, the human factors of the required motions must be considered.

Component mating is the act of bringing components together. Mating may be minimal, like setting one component on the flat surface of another, or it may require threading a fastener into a threaded hole. A term often synonymous with *mating* is *insertion.* During assembly some components are inserted in holes, others are placed on surfaces, and yet others are fitted over pins or shafts. In all these cases, the components are said to be inserted in the assembly, even though nothing may really be inserted, in the traditional sense of the word, but only placed on a surface.

DFA measures a product in terms of the efficiency of its overall assembly and the ease with which components can be retrieved, handled, and mated. A product with high assembly efficiency has a few components that are easy to handle and virtually fall together during assembly. Assembly efficiency can be demonstrated by considering the seat frames designed for the bicycle shown in the Preface. Figure 12.10 shows an old frame, which had nine separate components requiring 20 separate operations to put together. These included positioning and welding operations. This frame took 30 minutes to assemble. In contrast, the new frame (Fig. 12.11) was designed with assembly efficiency as a major engineering requirement. The resulting product has only four components, requiring eight operations and about eight

DESIGN FOR ASSEMBLY
INDIVIDUAL ASSEMBLY EVALUATION FOR ____________

EVALUATED BY ____________ DATE ______
REVIEWED BY ____________ DATE ______
TRIAL 01 02 03 04 05

OVERALL ASSEMBLY						*COMMENTS*
1 OVERALL PART COUNT MINIMIZED	POOR	FAIR	GOOD	VERY GOOD	OUTSTANDING	
2 MINIMUM USE OF SEPARATE FASTENERS	POOR	FAIR	GOOD	VERY GOOD	OUTSTANDING	
3 BASE PART WITH FIXTURING FEATURES (LOCATING SURFACES AND HOLES)	POOR	FAIR	GOOD	VERY GOOD	OUTSTANDING	
4 REPOSITIONING REQUIRED DURING ASSEMBLY SEQUENCE	TWO OR MORE REPOSITIONS		REPOSITION ONCE		NO REPOSITIONING	
5 ASSEMBLY SEQUENCE EFFICIENCY	POOR	FAIR	GOOD	VERY GOOD	OUTSTANDING	
PART RETRIEVAL						
6 CHARACTERISTICS THAT COMPLICATE HANDLING (TANGLING, NESTING, FLEXIBILITY) HAVE BEEN AVOIDED	NO PARTS	FEW PARTS	SOME PARTS	MOST PARTS	ALL PARTS	
7 PARTS HAVE BEEN DESIGNED FOR A SPECIFIC FEED APPROACH (BULK, STRIP, MAGAZINE)						
PART HANDLING						
8 PARTS WITH END-TO-END SYMMETRY	NO PARTS	FEW PARTS	SOME PARTS	MOST PARTS	ALL PARTS	
9 PARTS WITH SYMMETRY ABOUT THE AXIS OF INSERTION						
10 WHERE SYMMETRY IS NOT POSSIBLE PARTS ARE CLEARLY ASYMMETRIC						
PART MATING						
11 STRAIGHT LINE MOTIONS OF ASSEMBLY	NO PARTS	FEW PARTS	SOME PARTS	MOST PARTS	ALL PARTS	
12 CHAMFERS AND FEATURES THAT FACILITATE INSERTION AND SELF-ALIGNMENT						
13 MAXIMUM PART ACCESSIBILITY						
	TOTAL × 0	TOTAL × 2	TOTAL × 4	TOTAL × 6	TOTAL × 8	
					TOTAL SCORE	

NOTE:
EVALUATION SCORE TO BE USED ONLY TO COMPARE ONE ASSEMBLY TO ALTERNATE DESIGNS OF THE *SAME* ASSEMBLY

FIGURE 12.9
Design-for-assembly worksheet.

FIGURE 12.10
Old seat frame.

minutes to assemble. The saving in labor is obvious. Additionally, there are savings in component inventory, component handling, and dealings with component vendors.

Guidelines similar to those on the worksheet of Fig. 12.9 were used in the design of the new seat frame to make it efficient to assemble. The worksheet is designed to give an assembly efficiency score to each product evaluated. The score ranges from 0 to 104. The higher the score, the better the assembly. This score is used as a relative measure to compare alternative designs of the same product or similar products; the actual value of the score has no meaning. The design can be patched or changed on the basis of suggestions given in the guidelines and then reevaluated. The difference between the score of the original product and that of the redesign gives an indication of the improvement of assembly efficiency.

Although this technique is only applied late in the design process, when the product is so refined that the individual components and the methods of fastening are determined, its value can be appreciated much earlier in the design process. This is true because, after filling out the worksheet a few times, the designer develops the sense of what makes a product easy to assemble, knowledge that will have an effect on all future products.

Using ease of assembly as an indication of design quality makes sense only for mass-produced products, since the design-for-assembly guidelines encourage a few complex components. These types of components usually require expensive tooling, which can only be justified if spread over a large manufacturing volume.

Finally, the relationship between the cost of assembly to the overall cost of the product must be kept in mind when considering how much to modify a design

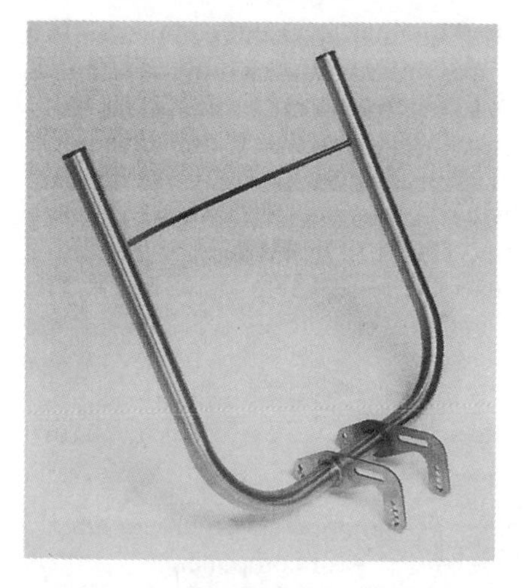

FIGURE 12.11
Redesigned seat frame.

according to these suggestions. In low-volume electromechanical products, the cost of assembly is only 1 to 5 percent of the total manufacturing cost. Thus, there is little payback for changing a design for easier assembly; the change will require extra design effort and may raise the cost of manufacturing, with little financial return.

Measures for each of the 13 design-for-assembly guidelines will be discussed in the sections below; the first section gives five guidelines, all concerned with the overall assembly efficiency; the succeeding three sections give design-for-assembly guidelines oriented toward the retrieval, handling, and mating of the individual components.

12.5.1 Evaluation of the Overall Assembly

GUIDELINE 1: OVERALL COMPONENT COUNT SHOULD BE MINIMIZED. The first measure of assembly efficiency is based on the number of components or subassemblies used in the product. The part count is evaluated by estimating the minimum number of components possible and comparing the design being evaluated to this minimum. The measure for this guideline is estimated in the following way:

a. Find the theoretical minimum number of components. Examine each pair of adjacent components in the design to see if they really should be separate components. Include fastening components such as bolts, nuts, and clips in this accounting. Assuming no production or material limitations: (1) Components must be separate if the design is to operate mechanically. For example, components that must slide or rotate relatively to each other must be separate components. However, if the relative motion is small, then elasticity can be built into the design to meet the need. This is readily accomplished in plastic components by using elastic hinges, thin sections of fatigue-resistant material that act as a one-degree-of-freedom joint. (2) Components must be separate if they must be made of different materials, for example, when one component is an electric or thermal insulator and another, adjacent component is a conductor. (3) Components must be separate if assembly or disassembly is impossible. (Note that the last word is "impossible," not "inconvenient.")

Thus each pair of adjacent components is examined to find if they absolutely need to be separate components. If they do not, then theoretically they can be combined into one component. After reviewing the entire product this way, we develop the theoretical minimum number of components. The Splashgard has a minimum of two separate components, and the seat frame used as an example above has 1. The actual number of components in the Splashgard is five: the keeper, a bolt, a nut, the clip, and the fender.

b. Find the improvement potential. To rate any product, we can calculate its improvement potential:

$$\text{Improvement potential} = \frac{\begin{pmatrix}\text{Actual number of}\\ \text{components}\end{pmatrix} - \begin{pmatrix}\text{Theoretical minimum number}\\ \text{of components}\end{pmatrix}}{\text{Actual number of components}}$$

c. Rate the product on the worksheet (Fig. 12.9).

- If the improvement potential is less than 10 percent, the current design is *outstanding*.
- If the improvement potential is 11 to 20 percent, the current design is *very good.*
- If the improvement potential is 20 to 40 percent, the current design is *good.*
- If the improvement potential is 40 to 60 percent, the current design is *fair.*
- If the improvement potential is greater than 60 percent, the current design is *poor.*

The improvement potential of the Splashgard is $(5-2)/5 = 60$ percent. In this case current design is fair.

As a product is redesigned, keep track of the actual improvement:

$$\text{Actual improvement} = \frac{\begin{pmatrix}\text{Number of components}\\ \text{in initial design}\end{pmatrix} - \begin{pmatrix}\text{Number of components}\\ \text{in redesign}\end{pmatrix}}{\text{Number of components in initial design}}$$

Typical improvement in the number of components in the range of 30 to 60 percent is realized by redesigning the product in order to reduce the component count.

To put this guideline in perspective, compare it with earlier phases of the design process. In the design philosophy of this text, the functionality of the product is broken down as finely as possible as a basis for the development of concepts (Chap. 7). We then used a morphology for developing ideas for each function. This can lead to poor designs, as can the effort to minimize the number of components. Consider the design of the common nail clipper (Fig. 12.12). If the assumption is made that all the functions are independent and that concepts are generated for each function, then the result, as seen in Fig. 12.13, is a disaster. Note that each function is mapped to

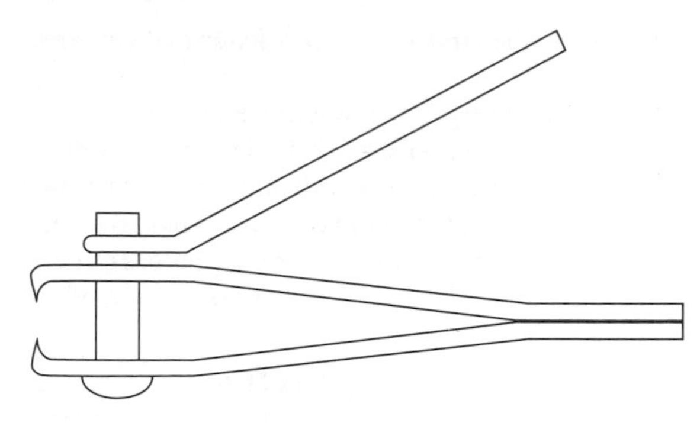

FIGURE 12.12
Common nail clipper.

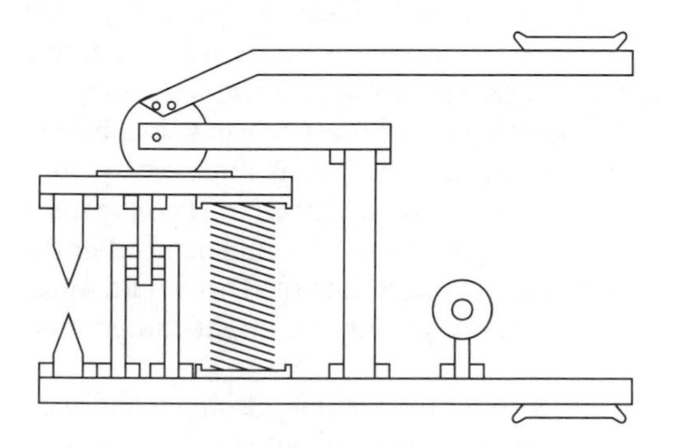

FIGURE 12.13
Nail clipper with one interface for each function. (Design developed by Karl T. Ulrich, Sloan School of Management, Massachusetts Institute of Technology.)

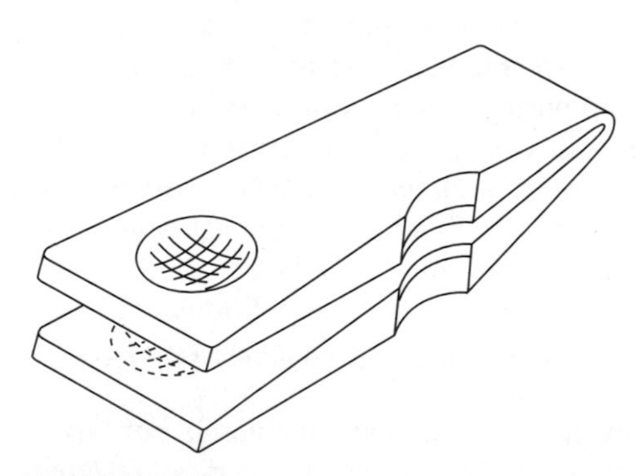

FIGURE 12.14
A one-piece nail clipper.

one or more interface. At the other extreme, the DFA philosophy leads to the product shown in Fig. 12.14.

Here, in evaluating the product for assembly, this guideline encourages lumping as many functions as possible into each component. This design philosophy, however, also has its problems. The cost of tooling (molds or dies) for the shapes that result from a minimized component count can be high—and that cost is not taken into account here. Additionally, tolerances on complex components may be more critical, and manufacturing variations might affect many functions that are now coupled.

GUIDELINE 2: MAKE MINIMUM USE OF SEPARATE FASTENERS. One way to reduce the component count is to minimize the use of separate fasteners. This is advisable for many reasons. First, each fastener used is one more component to handle, and there may be many more than one in the case of a bolt with its accompanying nut, flat washer, and lock washer. Each instance of component handling takes time, typically 10 seconds per fastener. Second, the total cost for fasteners is the cost of the components themselves as well as the cost of purchasing, inventorying, accounting for, and quality-controlling them. Third, fasteners are stress concentrators; they are points of potential structural failure in the design. For all these reasons, it is best to eliminate as many fasteners as possible from the design. This is more easily done on high-volume products, for which components can be designed to snap together, than on low-volume products or products utilizing many stock components. For the Splashgard we could eliminate the nut and bolt and use a snap fit. This would lower the part count, improving the design by 40 percent, although it might compromise the function of the keeper.

An additional point that should be considered in evaluating a design is how well the use of fasteners has been standardized. A good example of part standardization is the fact that almost everything on the Volkswagon Beetle, a car popular in the 1970s, can be fixed with a set of screwdrivers and a 13-mm wrench.

Finally, if the components fastened together must be taken apart for maintenance, use captured fasteners (fasteners that remain loosely attached to a component even when unfastened). Many varieties of captured fasteners are available, all designed so that they will not be misplaced during assembly or maintenance.

There are no general rules for the quality of a design in terms of the number of separate fasteners. Since the worksheet is just a relative comparison between two designs, an absolute evaluation is not necessary. Obviously, an outstanding design will have few separate fasteners, and those it does have will be standardized and possibly captured. Poor designs, on the other hand, require many different fasteners to assemble. If more than one-third of the components in a product are fasteners, the assembly logic should be questioned.

Figures 12.15 and 12.16 show some ideas for reducing the number of fasteners. In designing with injection-molded plastics, the best way to get rid of fasteners is through the use of snap fits. A typical cantilever snap is shown in Fig. 12.15*a*. Important considerations when designing snaps are the loads during insertion and when seated. During insertion the snap acts like a cantilever beam flexed by the amount of the insertion displacement. The major stress during insertion is therefore bending

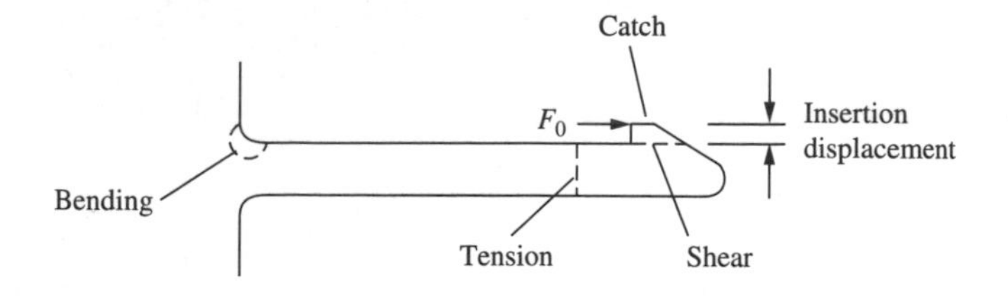

Cantilever snap

(*a*)

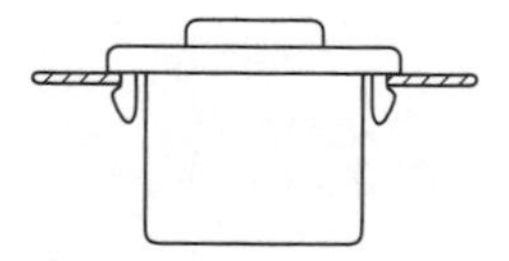

Undersized snap-fit lugs: Too short a bending length can cause breakage.

Properly sized snap-fit lugs: Longer lugs reduce stress.

(*b*)

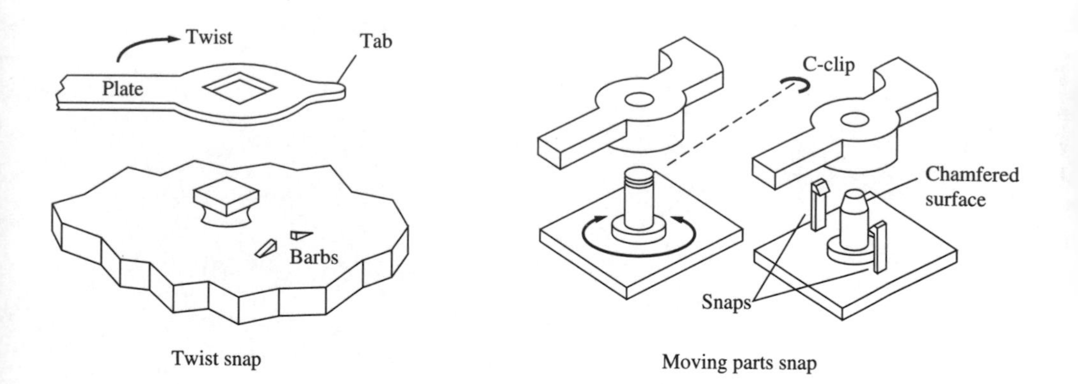

Twist snap

Moving parts snap

(*c*)

FIGURE 12.15
Snap-fastener design.

at the root of the beam. Thus it is important to have low stress concentrations at that point and to be sure that the snap can flex enough without approaching the elastic limit of the material (Fig. 12.15*b*). When seated, the snap's main load is the force F_0, the force holding the components together. It can cause crushing on the face of the catch, shear failure of the catch, and tensile failure of the snap body. (Think of the force flow here.)

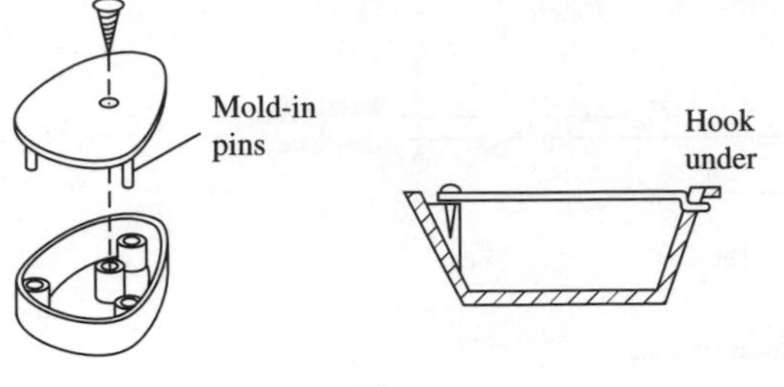

(*a*)

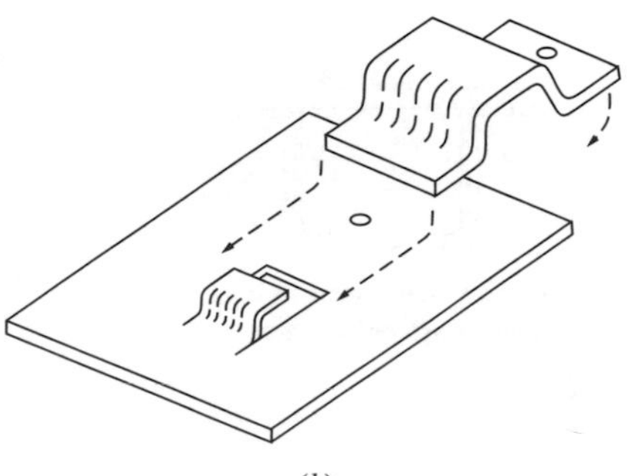

(*b*)

FIGURE 12.16
Single fastener examples.

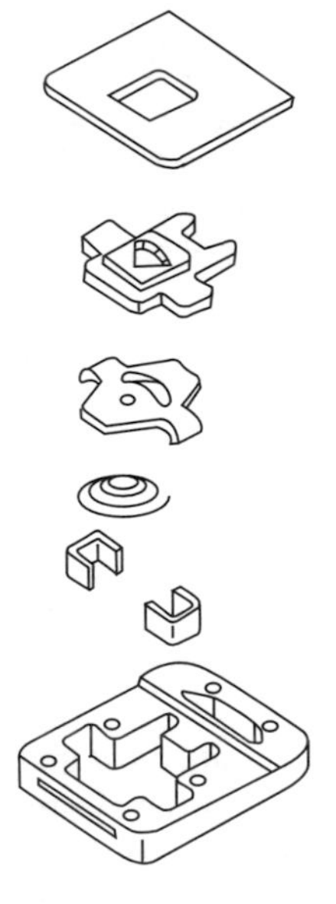

FIGURE 12.17
Meter assembly.

Additionally, design consideration must be given to unsnapping. If the device is ever to come apart for maintenance, then consider features that allow a tool or a finger to flex the snap while $F_0 = 0$. Additional snap configurations are shown in Fig. 12.15*c*. Note that each has one feature that flexes during insertion and another that takes the seated load.

Another way to reduce the number of fasteners is to use only one fastener and either pins, hooks, or other interference to help connect the components. The examples in Fig. 12.16 show both plastic and sheet-metal applications of this idea.

GUIDELINE 3: DESIGN THE PRODUCT WITH A BASE COMPONENT FOR LOCATING OTHER COMPONENTS. This guideline encourages the use of a single base on which all the other components are assembled. The base in Fig. 12.17 provides a foundation for consistent component location, fixturing, transport, orientation, and strength. The ideal design would be built like a layer cake, with each component or subassembly stacking on top of another one. Without this base to build on, assembly may consist of work on many subassemblies, each with its own fixturing and transport needs and final assembly requiring extensive repositioning and fixturing. The use of a single base component has shortened the length of some assembly lines by a factor of 2.

As with most of these measures, there are no absolute standards for determining an outstanding product and a poor one. Keep in mind that the rating on the worksheet is relative.

GUIDELINE 4: DO NOT REQUIRE THE BASE TO BE REPOSITIONED DURING ASSEMBLY. If automatic assembly equipment such as robots or specially designed component-placement machines are used during assembly, it is important that the base be positioned precisely. On larger products, repositioning may be time-consuming and costly. An outstanding design would require no repositioning of the base. A product requiring more than two repositionings is considered poor.

GUIDELINE 5: MAKE THE ASSEMBLY SEQUENCE EFFICIENT. If there are N components to be assembled, there are potentially $N!$ (N factorial) different possible sequences to assemble them. In reality, some components must be assembled prior to others; thus the number of possible assembly sequences is usually much less than $N!$. An efficient assembly sequence is one that

- Affords assembly with the fewest steps
- Avoids risk of damaging components
- Avoids awkward, unstable, or conditionally unstable positions for the product and the assembly personnel and machinery during assembly
- Avoids creating many disconnected subassemblies to be joined later

Since even a minor design change can alter the available choices in assembly sequence, it is important to consider the efficiency of the sequence during design. The technique described here will be demonstrated through a simple example, the assembly of a ballpoint pen (Fig. 12.18).

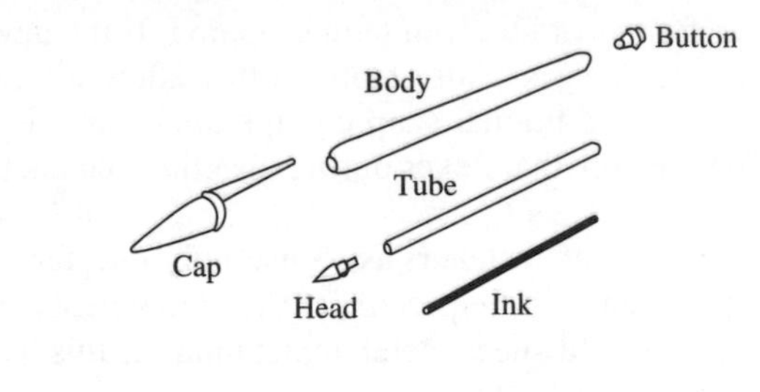

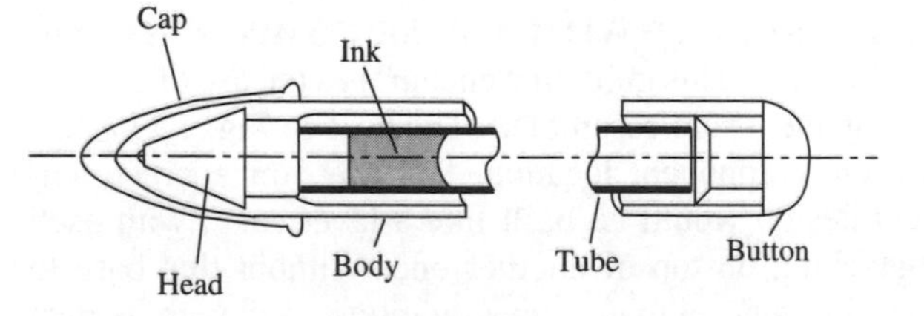

FIGURE 12.18
Ballpoint pen assembly.

a. List all the components and processes involved in the assembly process. Begin with a layout or assembly drawing of the product and a bill of materials. All components for the pen assembly are listed in Fig. 12.18. In some products the components to be assembled include subassemblies and processes—for example, the component called "ink" in the ballpoint pen includes the process of actually putting the ink in the tube. Additionally, some products require testing during the assembly process. These tests should also be included as components. Finally, fasteners should be lumped with the component they hold in place.

b. List the connections between components and generate a connections diagram. The connection diagram for the ballpoint pen is shown in Fig. 12.19. In this diagram the nodes represent the components and the links represent the connections. Connection diagrams can have loops. For example, the pen may have the button supporting the end of the tube, creating interface 6, a link between the tube and the

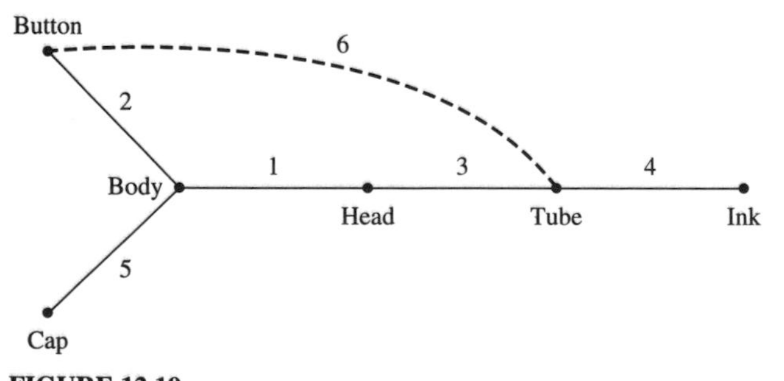

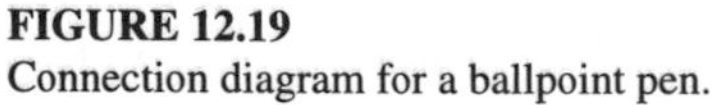
FIGURE 12.19
Connection diagram for a ballpoint pen.

button (shown as a dashed line in Fig. 12.19 and assumed not to exist throughout the remainder of this example).

c. Select a base component. The base component should be at one end of the connection diagram or be a large component. It should be the component that requires the least subassembly and allows assembly from the fewest directions. For the ballpoint pen, the options are the cap, the button, or the body. The cap requires subassembly of the head in the tube and is thus a poor candidate. The body requires assembly from two directions. The button may be the best base part, but it is hard to hold. Both the body and the button need to be further investigated.

d. Recursively add the next component. Add components to the base using the connection diagram as a guide. It is important to be aware of precedences; e.g., the tube must be on the head before the ink is installed. It is useful to list all precedences before starting this step. For the ballpoint pen the precedences are as follows:

Connection 3 must precede connection 4.
Connection 1 must precede connection 5.

e. Identify subassemblies. Subassemblies can be made of components that have a secure connection with each other, can be reoriented without falling apart, and have a simple connection with the other assembled components. Subassemblies should only be used if they simplify the process. For the pen the head, tube, and ink form a subassembly that simplifies assembly.

There are many potential assembly sequences for the ballpoint pen. One that is developed using the procedure above is

[2, [3, 4], 1, 5]

or

[button, body, [head, tube, ink], cap]

The first sequence lists the connections, and the second the components, in the order of assembly. The brackets denote subassemblies.

The process given here is very useful in evaluating the assembly sequence and determining the effects of design changes on the sequence. It also measures the efficiency of the assembly sequence. If all connections are made in a logical order, no subassemblies are generated, and no awkward connections made, then the efficiency is rated high; if the connection sequence cannot be accomplished, subassemblies are made, or awkward connections are needed, then the efficiency is low.

12.5.2 Evaluation of Component Retrieval

The measures associated with each guideline for retrieving components range from "all components" to "no components." If all components achieve the guideline, the quality of the design is high as far as component retrieval is concerned. Those components that do not achieve the guidelines should be reconsidered.

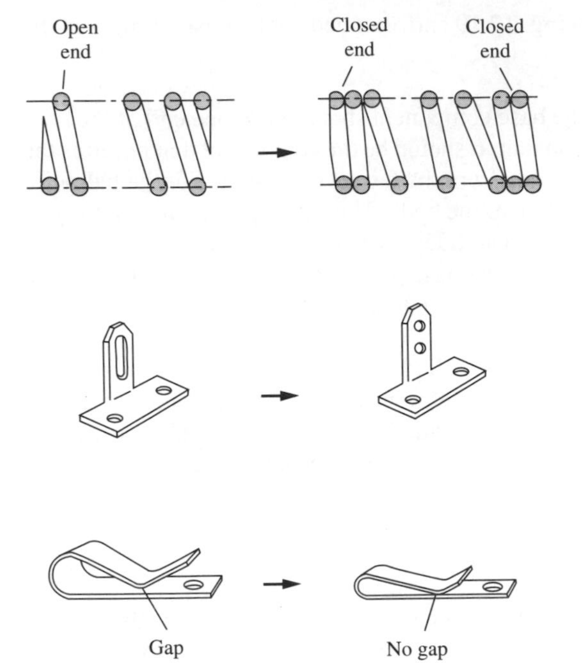

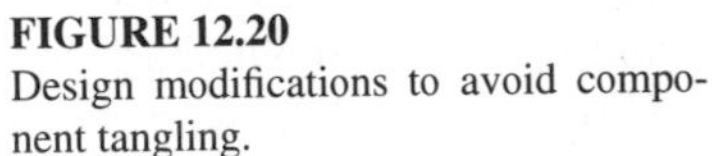
FIGURE 12.20
Design modifications to avoid component tangling.

GUIDELINE 6: AVOID COMPONENT CHARACTERISTICS THAT COMPLICATE RETRIEVAL. Three component characteristics make retrieval difficult: tangling, nesting, and flexibility. If components of the type shown in Fig. 12.20 column *a* are stored in a box or tray, they will be nearly impossible to pick up individually because they will become tangled. If the components are designed as shown in Fig. 12.21 column *b*, then they cannot tangle.

A second common problem that complicates retrieval is nesting, in which components jam inside each other (Fig. 12.21). There are two simple solutions for this problem: Either change the angle of the interlocking surfaces or add features that prevent jamming.

Finally, flexible components such as gaskets, tubing, and wiring harnesses are exceptionally hard components to retrieve and handle. When possible, make components as few, as short, and as stiff as possible.

GUIDELINE 7: DESIGN COMPONENTS FOR A SPECIFIC TYPE OF RETRIEVAL, HANDLING, AND INSERTION. Consider the assembly method of each component during design. There are three types of assembly systems: manual assembly, robot assembly, and special-purpose transfer machine assembly. In general, if the volume of the product is less than 250,000 annually, the most economic method of assembly is manual. For products that have a volume of up to 2 million annually, robots are generally best. Special-purpose machines are warranted only if the volume exceeds 2 million. Each of these systems has requirements for component retrieval, handling, and insertion. For example, components for manual assembly can be

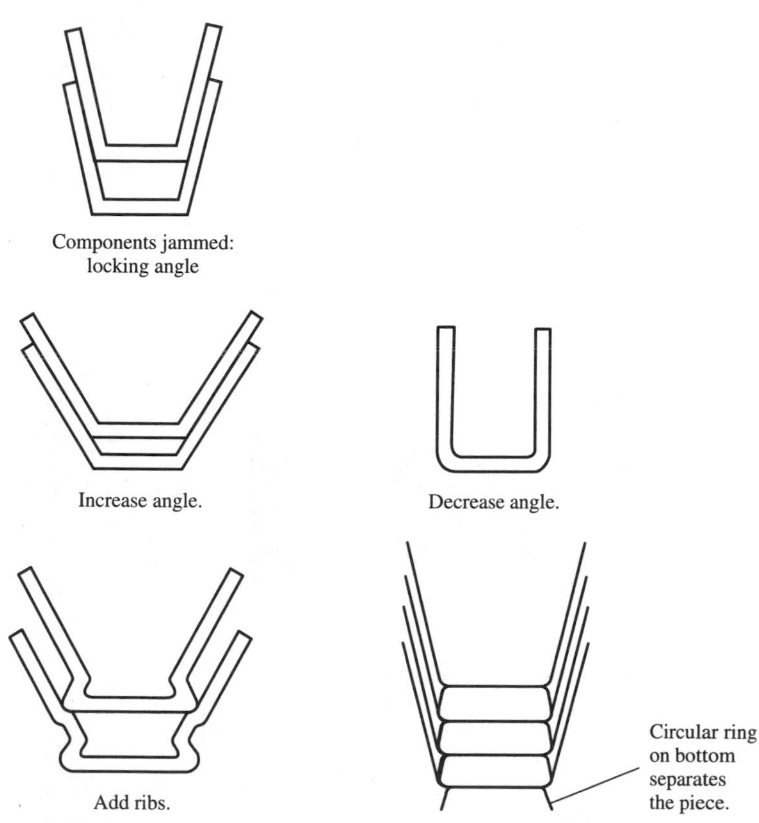

FIGURE 12.21
Design modifications to avoid jamming.

bulk-fed and must have features that make them easy to grasp. Robot grippers, on the other hand, may be fed automatically and can grasp a component externally, like a human; internally, with a suction cup on a flat surface; or with many other end effectors.

12.5.3 Evaluation of Component Handling

The following three design-for-assembly guidelines are all oriented toward the handling of individual components.

GUIDELINE 8: DESIGN ALL COMPONENTS FOR END-TO-END SYMMETRY. If a component can be installed in the assembly only in one way, then it must be oriented and inserted in just that way. The act of orienting and inserting the component takes time and either worker dexterity or assembly machine complexity. If assembly is to be done by a robot, for example, then having only one orientation for insertion may require the robot to be multiaxial. Conversely, if the component is spherical, then its orientation is of no consequence and handling is much easier. Most components in an assembly fall between these two extremes.

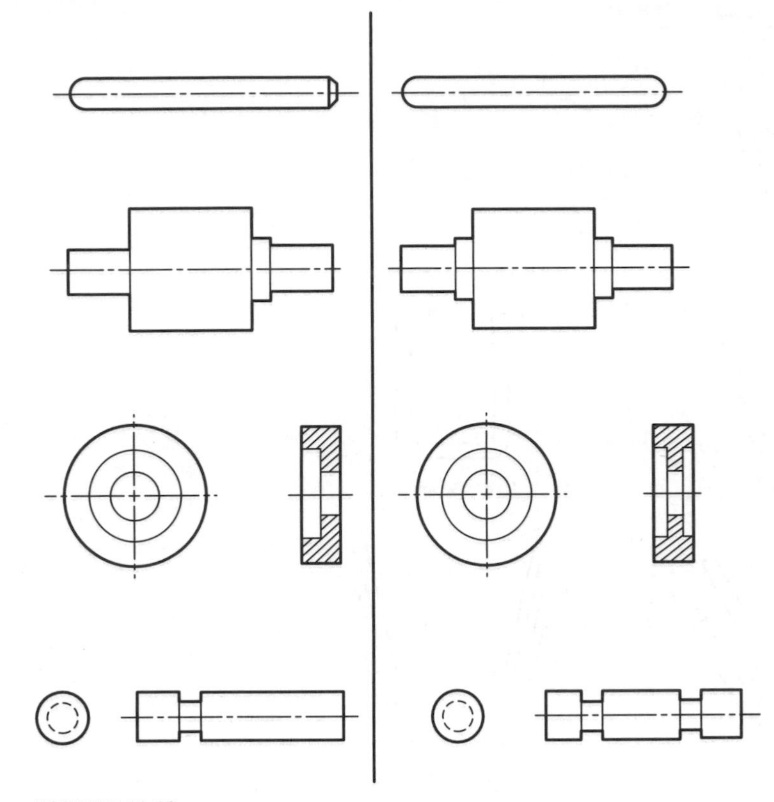

FIGURE 12.22
Modification of axisymmetric parts for end-to-end symmetry.

There are two measures of symmetry: end-to-end symmetry (symmetry about an axis perpendicular to the axis of insertion) and axis-of-insertion symmetry. (The latter is the focus of the next guideline and is not discussed here.) End-to-end symmetry means that a component can be inserted in the assembly either end first. Axisymmetric components that are intended to be inserted along their axes are shown in Fig. 12.22. Those in column *a* are designed to work in the design only if installed in one way. These same components are shown in column *b* modified so that they can be inserted either end first. In each case the asymmetrical feature has been replicated to make the component end-to-end symmetrical for ease of assembly.

Before modifying a component to meet this or similar guidelines, it is important to check the value of the modification. The cost of adding a feature may not improve its functionality for the assembler sufficiently to warrant the modification.

GUIDELINE 9: DESIGN ALL COMPONENTS FOR SYMMETRY ABOUT THEIR AXES OF INSERTION. Whereas the previous guideline called for end-to-end symmetry, a designer should also strive for rotational symmetry. The components in

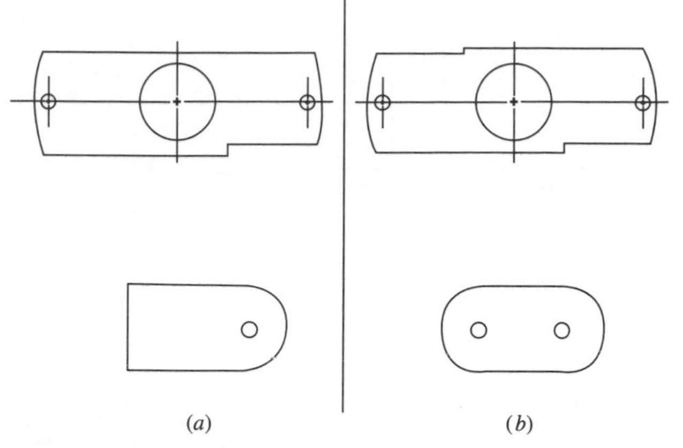

FIGURE 12.23
Modification of features for symmetry about the axis of insertion.

Fig. 12.22 are all axisymmetric if inserted in the direction of their centerline. In Fig. 12.23 the components in column *a* have only one orientation if they are inserted in the plane of the diagram. However, by adding a functionally useless notch (on the top component) or adding a hole and rounding an end (on the bottom component), we can give the components two orientations for insertion—a decided improvement.

In Fig. 12.24*a* the original design for the component fits only one way into the assembly. The addition of an opposing finger (Fig. 12.24*b*), which is useless functionally, gives the component two possible insertion orientations. Finally, modifying the component functions (Fig. 12.24*c*) can make the component axisymmetric. It is important to ask if the change in functionality is worth the gained ease of assembly. If not, then the asymmetry should be tolerated.

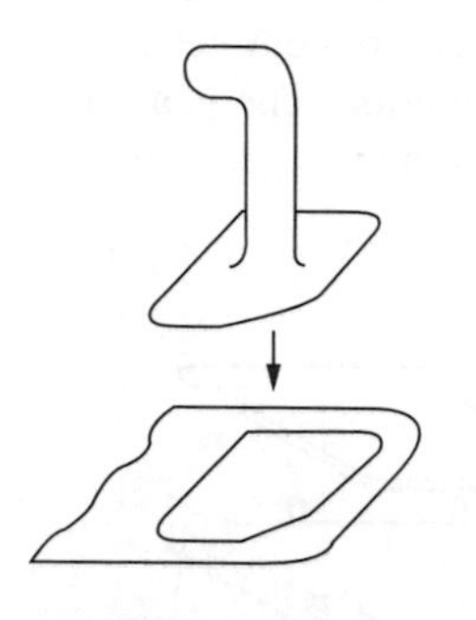

(*a*) The assembly fits together only one way.

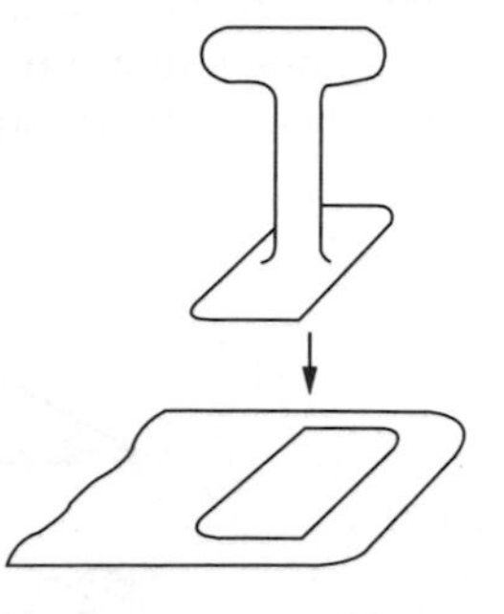

(*b*) Two possible directions of insertion.

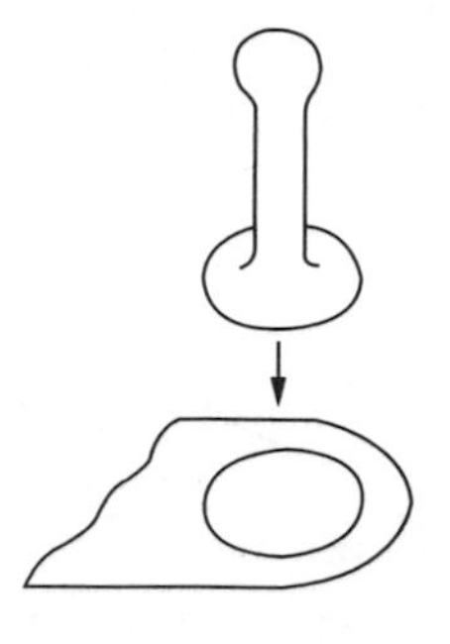

(*c*) 360° rotational symmetry.

FIGURE 12.24
Modification of a part for symmetry.

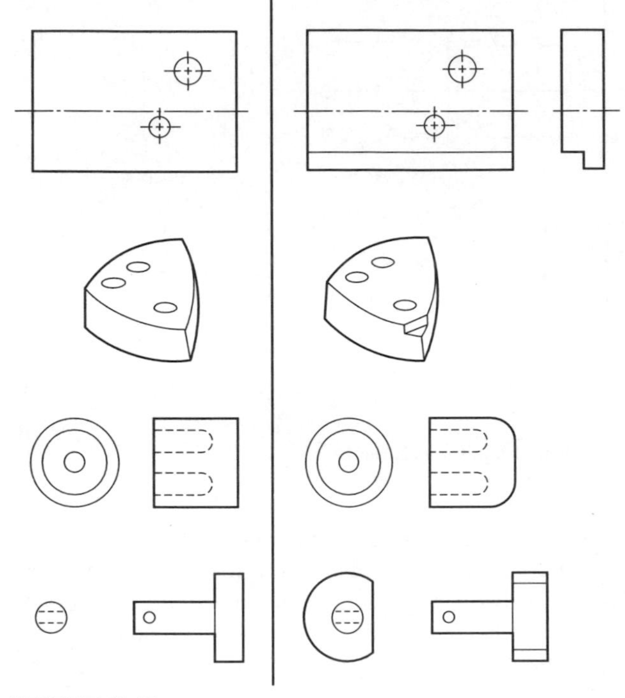

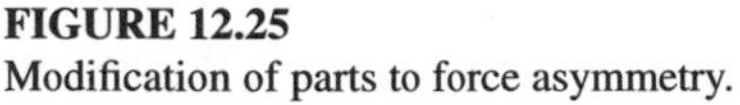

FIGURE 12.25
Modification of parts to force asymmetry.

GUIDELINE 10: DESIGN COMPONENTS THAT ARE NOT SYMMETRIC ABOUT THEIR AXES OF INSERTION TO BE CLEARLY ASYMMETRIC. The component in Fig. 12.24*a* is clearly asymmetric. If it were not asymmetric, the component could be inserted with the finger pointing the wrong way and, as a result, would not function as it was designed to. In Fig. 12.25 the four component designs of column *a* have been modified in column *b* to afford easy orientation. The goal of this guideline is to make components that can be inserted only in the way intended.

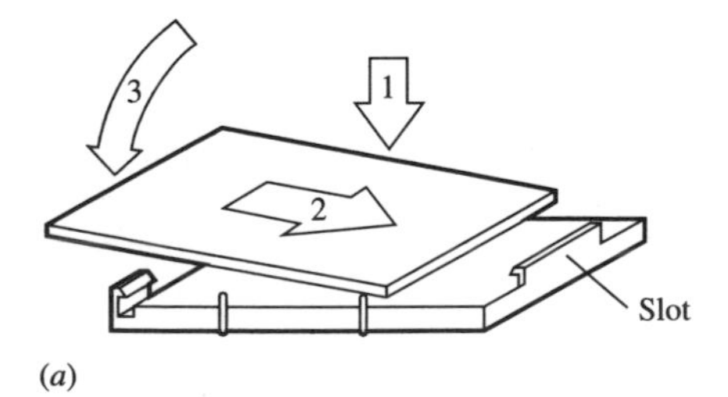

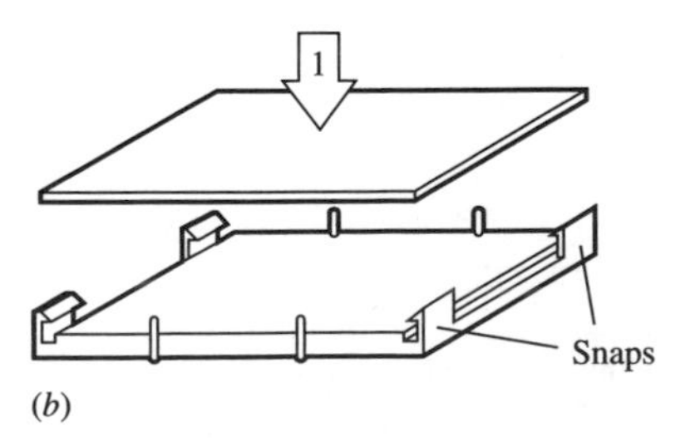

FIGURE 12.26
Example of one-direction assembly.

12.5.4 Evaluation of Component Mating

Finally, the quality of component mating should be evaluated. The following guidelines offer some design aids for improving assemblability.

GUIDELINE 11: DESIGN COMPONENTS TO MATE THROUGH STRAIGHT-LINE ASSEMBLY, ALL FROM THE SAME DIRECTION. This guideline, intended to minimize the motions of assembly, has two aspects: the components should mate through straight-line motion, and this motion should always be in the same direction. If both these corollaries are met, the assembly will then fall together from above. Thus, the assembly process will never require reorientation of the base nor any other assembly motion other than straight down. (Down is the preferred single direction, because gravity aids the assembly process.)

The components in Fig. 12.26*a* require three motions for assembly. This number has been reduced in Fig. 12.26*b* by redesigning the interface between the components. Note that the design in Fig. 12.16*b*, although improving the quality in terms of fastener use, has degraded the design in terms of insertion difficulty, again demonstrating that there are always tradeoffs to be considered in design.

GUIDELINE 12: MAKE USE OF CHAMFERS, LEADS, AND COMPLIANCE TO FACILITATE INSERTION AND ALIGNMENT. To make the actual insertion or mating of a component as easy as possible, each component should guide itself into place. This can be accomplished using three techniques. One common method is to use chamfers, or rounded corners, as shown in Fig. 12.27. Here the four components shown in column *a* are all modified with chamfers in column *b* to ease assembly.

In Fig. 12.28*a* the shaft has chamfers and still the disk is hard to align and press into its final position. This difficulty is alleviated by making part of the shaft a smaller diameter, allowing the disk to mate with the final diameter, as shown in column *b* of the figure. The lead section of the shaft has forced the disk into alignment with the final section. A similar redesign is shown in the lower component, where, in column *b*, by the time the shaft is inserted in the bearing from the right it is aligned properly.

Finally, component complicance, or elasticity, is used to ease insertion and also relax tolerances. The component mating scheme in colum *b* of Fig. 12.29 need not have high tolerance; even if the post is larger than the hole, the components will snap together.

GUIDELINE 13: MAXIMIZE COMPONENT ACCESSIBILITY. Whereas guideline 5 concerned itself with assembly sequence efficiency, this guideline is oriented toward sufficient accessibility. Assembly can be difficult if components have no clearance for grasping. Assembly efficiency is also low if a component must be inserted in an awkward spot.

Beside concerns for assembly, there is also maintenance to consider. To replace the fuses in one common computer printer, it is necessary to disassemble the entire machine. In both assembly and maintenance, tools are necessary and room must be

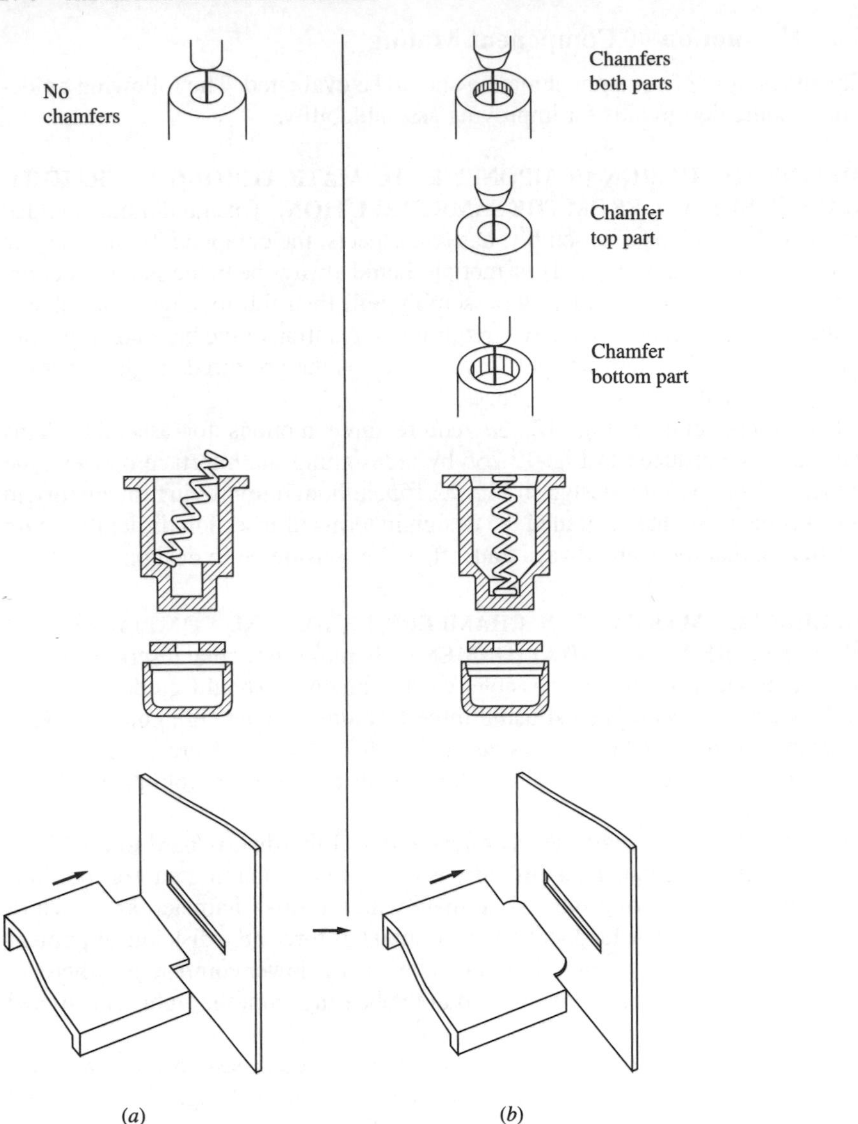

FIGURE 12.27
Use of chamfers to ease assembly.

allowed for the tools to mate with the components and to be manipulated. As shown in Fig. 12.30, sometimes simple design changes can make tool engagement and motion much easier.

12.6 DESIGN FOR RELIABILITY (DFR)

Reliability is a measure of how the quality of a product is maintained over time. Quality here is usually in terms of satisfactory performance under a stated set of operating

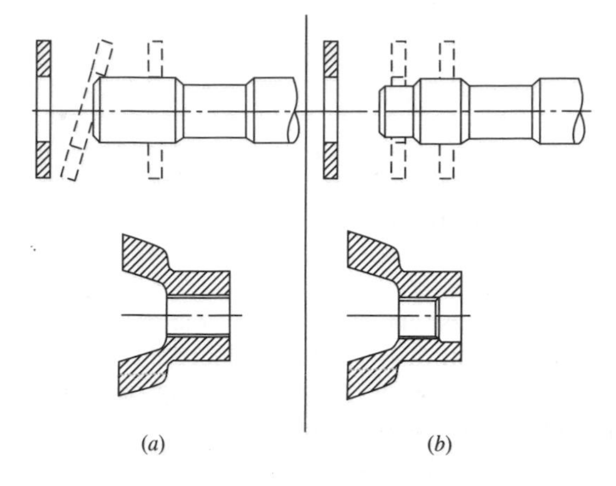

FIGURE 12.28
Use of leads to ease assembly.

conditions. Unsatisfactory performance is considered a *failure*, and so in calculating the reliability of a product we use a technique for identifying failure potential called *failure modes and effects analysis, FMEA*. This best practice is useful as a design evaluation tool and as an aid in hazard assessment, described in Section 8.7.2. (A failure can, but does not necessarily, present a hazard; it presents a hazard only if the consequence of its occurrence is sufficiently severe.) Traditionally, a *mechanical failure* is defined as any change in the size, shape, or material properties of a component, assembly, or system that renders the product incapable of performing its intended function. A failure may be the result of change in the hardware due to aging (for example, wear, material property degradation, or creep) or environmental conditions (for example, overloading, temperature effects, and corrosion). If deterioration or aging noises are taken into account, then the potential for mechanical failure is minimized (see Section 11. 6).

To use failure potential as a design aid, it is important to extend the definition of failure to include not only undesirable changes after the product is in service, but also design and manufacturing errors (for example, moving parts interfere, parts do not fit together, or systems do not meet engineering requirements).

Thus, a more general definition is the following: *a mechanical failure is any change or any design or manufacturing error that renders a component, assembly, or system incapable of performing its intended function.* Based on this definition,

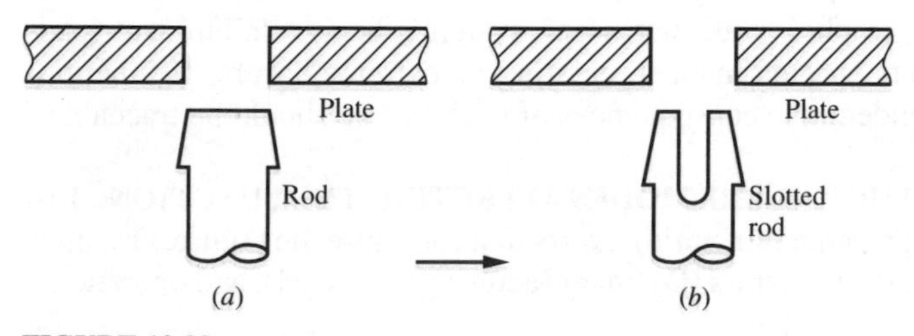

FIGURE 12.29
Use of compliance to ease assembly.

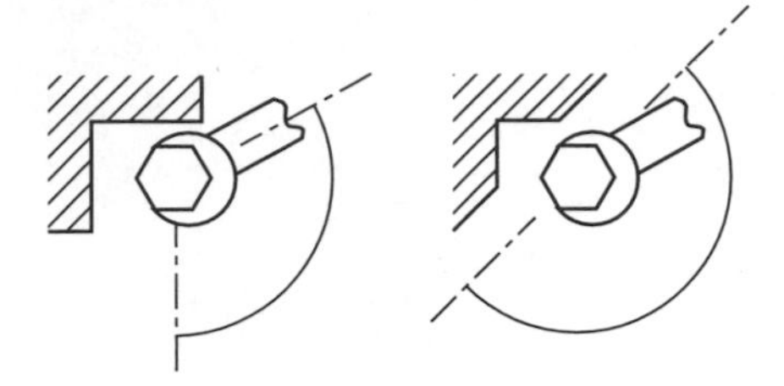

FIGURE 12.30
Modifications for tool clearance.

a failure has two attributes: the function affected and the source of the failure (i.e., the operational change or design or manufacturing error that produced the failure). Typical sources of failure or failure modes are wear, fatigue, yielding, jamming, bonding weakness, property change, buckling, and imbalance.

12.6.1 Failure-Potential Analysis

The failure-potential evaluation technique presented below can be used throughout the product development process and refined as the product is refined. The method aids in identifying where redundancy may be needed and in diagnosing failures after they have occurred. Failure analysis is the following four-step technique:

STEP 1: IDENTIFY THE FUNCTION AFFECTED. For each function identified in the evolution of the product, ask, "What if this function fails to occur?" If functional development has paralleled form development, this step is easy; the functions are already identified. However, if detailed functional information is not available, this step can be accomplished by listing all the functions of each component or assembly. For products being redesigned, the functions of a component or assembly are found by examining the connections or component interfaces and identifying the flow of energy, information, or materials through them. Additional considerations come from extending the basic question to read, "What if this function fails to occur at the right time?" "What if this function fails to occur in the right sequence?" or "What if this function fails to occur completely?"

STEP 2: IDENTIFY THE EFFECT OF FAILURE ON OTHER PARTS OF THE SYSTEM. What are the consequences on other parts of the system of each failure identified in Step 1? In other words, if this failure occurs, what else might happen? These effects may be hard to identify in systems in which the functions are not independent. Many catastrophes result when one system's benign failure overloads another system in an unexpected manner, creating an extreme hazard. If functions have been kept independent, the consequences of each failure should be traceable.

STEP 3: IDENTIFY THE FAILURE MODES AFFECTING THE FUNCTION. List the changes or the design or manufacturing errors that can cause the failure. Organize them into three groups: design errors (D), manufacturing errors (M), and operational changes (O).

STEP 4: IDENTIFY THE CORRECTIVE ACTION. For each design error listed in Step 3, note what redesign action should be taken to ensure that the error does not occur. The same is true for each potential manufacturing error. For each operational change, use the information generated to establish a clear way for the failure mode to be detected. This is important, as it is the basis for the diagnosis of problems when they do occur. For operational changes it may also be important to redesign the device so that the failure mode has a reduced effect on the function. This may include the addition of other devices (for example, fuses or filters) to protect the function under consideration; however, the failure potential of these added devices should also be considered. The use of redundant systems is another way to protect against failures. But redundancy might add other failure modes as well as increase costs.

12.6.2 Reliability

Once the different potential failures of the product have been identified, the reliability of the system can be found and rated in units of reliability called *mean time between failures (MTBF)*, or the average elapsed time between failures. MTBF data are generally accumulated by testing a representative sampling of the product. Often these data are collected by service personnel, who record the part number and type of failure for each component they replace or repair.

These data aid in the design of a new product. For example, a manufacturer of ball bearings collected data for many years. The data showed an MTBF of 77,000 hours for a ball bearing operating under manufacturer-specified conditions. On the average, a ball bearing would last 8.8 years (77,000/(365 × 24)) under normal operating conditions. Of course, a harsh environment or lack of lubrication would greatly reduce this lifetime. Often the MTBF value is expressesd as its inverse and called the *failure rate L*, the number of failures per unit time. Failure rates for common machine components are given in Table 12.2, where the failure rate for the ball bearing is 1/77,000, or 13 failures per 1 million hours.

TABLE 12.2
Failure rates of common components

Mechanical failures, per 10^6 hr		Electrical failures, per 10^6 hr	
Bearing		Meter	26
Ball	13	Battery	
Roller	200	Lead acid	0.5
Sleeve	23	Mercury	0.7
Brake	13	Circuit board	0.3
Clutch	2	Connector	0.1
Compressor	65	Generator	
Differential	15	AC	2
Fan	6	DC	40
Heat exchanger	4	Heater	4
Gear	0.2	Lamp	
Pump	12	Incandescent	10
Shock absorber	3	Neon	0.5
Spring	5	Motor	
Valve	14	Fractional hp	8
		Large	4
		Solenoid	1
		Switch	6

The actual reliability of a component is determined from the failure rate information. Assuming that the failure rate is constant over the life of the component—which is generally true for all but the initial (infant mortality) and the final (wear-out) periods—the reliability is defined as

$$R(t) = e^{-Lt}$$

where R, the reliability, is the probability that the component has not failed. For the ball bearing,

$$R(t) = e^{-0.000013t}$$

with t in hours. Thus:

t, hours	R
0	1.000
100	0.999
1000	0.987
8760 (1 year)	0.892
10,000	0.878
43,800 (5 years)	0.566

If 1000 ball bearings are tested, it would be expected that 892 of them would still be operating a year later within specifications.

What if there are four ball bearings in a product and the product will fail if any one bearing fails? The total reliability of that device is the product of the reliabilities of all its components (this is often called *series reliability*):

$$R_{\text{product}} = R_{\text{bearing 1}} \cdot R_{\text{bearing 2}} \cdot R_{\text{bearing 3}} \cdot R_{\text{bearing 4}}$$

Because of the exponential nature of the definition of reliability, the failure rate for that device would be

$$L_{\text{product}} = L_{\text{bearing 1}} + L_{\text{bearing 2}} + L_{\text{bearing 3}} + L_{\text{bearing 4}}$$

For the product with four bearings, $L = 4 \cdot 0.000013 = 0.000052$. Thus, after one year, $R = 0.634$; about one-third of the products will have had a bearing failure.

There are essentially two ways to increase reliability. First, decrease the failure rate. This is accomplished by lowering the bearing's load or by decreasing its rotation rate. A second way to increase reliability is through redundancy, often called *parallel reliability*. For redundant systems, the failure rate is

$$L = \frac{1}{1/L_1 + 1/L_2 + \cdots}$$

Thus, if a ball bearing and a sleeve bearing are designed into the product so that either can carry the applied load, then

$$L = \frac{1}{1/0.000013 + 1/0.000023} = 8.3 \text{ failures}/10^6 \text{ hr}$$

With this technique, reliability evaluations can also be made on complex systems. A model of the failure modes and the MTBF for each of them is needed to accomplish such an evaluation.

12.7 DESIGN FOR TEST AND MAINTENANCE (DFTM)

Testability refers to the ease with which the performance of critical functions is measured. For instance, in the design of VLSI chips, circuits are included on the chip that allow critical functions to be measured. Measurements can be made during manufacturing to ensure that no errors are built into the chip. Measurements can also be made later in the life of the chip to diagnose failures.

Adding structure in this way, to make testability easier, is often impossible in mechanical products. However, if the technique developed in the previous section for identifying failures is extended, at least some measure of the testability of the product can be realized. For instance, Step 3 of the failure potential evaluation technique required the listing of errors that can cause each failure. An additional step here would address testability:

STEP 3A: FOR EACH ERROR, IS IT POSSIBLE TO IDENTIFY THE PARAMETERS THAT COULD CAUSE THE FAILURE? If there are a significant number of cases in which the parameters cannot be measured, there is a lack of testability in the product.

There are no firm guidelines in developing an acceptable level of testability. The designer should ensure, however, that the critical parameters that affect the critical functions can be tested. In this way, the ability to diagnose manufacturing problems and failures when they occur is increased.

The terms *maintainability, serviceability,* and *repairability* are often used interchangeably to describe the ease of diagnosing and repairing a product. During the 1980s a dominant philosophy was to design products that were totally disposable or composed of disposable modules that could be removed and replaced. These modules often contained still-functioning components along with those that had failed. The structure of the module forced replacement of both good and bad components. This philosophy was characteristic of the "throwaway" attitude of the time, and products designed during this period were often easy to replace and hard to repair. A different philosophy is to design products that are easy to diagnose, disassemble, and repair at any level of function. As discussed above, designing diagnosability into a mechanical product is possible, but it takes extra effort and may be of questionable value. This also applies to designing a product that is easy to disassemble and repair. Since the guidelines given for the design-for-assembly technique do not lead to a product that is easy to disassemble, special care must be taken to ensure that, if desired, the snap fits can be unsnapped and that the disassembly sequence has been considered with as much care as the assembly sequence. Further, the ability to disassemble a product is also important if the product is to be recycled at the end of its useful life. This topic is discussed in the next section.

12.8 DESIGN FOR THE ENVIRONMENT

Design for the environment is often called green design, environmentally conscious design, life-cycle design, or design for recyclability. Treating environmental

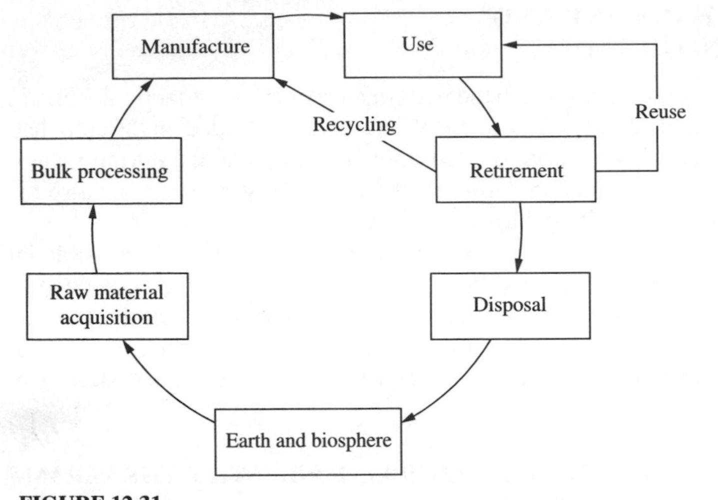

FIGURE 12.31
Green design life cycle.

concerns as important requirements in the design process began in the 1970s. It was not until the 1990s that it became an important issue in the design community. The major consideration of design for the environment is seen in Figure 12.31. Here the arrows represent materials that are taken from the earth or the biosphere and ultimately returned to it. In this figure all the major green design issues are considered.

When a product's useful life is over, one of three things happens to its components. They are either disposed of, reused, or recycled. For many products there is no thought given beyond disposal. However, in 1995, 94 percent of all cars and trucks scrapped in the United States were dismantled and shredded, and 75 percent of the content by weight was recycled. Where, in the 1970s and 1980s, there was design emphasis on disposable products, more and more industries are now trying to design in the ability to recycle or reuse parts of retired products.

For example, even though the single-use camera appears to be disposable after use, Kodak has recycled 41 million of its cameras, or 75 percent of those sold. Likewise, Xerox reuses or recycles 97 percent of parts and assemblies from the toner cartridges it manufactures.

This attention to the entire product life cycle is fueled by economics, customer expectation, and government regulation. First, it is becoming less expensive to recycle some materials than it is to pay the expense of processing new raw materials. This is especially true if the product is designed so that it is easily disassembled into components made of a single material. Expense increases if materials are difficult to separate or if one material contaminates another, adversely affecting its material properties. Further, the realization that the resources of raw materials are limited has only recently dawned on many engineers and consumers.

Second, consumers are increasingly more environmentally conscious and aware of the value of recycling. Thus, companies that pollute, generate excessive waste, or produce products that clearly have adverse effects on the environment are looked down on by the public.

Finally, government regulation is forcing attention on the environment. In Germany manufacturers are responsible for *all* the packaging they create and use. They must collect and recycle it. Further, Mercedes and BMW are designing their new cars so that they, too, can be collected and recycled. European Union laws are forcing this corporate responsibility for the entire life of the product.

In evaluating a product for its "greenness," the following guidelines help ensure that environmental design issues have been addressed. The guidelines serve to compare two designs as do the design for assembly measures in Section 12.5.

GUIDELINE 1: BE AWARE OF THE ENVIRONMENTAL EFFECTS OF THE MATERIALS USED IN PRODUCTS. In Fig. 12.31 every step requires energy, produces waste products, and may deplete resources. Although it is not realistic for the design engineer to know the environmental details of every material used in a product, it is important to know about those materials that may have high environmental impact.

GUIDELINE 2: DESIGN THE PRODUCT WITH HIGH SEPARABILITY. The guidelines for design for disassembly are similar to those for design for assembly. Namely, a product is easy to disassemble if fewer components and fasteners are used, if they come apart easily, and if the components are easy to handle. Other aids for high separability are the following:

- Make fasteners accessible and easy to release
- Avoid laminating dissimilar materials
- Use adhesives sparingly and make them water soluble if possible
- Route electrical wiring for easy removal

One clear measure of separability is the percentage of material that is easily isolated from other materials.

If some of the components are to be reused, the designer must consider disassembly, cleaning, inspection, sorting, upgrading, renewal, and reassembly.

GUIDELINE 3: DESIGN COMPONENTS THAT CANNOT BE REUSED SO THAT THEY CAN BE RECYCLED. One design goal is to use only recyclable materials. Automobile manufacturers are striving for this goal. In recycling there are five steps: retrieval, separation, identification, reprocessing, and marketing. Of these five, the design engineer can have the most influence on the separation and identification. Separation was addressed in item 2 above. Identification means to be able to tell after disassembly exactly what material was used in the manufacture of each component. With few exceptions, it is difficult to identify most materials without laboratory testing. Identification is made easier with the use of standard symbols, such as those used on plastics that identify polymer type.

GUIDELINE 4: BE AWARE OF THE ENVIRONMENTAL EFFECTS OF THE MATERIAL NOT REUSED OR RECYCLED. Currently 18 percent of the solid waste in landfills is plastic and 14 percent is metal. All of this material is reusable or recyclable. If a product is not designed to be recycled or reused, it should at least be degradable. The designer should be aware of the percentage of degradable material in a product and the time it takes this material to degrade.

12.9 SUMMARY

* Cost estimation is an important part of the product evaluation process.
* Features should be judged on their value—the cost for a function.
* Design for manufacture focuses on the production of components.
* Design for assembly is a method for evaluating the ease of assembly of a product. It is most useful for high-volume products that have molded components. Thirteen guidelines are given for this evaluation technique.
* Functional development gives insight into potential failure modes. The identification of these modes can lead to the design of more reliable and easier-to-maintain products.
* Design for the environment emphasizes concern for energy, pollution, and resource conservation in processing raw materials for products. It also emphasizes concern for recycling, reuse, or disposal of the product after its useful life is over.

12.10 SOURCES

Boothroyd, G., and P. Dewhurst: *Product Design for Assembly,* Boothroyd and Dewhurst Inc., Wakefield, R.I., 1987. Boothroyd and Dewhurst have popularized the concept of DFA. The range of their tools is much broader than that of those presented here.

Chow, W. W-L.: *Cost Reduction in Product Design,* Van Nostrand Reinhold, New York, 1978. An excellent book that gives many cost-effective design hints, written before the term *concurrent design* became popular yet still a good text on the subject. The title is misleading; the contents of the book are a gold mine for the designer engineer.

Life Cycle Design Manual: Environmental Requirements and the Product System, EPA/600/R-92/226, United States Environmental Protection Agency, Jan. 1992. A good source for design for the environment information.

Michaels, J. V., and W. P. Wood: *Design to Cost,* Wiley, New York, 1989. A good text on the management of costs during design.

Nevins, J. L., and D. E. Whitney: *Concurrent Design of Products and Processes,* McGraw-Hill, New York, 1989. This is a good text on concurrent design from the manufacturing viewpoint; a very complete method for evaluating assembly order appears in this text.

Rivero, A., and E. Kroll: "Derivation of Multiple Assembly Sequences from Exploded Views," *Advances in Design Automation,* ASME DE-Vol. 2, American Society of Mechanical Engineers—Design Engineering, Minneapolis, MN, 1994, pp. 101–106. More guidance on determining the assembly sequence.

Trucks, H. E.: *Designing for Economical Production,* 2d ed., Society of Manufacturing Engineers, Dearborn, Mich., 1987. This is a very concise book on evaluating manufacturing techniques. It gives good cost-sensitivity information.

12.11 EXERCISES

12.1. For the product developed in response to the design problem begun in Exercise 4.1, estimate material costs, manufacturing costs, and selling price. How accurate are your estimates?

12.2. For the redesign problem begun in Exercise 4.2, estimate the changes in selling price that result from your work.

The following two problems assume that a cost estimation computer program is available or that a vendor can help with the estimates.

12.3. Estimate the manufacturing cost for a simple machined component:
(a) Compare the costs for manufacturing volumes of 1, 10, 100, 1000, and 10,000 pieces with an intermediate tolerance and surface finish. Explain why there is a great change between 1 and 10 and a small change between 1000 and 10,000 pieces.
(b) Compare the costs for fit, intermediate, and rough tolerances with a volume of 100 pieces.
(c) Compare the costs of manufacturing the component out of various materials.

12.4. Estimate the manufacturing cost for a plastic injection-molded component:
(a) Compare the costs for manufacturing volumes of 100, 1000, 10,000, and 100,000. The tolerance level is intermediate, and surface finish is not critical.
(b) Compare the cost for a change in tolerance.
(c) Why does changing the material have virtually no effect on cost at low plastic-injection volume (i.e., 100 pieces)?

12.5. Perform a design-for-assembly evaluation for one of the following devices. Based on the results of your evaluation, propose product changes that will improve the product. Be sure that your proposed changes do not affect the function of the device. For each change proposed, estimate its "value."
(a) A simple toy (fewer than 10 parts)
(b) An electric iron
(c) A kitchen mixing machine or food processor
(d) A Walkman-type radio, cassette, or disk player
(e) The product resulting from the design problem (Exercise 4.1) or the redesign problem (Exercise 4.2)

12.6. For the device chosen in the previous problem, perform a failure mode and effects analysis.

12.7. For one of the products in Exercise 12.5, evaluate it for disassembly, reuse, and recycling.

CHAPTER 13

LAUNCHING THE PRODUCT

13.1 INTRODUCTION

This chapter discusses issues that occur at the end of the design process. In many ways this chapter is a partner to Chap. 9, which introduced the product design phase of the process. Even if the techniques presented in the ensuing chapters have led to the development of a set of detail and assembly drawings and a bill of materials, the process is not yet complete. We must still finalize all the documentation and pass a final design review before launching the product into the marketplace.

This chapter is concerned with the final documents that represent the product—specifically, design records; patent applications; assembly, quality control, and quality assurance information; and installation, operation, maintenance, and retirement instructions. Another type of documentation that is usually generated after the product is released for production is the engineering change notice. This too is discussed.

Figure 13.1 is a copy of Fig. 4.1. It depicts the entire design process. The last design review shown in the figure, made when the product is complete and all its documentation is in order, releases the product for production. Just because a product has been refined to this point does not mean that it will be produced. An engineer can spend many years designing a product only to have the project canceled prior to production release. However, production should be approved if the customer requirements have been met with a quality design in a timely and cost-effective manner. The techniques, or best practices, given in this book have focused on aiding this positive outcome.

13.2 DESIGN RECORDS

In the previous chapters many design best practices were introduced to aid in the development of a product. The documentation of the results of these techniques,

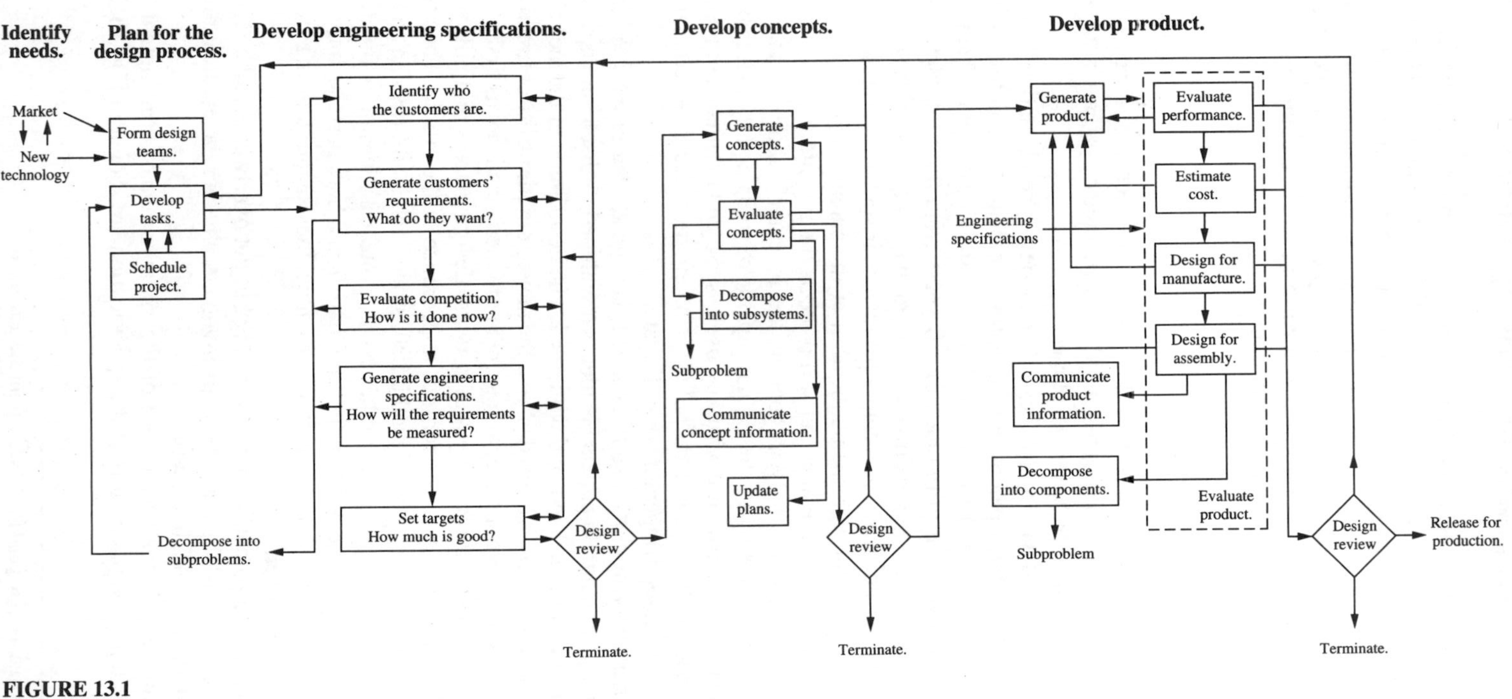

FIGURE 13.1
The product design process.

along with the personal notebooks of the design team members, and the drawings and the bill of materials, constitutes a record of a product's evolution. If all the techniques were used, the documentation generated would have been patched and updated many times. Additionally, summaries of the results of the design review meetings would also exist. All this information constitutes a complete record of the design process. Most companies archive this information for use in patent disputes or litigation or simply as a history of the evolution of the product.

13.3 PATENT APPLICATIONS

In Chap. 7 patent literature was used as a source of conceptual ideas. During the evolution of a product, new ideas, ones not already covered by patents, are sometimes generated and a patent application developed.

Just about any device or process that is new, useful, and not obvious is patentable. In obtaining a patent, the inventor is essentially entering into a contract with the U.S. government, as provided for in the Constitution. The inventor is granted an exclusive right to the idea for 20 years from the filing date.[1] In many cases the contract is between the inventor's employer—called the "assignee"—and the government. In return for receiving a patent, the inventor must make full public disclosure of the idea through the publication of the patent. This system allows all of society to benefit from the idea and still protects the inventor.

In reality, a patent is only as good as the inventor's ability to enforce it. In other words, if the holder of a patent is not capable of suing a party that is infringing on that patent, then the patent is virtually useless. Since lawsuits can be extremely expensive, an individual or company holding a patent must decide whether it is better business to allow the infringement or to litigate.

Whether applying for a patent or not, it is a good idea to write a *disclosure* whenever a new, potentially patentable idea is developed. A disclosure is description of the idea that is signed, dated, and witnessed. The description of the idea should be in terms of specific claims. Claims describe the new or unique utility or function of the device. A disclosure serves as a legal statement of when the idea was first conceived.

There are essentially two types of patents: design patents and utility patents. *Design patents* relate only to the form or appearance of the article. Thus there are many design patents for the basic toothbrush, since each toothbrush has a different appearance. *Utility patents*, on the other hand, protect utilitarian or functional aspects of the idea. The patent discussion in Chap. 7 and here pertains only to utility patents. Utility patents are given for processes, machines, manufacturing techniques, or composition of matter.

Applying for a patent is time-consuming but not overly expensive. The procedure outlined in Fig. 13.2 shows the steps involved, whether the application is done by an individual or through a lawyer.

The first step is the preparation of the *specification*, the document that becomes the patent. The basic parts of the specification are shown in Fig. 13.2. When

---

[1] Prior to 1995 patents were good for 17 years from the date of issue.

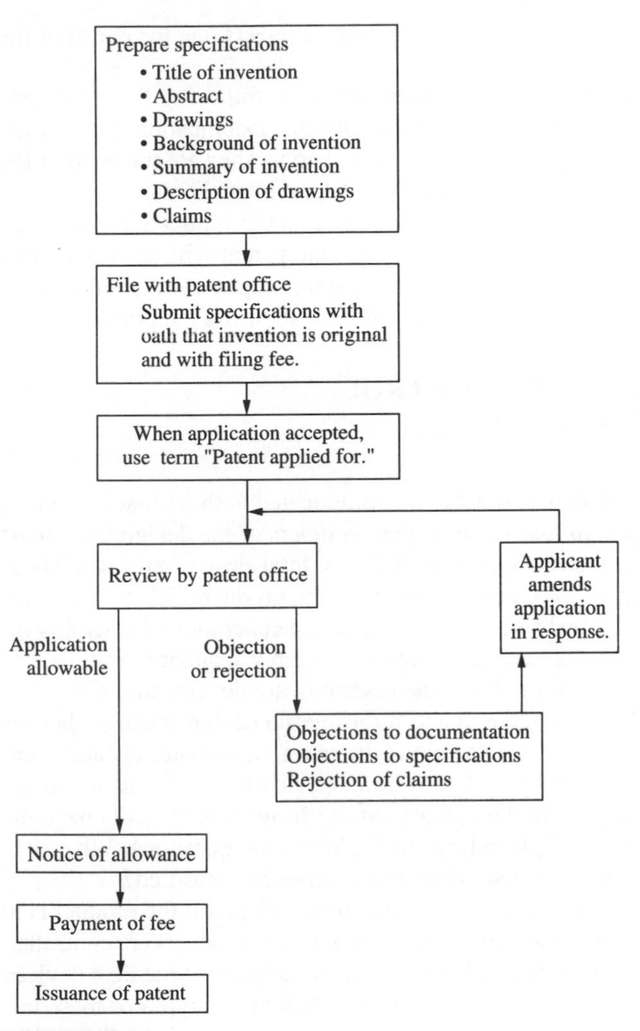

FIGURE 13.2
Patent application procedure.

preparing a specification, it is best to seek help from a patent lawyer or reference a book with details for writing claims and the proper patent format. Within three months of receiving the patent application, the patent office will acknowledge its submission and accept the application for consideration. The term "patent applied for" can be used after the specification is accepted. (The term "patent pending," often seen on products, has no legal meaning.)

The second phase of the patent process begins after the application is filed. The patent office assigns the application to an examiner familiar with the state of the patent art in the area of the application. This examiner then reviews the application and searches the patent literature. This initial study of the application and the subsequent iteration with the applicant takes months or even years. Seldom does the examiner accept the first application outright. It is more likely that he or she objects to the document itself or to the specification or rejects one (or more) of the claims.

Problems with the document usually stem from not following the rigid style required of patent text and drawings. Problems with the specification concern the content of the claims. Since the claims define the invention, they are the heart of the patent. All the other information is there simply to support the claims.

After the applicant and the patent examiner agree on the application, the patent office issues a notice of allowance. This means that the patent will be issued upon payment of an issue fee. About 65 percent of the patents applied for are ultimately granted. Virtually all of these have been greatly modified during the process.

13.4 ASSEMBLY, QUALITY CONTROL, AND QUALITY ASSURANCE DOCUMENTATION

Other information besides drawings must be communicated to those manufacturing and assembling the product. In many companies members of the design team must develop a set of assembly instructions as part of the total design package. These instructions spell out, step by step, how to assemble the product. This is necessary whether the assembly is done by hand or by machine. The generation of assembly instructions, while tedious, is enlightening in that the assembly sequence (Sec. 12.5.1) is refined, and jigs and fixtures for holding the assembly are developed.

Even if quality has been a major concern during the design process, there is still a need for quality control (QC) inspections. Incoming raw materials and manufactured components and assemblies should be inspected for conformance to the design documentation. The industrial engineers on the design team usually have the responsibility to develop the QC procedures that address the questions, What is to be measured? How will it be measured? How often will it be measured?

Quality assurance (QA) documentation must be developed if the product is to be regulated by government standards. For example, medical products are controlled by the Food and Drug Administration (FDA), and manufacturers of medical devices must keep a detailed file of quality assurance information on the types of materials and processes used in their products. FDA inspectors can come on site without prior notification and ask to see this file.

13.5 INSTALLATION, OPERATION, MAINTENANCE, AND RETIREMENT INSTRUCTIONS

Since the design team is responsible for the entire product life cycle, they must prepare documentation to support the product through to its retirement. This requires the development of instructions for the product's installation, maintenance, and final disposal.

INSTALLATION INSTRUCTIONS. These include instructions for unpacking the product and making the necessary connections for power, support, and environmental control. Instructions for initial start-up and testing may also be included.

OPERATION INSTRUCTIONS. These include instructions on how to operate the device over the normal range of activity. Various modes—start-up, standby, emergency operation, and shutdown—may be described. Instructions on how to determine when the equipment is failing may also be included.

MAINTENANCE INSTRUCTIONS. Diagnostic and maintenance procedures, which were considered during product design, must be documented so that preventive maintenance, failure analysis, and rapid repair can be accomplished.

RETIREMENT INSTRUCTIONS. Instructions on how to decommission the device and dispose of it as developed in Sec. 12.8 are included here.

13.6 ENGINEERING CHANGES

Although this book encourages change early in the design process, change may still occur after the product is released to production. Changes are caused by correction of a design error, a change in the customers' requirements necessitating the redesign of part of the product, or a change in material or manufacturing method. To make a change in an approved configuration, an *engineering change notice (ECN)*, also called *engineering change order (ECO)*, is required. An ECN is an alteration to an approved set of final documents and thus needs approval itself. As shown in the example in Fig. 13.3, an ECN must contain at least the following information:

- Identification of what needs to be changed. This should include the part number and name of the component and reference to the drawings that show the component in detail or assembly.
- Reason(s) for the change.
- Description of the change. This includes a drawing of the component before and after the change. Generally these drawings are only of the detail affected by the change.
- List of documents and departments affected by the change. The most important part of making a change is to see that all pertinent groups are notified and all documents updated.
- Approval of the change. As with the detail and assembly drawings, the changes must be approved by management.
- Instruction about when to introduce the change—immediately (scrapping current inventory), during the next production run, or at some other milestone.

13.7 VENDOR RELATIONS

Very few products are made solely in-house. In fact, many companies make no components themselves and only specify, assemble, sell, or distribute what others make. Thus for most companies relations with their vendors are crucial. Prior to 1980 large companies might have had thousands of vendors, each chosen for its low bid to make a component or assembly. These companies realized, however, that this was a poor

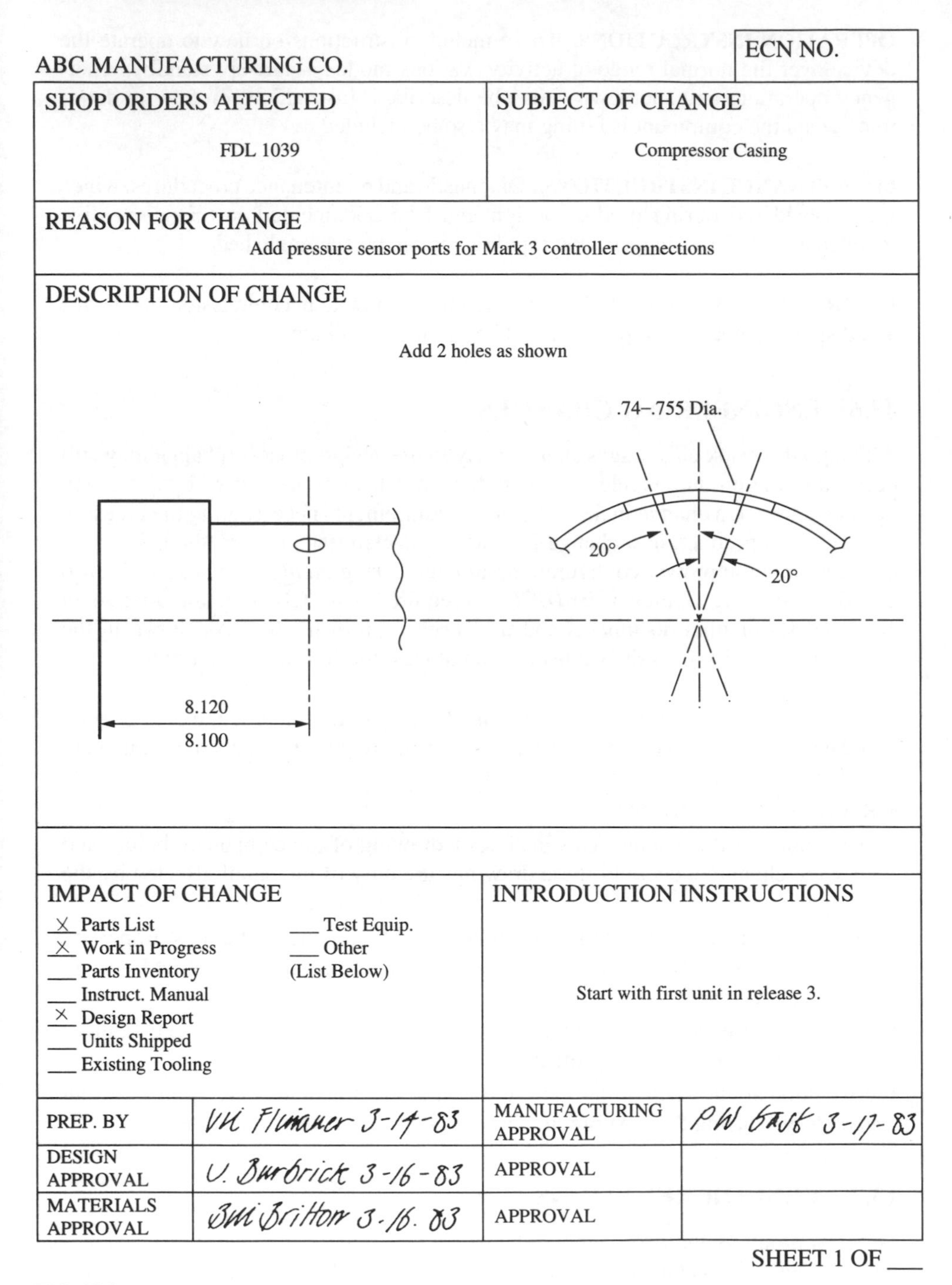

ABC MANUFACTURING CO. | ECN NO.

SHOP ORDERS AFFECTED	SUBJECT OF CHANGE
FDL 1039	Compressor Casing

REASON FOR CHANGE

Add pressure sensor ports for Mark 3 controller connections

DESCRIPTION OF CHANGE

Add 2 holes as shown

IMPACT OF CHANGE		INTRODUCTION INSTRUCTIONS
X Parts List	___ Test Equip.	
X Work in Progress	___ Other	
___ Parts Inventory	(List Below)	
___ Instruct. Manual		Start with first unit in release 3.
X Design Report		
___ Units Shipped		
___ Existing Tooling		

PREP. BY	VH Flimauer 3-14-83	MANUFACTURING APPROVAL	PW Gaust 3-17-83
DESIGN APPROVAL	U. Burbrick 3-16-83	APPROVAL	
MATERIALS APPROVAL	Bill Britton 3-16-83	APPROVAL	

SHEET 1 OF ___

FIG. 13.3
Engineering change notice.

way to do business, because the cheapest components were not always of the highest quality even if they met the specifications.

By the 1990s the philosophy of including vendors on the design team from the beginning of the project had evolved. This allows for fewer vendors, empowers the vendors as "stake holders" in the product, and requires that organized design processes be used. Making this shift from low bidder to virtual partner required changing the philosophy of vendor relationships. Most companies reduced their number of vendors by an order of magnitude. Many now only use vendors from a small, select list. In some cases the product manufacturing company has a financial interest in the vendor, or vice versa.

One important standard that has come out of this change of philosophy is ISO 9000. This international standard specifies that a company must have a documented process for designing and manufacturing a product that it supplies to another company. The standard does not specify the process exactly but only that it exists, is documented, and is maintained. If a company follows the process shown in Figure 13.1, it can gain ISO 9000 certification. To gain certification, a company has to present its design process documentation to an accredited inspector. An important point is that the inspector is not looking for a specific design process, only documentation of the process followed by the company. The idea behind ISO 9000 certification is that, if a company is going to take the time to study its process and document it, it will understand the strengths and weaknesses of the process and will work to make it stronger. In other words, ISO 9000 encourages companies to implement the ideas presented in this book to improve their product development process without dictating one specific process.

13.8 SOURCES

Burgess, J. A.: *Design Assurance for Engineers and Managers,* Marcel Dekker, New York, 1984. A very complete and well-written book on the development and control of engineering documentation.

Kivenson, G.: *The Art and Science of Inventing,* Van Nostrand Reinhold, New York, 1977. Good overview of patents and patent applications; however, it is out of date on application details.

ISO 9001: Quality Systems—Model for Quality Assurance in Design/Development, Production, Installation and Servicing, International Organization for Standardization, 1987. ISO 9000 is made up of three parts. The first one, ISO 9001, is the most important for design engineers.

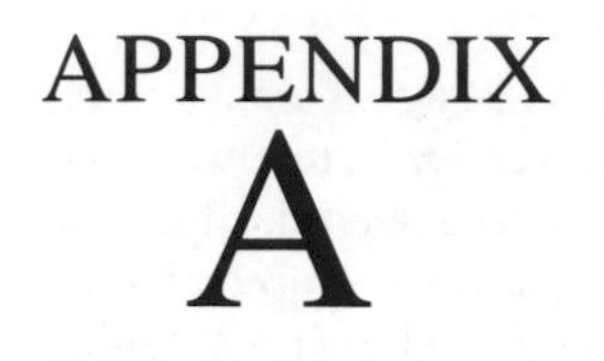

APPENDIX A

PROPERTIES OF 25 MATERIALS MOST COMMONLY USED IN MECHANICAL DESIGN

A.1 INTRODUCTION

There are literally an infinite number of materials available for use in products. In addition, it is now possible to actually design materials for a specific use. There is no way a design engineer can have knowledge of all these materials; however, all design engineers should be familiar with the materials that are the most available and the most commonly used in product design. Because these same materials are representative of a broad spectrum of materials, the design engineer can use his or her knowledge about them to communicate with materials engineers about other, less common materials.

In addition to the important properties of the 25 most used materials, this appendix also contains a list of the specific materials used in many common items. During material selection it is vital to know what materials have been used for similar applications in the past; this list provides a source of such information.

This appendix concludes with an extensive bibliography; the publications listed there are a source for information beyond the basic data presented here.

A.2 PROPERTIES OF THE MOST COMMONLY USED MATERIALS

The following 25 materials are those most commonly used in the design of mechanical products; in themselves they represent the broad range of other materials.

Steel and irons
1. 1020
2. 1040
3. 4140
4. 4340
5. S30400
6. S316
7. 01 tool steel
8. Gray cast iron

Aluminum and copper alloys
9. 2024
10. 3003 or 5005
11. 6061
12. 7075
13. C268

Other metals
14. Titanium 6-4
15. Magnesium AZ63A

Plastics
16. ABS
17. Polycarbonate
18. Nylon 6/6
19. Polypropylene
20. Polystyrene

Ceramics
21. Alumina
22. Graphite

Composite materials
23. Douglas fir
24. Fiberglass
25. Graphite/epoxy

The properties of these 25 materials, given in Figs. A.1–A.12, are the properties most commonly needed for design purposes.[1] Other properties can be found in the references at the end of this appendix. The properties are given as ranges, since they will depend on specific heat treatment (metals) and additives (plastics).

---

[1]An excellent book with more properties presented in this manner and a computer program to support material selection is M. F. Ashby, *Materials Selection in Mechanical Design,* Pergamon Press, Oxford, U.K.

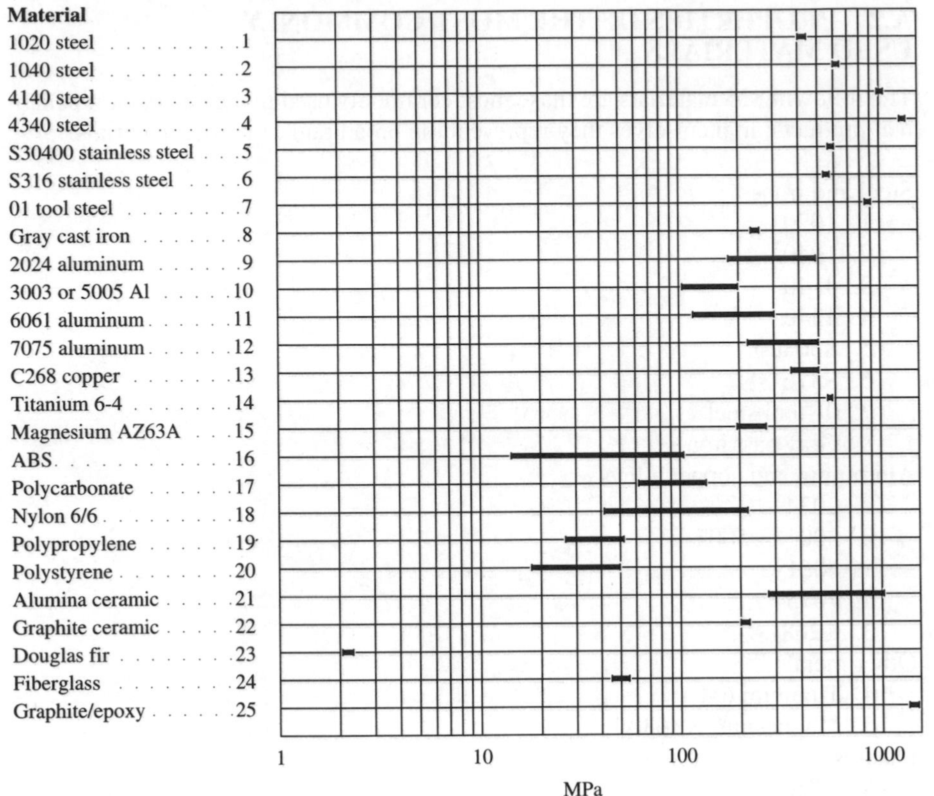

Note: Longitudinal value for graphite/epoxy.
1 MPa = 144.7 psi.

FIGURE A.1
Tensile strength.

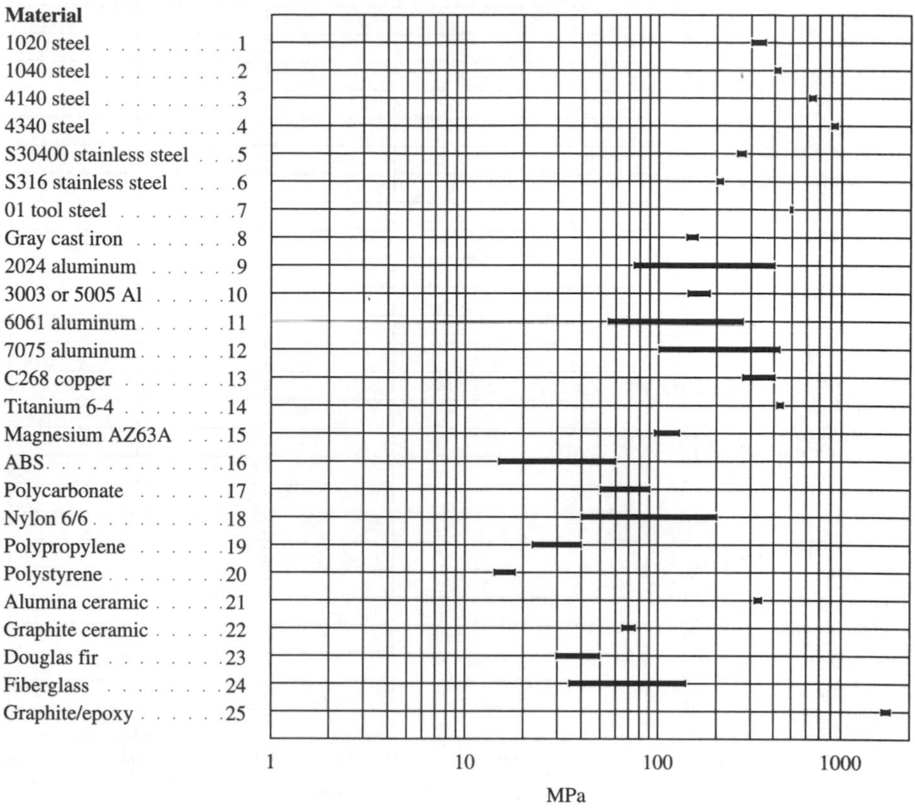

Material

1020 steel 1
1040 steel 2
4140 steel 3
4340 steel 4
S30400 stainless steel . . .5
S316 stainless steel 6
01 tool steel 7
Gray cast iron 8
2024 aluminum 9
3003 or 5005 Al 10
6061 aluminum 11
7075 aluminum 12
C268 copper 13
Titanium 6-4 14
Magnesium AZ63A 15
ABS 16
Polycarbonate 17
Nylon 6/6 18
Polypropylene 19
Polystyrene 20
Alumina ceramic 21
Graphite ceramic 22
Douglas fir 23
Fiberglass 24
Graphite/epoxy 25

1 10 100 1000

MPa

Note: 1 MPa = 144.7 psi.

FIGURE A.2
Yield strength.

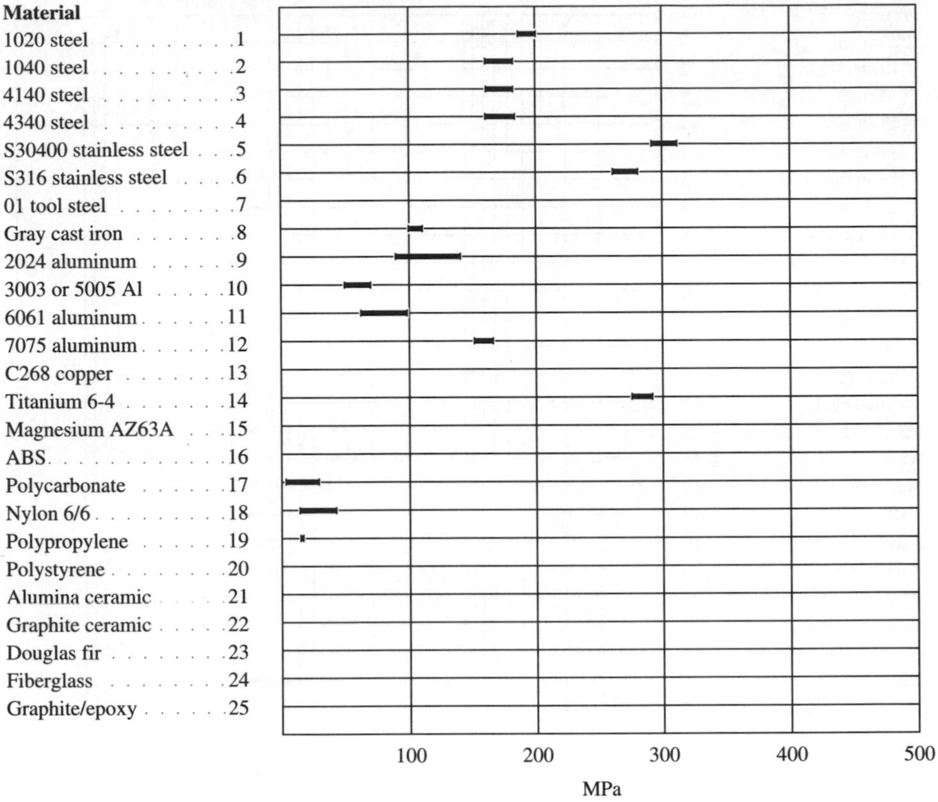

Note: Some materials do not have an endurance limit.
1 MPa = 144.7 psi.

FIGURE A.3
Endurance limit.

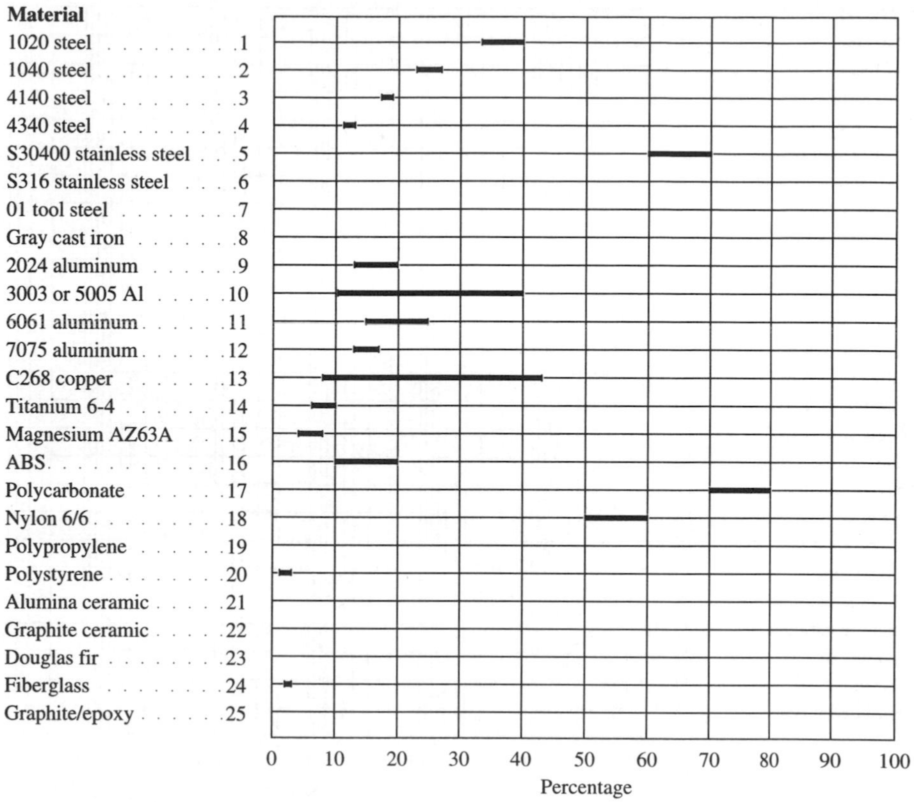

Material

1020 steel 1
1040 steel 2
4140 steel 3
4340 steel 4
S30400 stainless steel . . . 5
S316 stainless steel 6
01 tool steel 7
Gray cast iron 8
2024 aluminum 9
3003 or 5005 Al 10
6061 aluminum 11
7075 aluminum 12
C268 copper 13
Titanium 6-4 14
Magnesium AZ63A . . . 15
ABS 16
Polycarbonate 17
Nylon 6/6 18
Polypropylene 19
Polystyrene 20
Alumina ceramic 21
Graphite ceramic 22
Douglas fir 23
Fiberglass 24
Graphite/epoxy 25

0 10 20 30 40 50 60 70 80 90 100
Percentage

Note: Data unavailable for some materials.
Elongation in plastics depends on filler materials.

FIGURE A.4
Elongation.

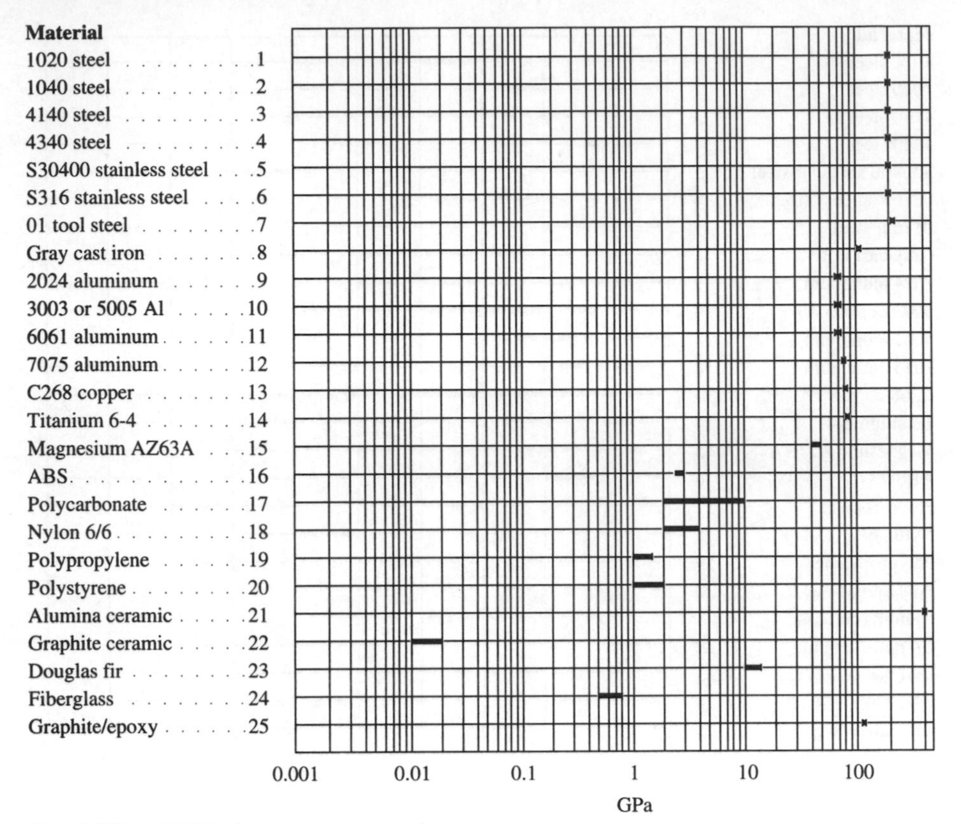

Note: 1 GPa = 144.7 kpsi.

FIGURE A.5
Modulus of elasticity.

Material
1020 steel 1
1040 steel 2
4140 steel 3
4340 steel 4
S30400 stainless steel 5
S316 stainless steel 6
01 tool steel 7
Gray cast iron 8
2024 aluminum 9
3003 or 5005 Al 10
6061 aluminum 11
7075 aluminum 12
C268 copper 13
Titanium 6-4 14
Magnesium AZ63A 15
ABS 16
Polycarbonate 17
Nylon 6/6 18
Polypropylene 19
Polystyrene 20
Alumina ceramic 21
Graphite ceramic 22
Douglas fir 23
Fiberglass 24
Graphite/epoxy 25
0
2
4
6
8
10
g/cc

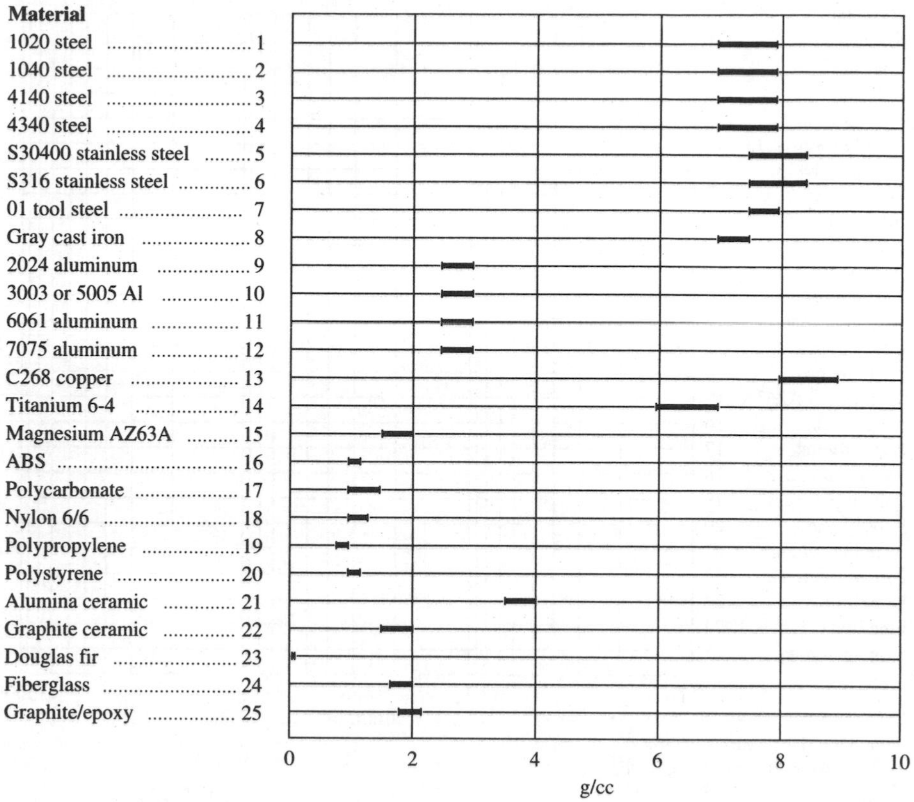

Note: 1g/cc = 0.036 1b/in^3.

FIGURE A.6
Density.

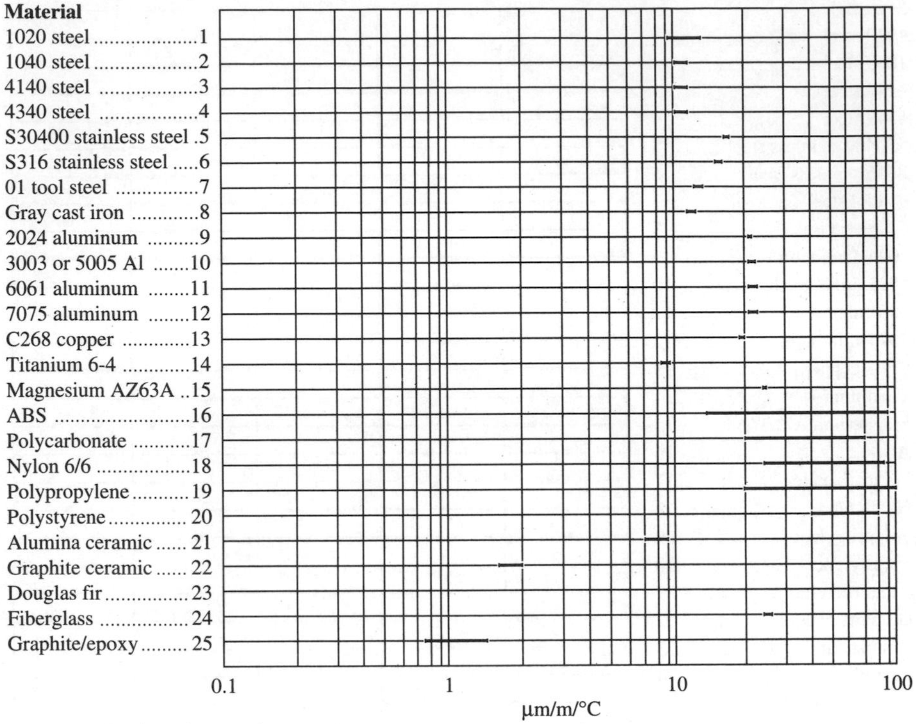

Note: Douglas fir varies greatly.
1 μm/m/°C = 0.55 μin/in/°F.

FIGURE A.7
Coefficient of thermal expansion.

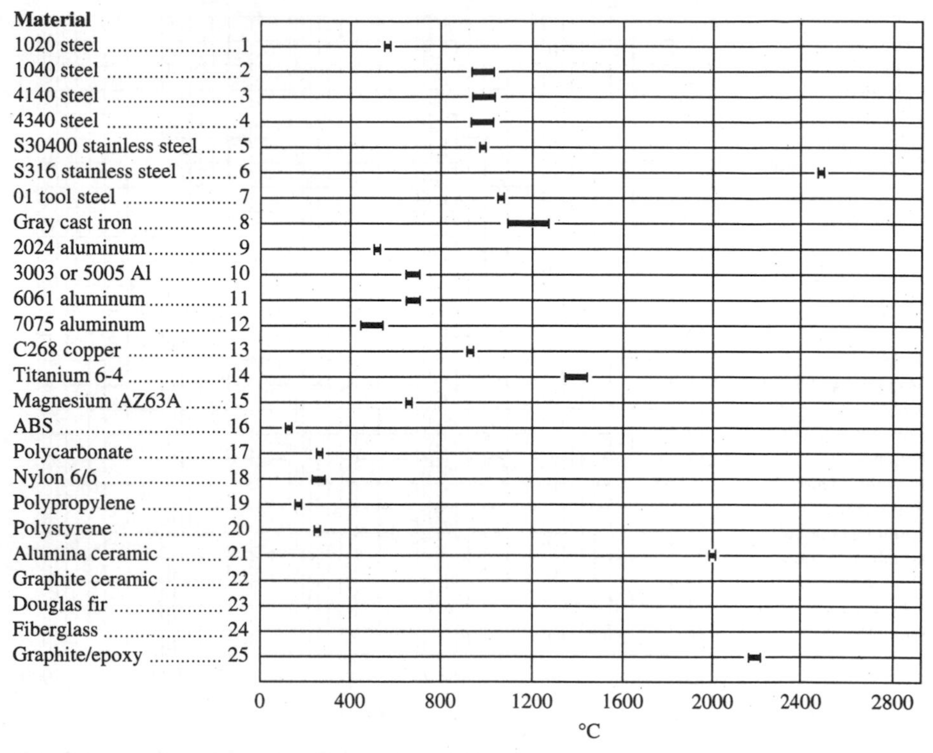

Note: Data unavailable for graphite, Douglas fir, and Fiberglass.
°F = 32.2 + (9/5)°C.

FIGURE A.8
Melting temperature.

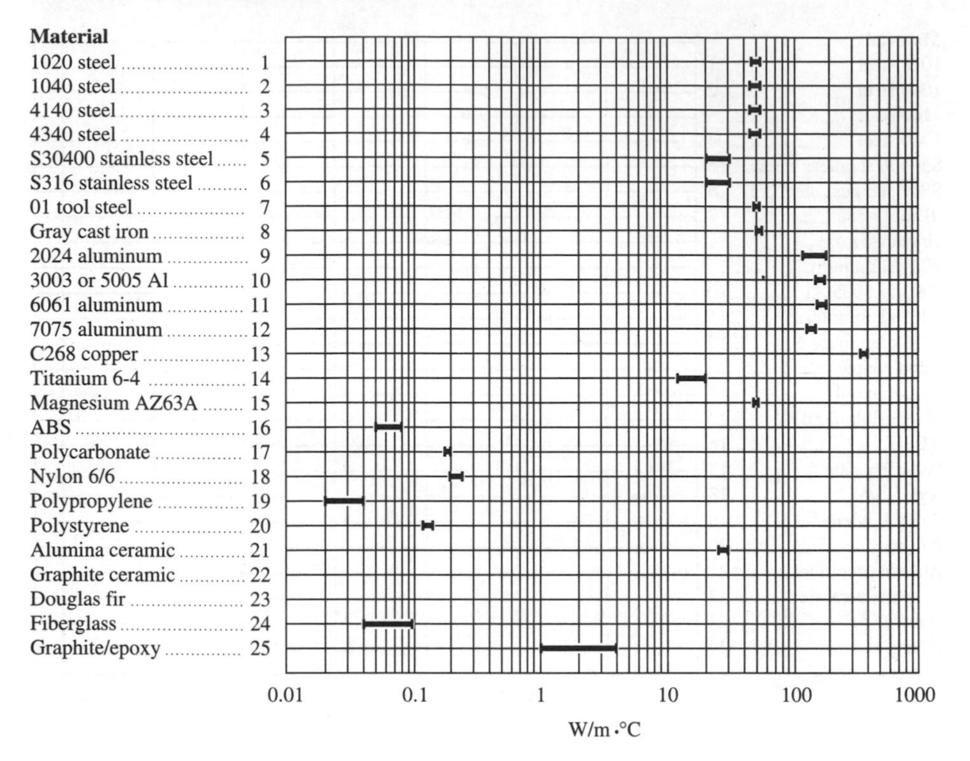

Note: Data unavailable for some materials.
W/m·°C = 0.57 Btu/h·ft·°F.

FIGURE A.9
Thermal conductivity.

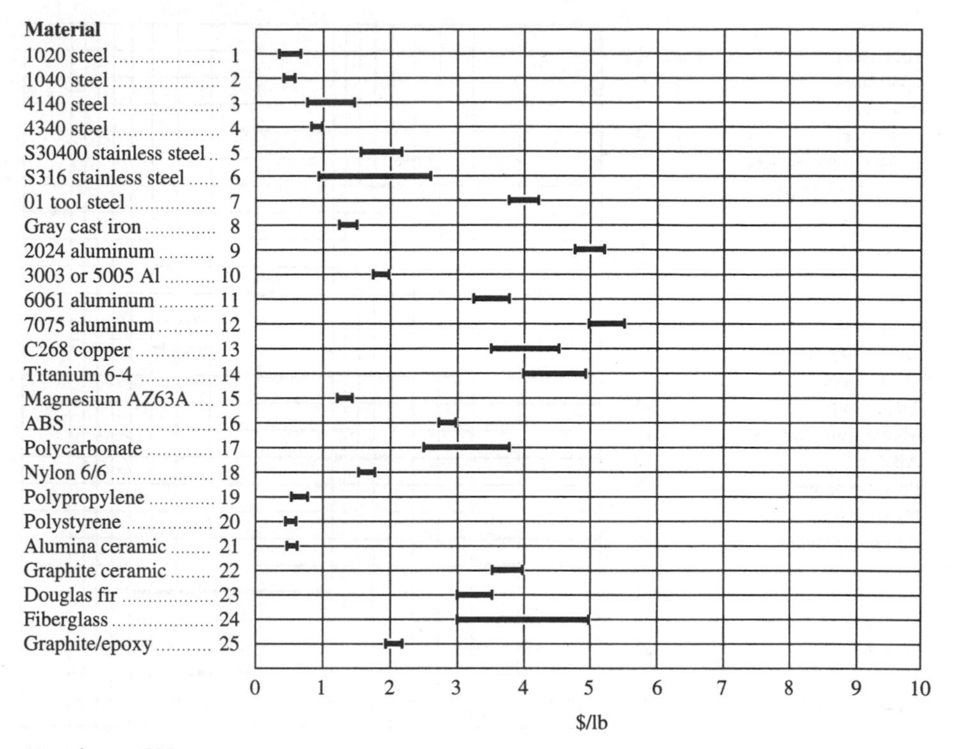

Note: $ year 1990.

FIGURE A.10
Cost per pound.

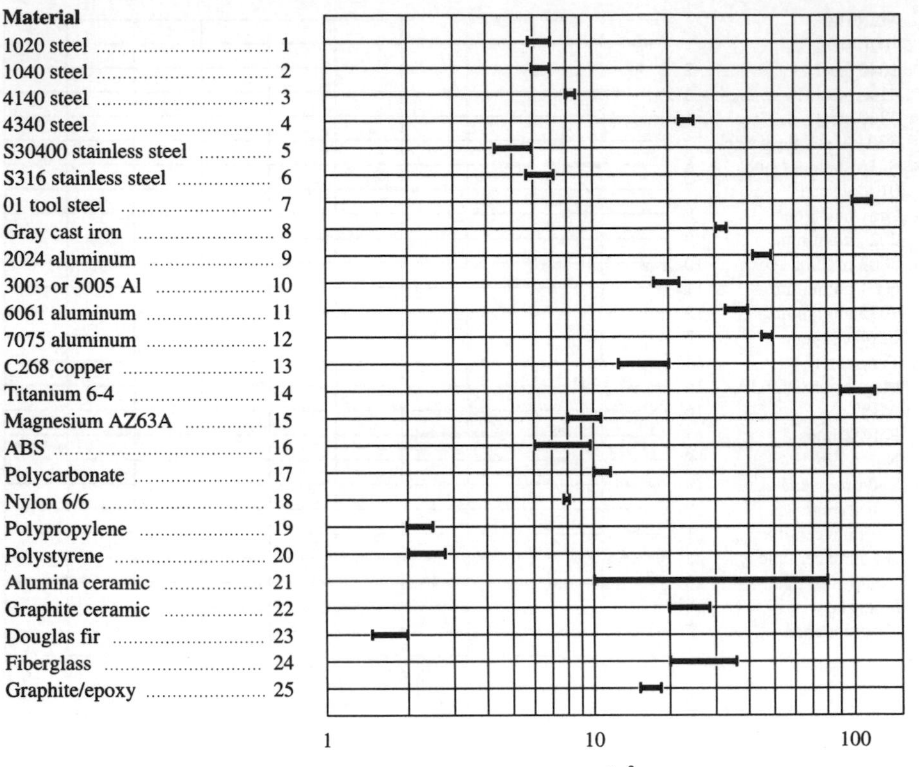

Note: $ year 1990.

FIGURE A.11
Cost per unit volume.

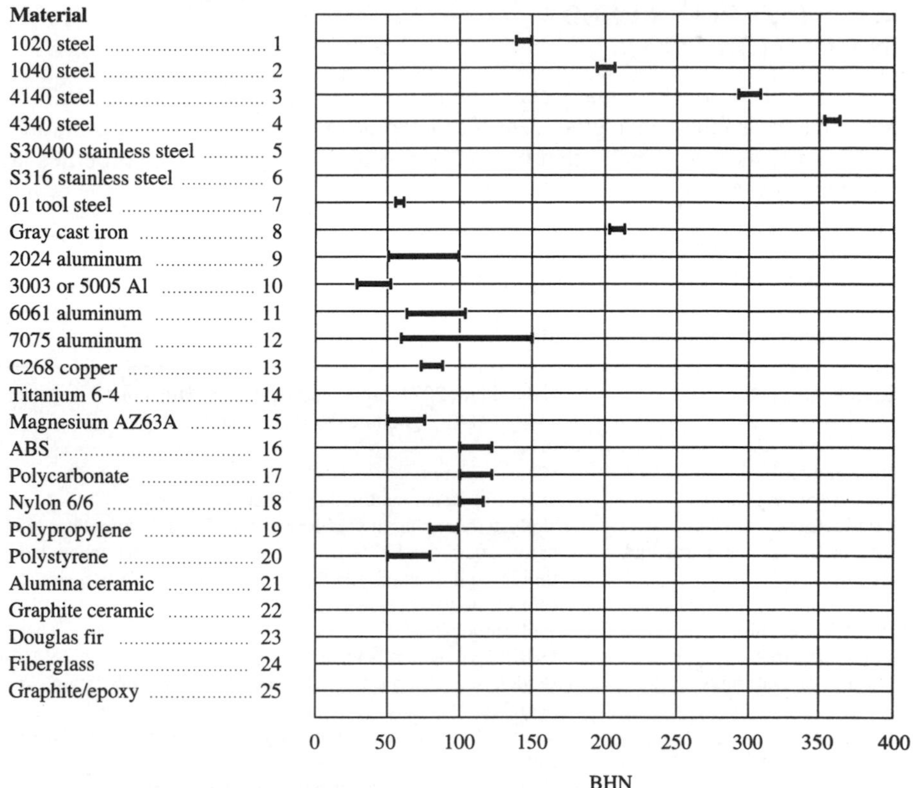

Note: Data unavailable for some materials.

FIGURE A.12
Hardness.

A.3 MATERIALS USED IN COMMON ITEMS

The materials from which many components are commonly made are listed in Table A.1. This list is intended to give ideas for materials to be specified in a new product. In the list an asterisk denotes that the material is one of the 25 whose properties are given in Figs. A.1–A.12.

TABLE A.1
Materials used in common components

Component	Materials
Aircraft components	*Aluminum 2024, 5083, *6061, *7075; *titanium 6-4; *steel 4340, *4140; *graphite/epoxy; polyamide-imide, maraging steel
Auto engines	*Gray cast iron
Auto instrument panels	*Polypropylene; modified polyphenylene oxide
Auto interiors	*Polystyrene; polyvinyl chloride; *ABS; *polypropylene
Auto bodies, panels, parts, and coverings	*Steel 1020, *1040; *ABS; *alumina *nylon; polyphenylene sulfide; aluminum 5083; *stainless steel 316; *fiberglass/epoxy; *polycarbonate
Auto taillight lenses	*polycarbonate
Automotive trim	*ABS; styrene acrylonitrile; *polypropylene
Bearings and bushings	*Nylon; acetal; bronze; Teflon; beryllium copper; *stainless steel 316
Boat hulls	Styrene acrylonitrile; *aluminum 6061; polyethylene; *fiber-glass/epoxy
Bottles	*Polycarbonate; thermoplastic polyester; high-density polyethylene; *polypropylene; polyvinyl chloride; PET
Battery cases	*Polystyrene
Business machine cases	*ABS; *polystyrene; *polycarbonate
Buttons	Melamine; urea
Cams	*Nylon; *aluminum 6061
Cabinets and housings	*ABS; acetal; *polycarbonate; polysulfone; *nylon; *polypropylene; *polystyrene; polyvinyl chloride; *steel 1040
Compact discs	*Polycarbonate
Computer casings	*ABS
Conveyor chains	Acetal; *steel 1040,
Cryogenic parts	Boron/aluminum; *graphite/epoxy; aluminum 5086
Dies	*01 tool steel; cemented carbide
Electric connectors	Polysulfone; phenolic; *nylon; polyethersulfone (PES); *ABS
Fan blades	*Nylon; *steel 4340
Forgings	*Steel 1020, *4340; copper alloys; aluminum alloys
Fixtures	*01 and A2 tool steels; epoxy; *aluminum 6061
Gears	Acetal; *Gray cast iron; *steel 1020, *4340; *nylon; *polycarbonate; polyamide; filled phenolic; aluminum bronze
Handles and knobs	Phenolic; melamine; urea; nylon*; ABS*; acetal
Helmets	*Polycarbonate; *ABS
Heat exchangers	Stainless steel
Hoses	*Nylon; *polycarbonate
House siding and gutters	*Polypropylene
Keys	*Steel 1020, *4140

TABLE A.1
Materials used in common components (*continued*)

Component	Materials
Levers and linkages	Acetal; *steel 4140, *4340
Machine bases	*Gray cast iron; Steel 1020; ductile iron
Marine parts and instruments	Styrene acrylonitrile; *steel 1020; aluminum 5083, *6061; *polycarbonate; *titanium 6-4
Microwave cookware	Thermoplastic polyester; *polycarbonate; *polypropylene
Molded containers	*ABS; *polypropylene; *polycarbonate; polystyrene; polyvinyl chloride; vitreous graphite
Pipes	Polyvinyl chloride; *copper
Screws and bolts	*Steel 1020, *1040, *4140; acetal; *nylon
Shafts	*Steel 1020, *1040, *4140, *4340
Springs	*Steel 1080, *4140, 6250; stainless steel; beryllium copper; *nylon; titanium 6-4; maraging steel; phosphor bronze; acetal; nylon
Structural components	*Cast iron; *Douglas fir; *alumina; *aluminum 2024, *6061, *7075; *steel 1020, *4140
Storage boxes	Polystyrene*
Switches and wire jacketing	Fluoropolymers; thermoplastic polyester; modified polyphenylene oxide; *nylon; copper C11400
Telephone cases	*ABS
Toys (plastic)	*ABS; high- and low-density polyethylene; *polypropylene; polyvinyl chloride
Utensils	*Aluminum 3003; *polycarbonate; polyphenylene sulfide; polytetrafluoroethylene
Valves	Acetal; polyamide-imide; *alumina; *nylon; *aluminum; bronze

A.4 SOURCES

The following books have proven to be good sources of material property data.

Steels

Harvey, P. D.: *Engineering Properties of Steels,* American Society for Metals, Metals Park, Ohio, 1982.

Juvinall, R. C.: *Fundamentals of Machine Component Design,* Wiley, New York, 1983.

American Society for Metals: *Properties and Selection: Iron, Steels and High Performance Alloys,* 10th ed., vol. 1 of *Metals Handbook,* ASME, Cleveland, Ohio, 1990.

——: *Properties and Selection: Stainless Steels, Tool Steels, and Special Purpose Metals,* vol. 3 of *Metals Handbook,* ASME, Cleveland, Ohio, 1979.

Peckner, D., and I. M. Bernstein: *Handbook of Stainless Steels,* McGraw-Hill, New York, 1977.

Other metals

Copper Development Association: *Standards Handbook: Copper-Brass-Bronze, Wrought Mill Products,* part 2, Copper Development Association, New York, 1968.

Mantell, C. L. (ed.): *Engineering Materials Handbook,* McGraw-Hill, New York, 1958.

Materials Engineering: Material Selector, Penton, Cleveland, Ohio, 1989.

Properties and Selection: Non-ferrous Alloys and Special Purpose Materials, vol. 2 of *Metals Handbook,* ASME, Cleveland, Ohio, 1990.

Ross, R. B.: *Metallic Materials Specification Handbook,* 4th ed., Chapman & Hall, London, 1992.

Plastics

Flick, E. W.: *Engineering Resins: An Industrial Guide,* Noyes, Park Ridge, N.J., 1988.
Harper, C. A.: *Handbook of Plastics and Elastomers,* McGraw-Hill, New York, 1975.
Juran, R.: *Modern Plastics Encyclopedia 96,* vol. 72, no. 12, McGraw-Hill, New York, 1996.
Lubin, G.: *Handbook of Composites,* Van Nostrand Reinhold, New York, 1982.
Michael, J. H.: *The International Plastics Selector: Extruding and Molding Grades,* Cordura, La Jolla, Calif., 1979.
Perry, R. H. (ed.): *Engineering Manual,* McGraw-Hill, New York, 1967.
Pethrick, R. A.: *Polymer Yearbook 6,* Harwood Academic, New York, 1990.

Others

Avallone, E. A., and J. T. Baumeister: *Mark's Standard Handbook for Mechanical Engineers,* McGraw-Hill, 1992.
———: *Standard Handbook for Mechanical Engineers,* McGraw-Hill, New York, 1978.
Bever, M. B. (ed.): *Encyclopedia of Materials Science and Engineering,* MIT Press, Cambridge, Mass., 1986.
Budinski, K. G.: *Engineering Materials, Properties and Selection,* Prentice Hall, Englewood Cliffs, N.J., 1979.
Callister, W. D.: *Materials Science and Engineering,* Wiley, New York, 1985.
Ceramic Source, vol. 4, American Ceramic Society, Columbus, Ohio, 1989.
Chew, R. Z.: *Materials Engineering,* Penton, Cleveland, Ohio, 1988.
Clark, A. F., and R. P. Reed: *Materials at Low Temperatures,* American Society for Metals, Metals Park, Ohio, 1983.
Gere, J. M., and S. P. Timoshenko: *Mechanics of Materials,* 3d ed., PWS-Kent, Boston, 1990.
Higdon, A.: *Mechanics of Materials,* 4th ed., Wiley, New York, 1985.
Horton, H. L., and E. Oberg: *Machinery's Handbook,* vol. 25, Industrial Press, New York, 1996.
Kutz, M.: *Mechanical Engineer's Handbook,* Wiley-Interscience, New York, 1986.
Parker, E. R.: *Material Data Book for Engineers and Scientists,* McGraw-Hill, New York, 1967.
Shaw, K.: *Refractories and Their Uses,* Wiley, New York, 1972.
Summitt, R., and A. Sliker: *Handbook of Material Science,* vol. 4, CRC Press, 1980.
U. S. Forest Products Laboratory: *Wood Engineering Handbook,* Prentice Hall, Englewood Cliffs, N.J., 1982.

APPENDIX B

NORMAL PROBABILITY

B.1 INTRODUCTION

In this book we consider all data distributions to be normal (gaussian) distributions. This assumption is fairly accurate for most considerations in mechanical design.

To demonstrate the normal distribution, consider the data in Fig. B.1. This plot is a histogram of the ultimate strength as measured for 913 samples of 1035 steel. The data are plotted to the nearest 1 kpsi. The rounded values are given in the table below:

Ultimate strength, kpsi	Number of occurrences	Ultimate strength, kpsi	Number of occurrences
75	4	86	86
76	10	87	93
77	6	88	66
78	19	89	66
79	21	90	67
80	24	91	39
81	46	92	21
82	57	93	24
83	74	94	10
84	85	95	11
85	82	Total	913

These data are but a sample of an entire population. The entire population is all the possible samples of 1035 steel. The object is to use the statistics of this sample to infer something about the statistics of the entire population.

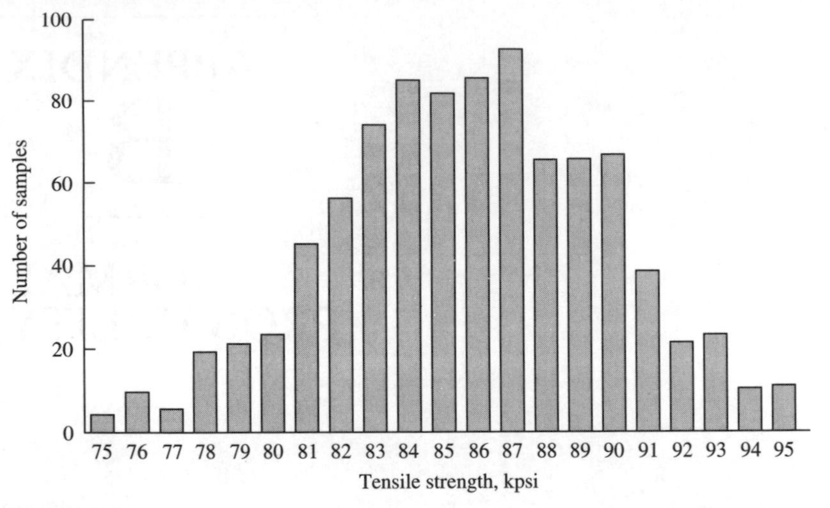

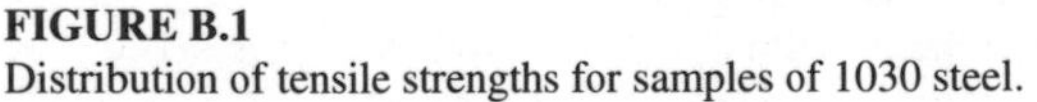
FIGURE B.1
Distribution of tensile strengths for samples of 1030 steel.

The data plotted in Fig. B.1 are replotted in Fig. B.2 on normal distribution paper. (Paper for this type of plot is commonly available.) Each ultimate strength value is plotted versus the percentage of the values that are less than it. Consider, for example, $S_u = 79$ kpsi: there are 39 values (4 + 10 + 6 + 19), or 39/913 = 4.3 percent of the total samples, that are less than 79 kpsi. The fact that the data are well fit by a straight line implies that their distribution can be modeled by a normal distribution.

Since any straight line can be characterized by two parameters, these data can be also. These two parameters are the *sample mean*, $\overline{x}$, and the *sample standard deviation*, σ, and they are defined as

$$\overline{x} = \frac{1}{N} \sum_{i=1}^{N} x_i$$

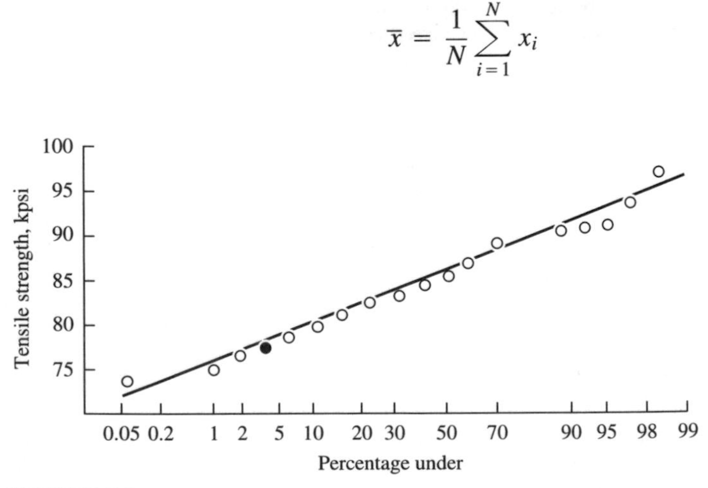

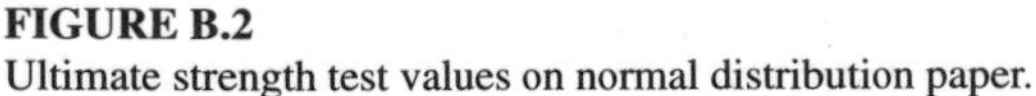
FIGURE B.2
Ultimate strength test values on normal distribution paper.

$$\sigma = \sqrt{\frac{1}{N-1}\sum_{i=1}^{N}(x_i - \overline{x})^2}$$

where the x_i are the data values S_u and N is the total number of data points (913). For the data here, $\overline{x} = 86$ kpsi and $\sigma = 4$ kpsi (both rounded to the accuracy of the input data). Another term that needs to be defined before we continue is the *sample variance*, which is the square of the standard deviation.

The three statistics defined above give only information about the sample. From the experimental data, much can be learned about the probability of finding the ultimate strength for a specific sample in a range between two values. For example, the probability of the ultimate strength being greater than or equal to 80 kpsi and less than 85 kpsi is

$$\Pr(80 \le S_u < 85) = \frac{24 + 46 + 57 + 74 + 85}{913} = 31.3 \text{ percent}$$

Also, the probability of being within two standard deviations of the mean value ($86 - 2 \times 4 = 78$ and $86 + 2 \times 4 = 94$) is

$$\Pr(78 \le S_u < 94) = 95.29 \text{ percent}$$

In both the examples above the probability was given by

$$\Pr(a \le x < b) = \frac{1}{N}\sum_{i=a}^{b-1} x_i$$

where the summation limits reflect the inequality or equality of the probability bound.

Leaving the sample data for a moment, we will develop the population normal distribution and then compare it with the sample data. The normal distribution is a continuous distribution, whereas the sample data were treated as discrete (they were rounded to the nearest kpsi). The distribution of the population is based on two parameters: the mean, μ, and the standard deviation, s, and is defined as

$$\Pr(a \le x < b) = \int_a^b \frac{e^{-(1/2)[(x-\mu)/s]^2}\,dx}{\sqrt{2\pi}s}$$

The integration implied here is seldom performed, as the variable x is normalized by defining $x' = (x - \mu)/s$. This normalization transforms the mean to 0 and the standard deviation to 1 for the new variable x'. The equation for the normal distribution can then be rewritten as

$$\Pr(a \le x < b) = \int_{(a-\mu)/s}^{(b-\mu)/s} \frac{e^{-(1/2)x'}\,dx'}{\sqrt{2\pi}}$$

and a table such as Table B.1 used to find the probability. For example, assume that the population mean and standard deviation are the same as in the sample for

TABLE B.1
Percentiles of the normal distribution: probability of obtaining a value less than or equal to x' where $x' = (x - \mu)/s$

x'	.00	.01	.02	.03	.04	.05	.06	.07	.08	.09
0.0	.5000	.5040	.5080	.5120	.5160	.5199	.5239	.5279	.5319	.5359
0.1	.5398	.5438	.5478	.5517	.5557	.5596	.5636	.5675	.5714	.5753
0.2	.5793	.5832	.5871	.5910	.5948	.5987	.6026	.6064	.6103	.6141
0.3	.6180	.6217	.6255	.6293	.6231	.6368	.6406	.6443	.6480	.6517
0.4	.6554	.6591	.6628	.6664	.6700	.6736	.6772	.6808	.6844	.6879
0.5	.6915	.6950	.6985	.7019	.7054	.7088	.7123	.7157	.7190	.7224
0.6	.7257	.7291	.7324	.7357	.7389	.7422	.7454	.7486	.7517	.7549
0.7	.7579	.7611	.7642	.7673	.7704	.7734	.7764	.7794	.7823	.7852
0.8	.7881	.7910	.7939	.7967	.7995	.8023	.8051	.8078	.8106	.8133
0.9	.8159	.8186	.8212	.8238	.8264	.8289	.8315	.8340	.8365	.8389
1.0	.8413	.8438	.8461	.8485	.8508	.8531	.8554	.8577	.8599	.8621
1.1	.8643	.8665	.8686	.8708	.8729	.8749	.8770	.8790	.8810	.8830
1.2	.8849	.8869	.8888	.8907	.8925	.8944	.8962	.8980	.8997	.9015
1.3	.9032	.9049	.9066	.9082	.9099	.9115	.9131	.9146	.9162	.9177
1.4	.9192	.9207	.9222	.9236	.9251	.9265	.9279	.9292	.9306	.9319
1.5	.9332	.9345	.9356	.9370	.9382	.9394	.9406	.9418	.9429	.9441
1.6	.9452	.9463	.9474	.9484	.9495	.9505	.9515	.9525	.9535	.9545
1.7	.9554	.9564	.9573	.9582	.9591	.9599	.9608	.9616	.9625	.9633
1.8	.9641	.9649	.9656	.9664	.9671	.9678	.9689	.9693	.9699	.9706
1.9	.9713	.9719	.9726	.9732	.9738	.9744	.9750	.9756	.9761	.9767
2.0	.9772	.9778	.9783	.9788	.9793	.9798	.9803	.9808	.9811	.9816
2.1	.9820	.9825	.9829	.9833	.9837	.9841	.9845	.9849	.9853	.9856
2.2	.9860	.9863	.9867	.9870	.9874	.9877	.9880	.9883	.9886	.9889
2.3	.9882	.9895	.9897	.9900	.9903	.9905	.9907	.9910	.9912	.9915
2.4	.9917	.9920	.9921	.9924	.9926	.9928	.9929	.9931	.9933	.9935
2.5	.9937	.9938	.9940	.9941	.9943	.9944	.9946	.9947	.9949	.9950
2.6	.9951	.9953	.9954	.9955	.9957	.9958	.9959	.9960	.9961	.9962
2.7	.9963	.9964	.9965	.9966	.9967	.9968	.9969	.9970	.9971	.9971
2.8	.9972	.9973	.9974	.9974	.9975	.9976	.9976	.9977	.9977	.9978
2.9	.9979	.9979	.9980	.9981	.9981	.9982	.9982	.9983	.9983	.9984
3.0	.9984									

the 1035 steel, 86 and 4 kpsi. Obviously $\Pr(S_u \leq 86 \text{ kpsi})$ is 50 percent. This can be found by normalizing the value 86 by subtracting the mean and dividing by the standard deviation, which results in 0. The upper leftmost entry in the table, $x' = 0$, gives 0.5000 or 50 percent. If the probability of the ultimate strength being less than the mean plus 1 standard deviation is desired, then the value of x is 90 (86 + 4) and so $x' = 1.000$, resulting in $\Pr(S_u < 90 \text{ kpsi}) = 84.13$ percent as marked in the table. Obviously, the probability of the ultimate strength being greater than 90 kpsi is $1 - 0.8413$, or $\Pr(S_u > 90 \text{ kpsi}) = 15.87$ percent.

The probability of the ultimate strength being greater than 78 kpsi is a little more difficult to find. For $x = 78$, $x' = -2.0$. Since the distribution is symmetrical about the mean value, this can be treated as the same problem as finding the probability that $x' < 2.0$ and subtracting it from 1. From Table B.1, $\Pr(x' < 2) = 97.72$ percent; therefore $\Pr(S_u \leq 78 \text{ kpsi}) = 2.28$ percent $(1 - 97.72)$.

The probability that S_u is between ± 2 standard deviations of the mean can be found by taking the probability that $x' \leq 2$, which is 0.9772, and subtracting the probability that it is less than -2, so $0.9772 - 0.0228 = 0.9544$, or $\Pr(78 \leq S_u < 94 \text{ kpsi}) = 95.44$ percent. This compares well with the value found for the sample of 95.29. For ± 1 standard deviation, the result is 68.26 percent, and for three standard deviations it is 99.68 percent. This last value is what is generally assumed to be the limit on dimensional tolerances.

One last example of the use of normal distributions. This one is about dimensions. Say a dimension on a drawing is given as 4.000 ± 0.008 cm. From a statistical viewpoint the dimension is the mean value of all the samples to be manufactured; the tolerance represents ± 3 standard deviations from the mean. This implies that 99.68 percent of all samples should have dimensions within the tolerance range. Modifying this example, consider the dimension $4.000 + 0.008 - 0.016$ cm. The nominal value is 4.000 cm; however, the mean value is 3.996 cm. Also, the standard deviation of the dimension is $(0.008 - (-0.016))/6 = 0.004$ cm and the variance is $1.6 \times 10^5 \text{ cm}^2$. From normal distribution tables, the above data allow the calculation of other statistics. For example; 68.3 percent of the samples should be between 3.992 and 4.000 cm (± 1 standard deviation), 84.1 percent should be less than 4.000 cm (less than the mean plus 1 standard deviation), and 95.44 percent should be between 3.988 and 4.004 cm (± 2 standard deviations).

B.2 OTHER MEASURES

Besides the mean and standard deviation, other measures for the normal distribution are often used. For example, statistics on human size or strength measurements are often given in terms of statistics for the 5th and 95th percentiles. The stature for a 5th-percentile man is 64.1 in., and that for a 95th-percentile man is 73.9 in. From these, the mean, 50th percentile, is 69 in. Additionally, since the 95th percentile is 1.6045 standard deviations from the mean (see Table B.1) and the difference between the mean and 73.9 is 4.9, the standard deviation is $4.9/1.6045 = 3.05$ in.

Finally, the statistics for the sum and difference of variables with normal distributions are easily found. If x_1, x_2, and x_3 are all normally distributed with means μ_1, μ_2, and μ_3 and standard deviations s_1, s_2, and s_3 and if $y = x_1 + x_2 - x_3$, then the mean value of y is $\overline{y} = \mu_1 + \mu_2 - \mu_3$ and $s_y = (s_1^2 + s_2^2 + s_3^2)^{1/2}$. Note that the mean values are just the sums and differences, whereas all the signs on the standard deviations are positive. These formulas are readily extended to any number of terms, as shown in Eq. 11.3 and 11.4.

B.3 SOURCES

Haugen, E. B.: *Probabilistic Mechanical Design,* Wiley-Interscience, New York, 1980.
Siddal, J. N.: *Probabilistic Engineering Design,* Marcel Dekker, New York, 1983.

APPENDIX C

THE FACTOR OF SAFETY AS A DESIGN VARIABLE

C.1 INTRODUCTION

The factor of safety is a factor of ignorance. If the stress on a part at a critical location (the applied stress) is known precisely, if the material's strength (the allowable strength) is also known with precision, and the allowable strength is greater than the applied stress, then the part will not fail. However, in the real world, all of the aspects of the design have some degree of uncertainty, and therefore a fudge factor, a factor of safety, is needed. A factor of safety is one way to account for the uncontrollable noises that were discussed in Chap. 11.

In practice the factor of safety is used in one of three ways. (1) It can be used to reduce the allowable strength, such as the yield or ultimate strength of the material, to a lower level for comparison with the applied stress; (2) it can be used to increase the applied stress for comparison with the allowable strength; or (3) it can be used as a comparison for the ratio of the allowable strength to the applied stress. We apply the third definition here, but all three are based on the simple formula

$$\mathrm{FS} = \frac{S_{\mathrm{al}}}{\sigma_{\mathrm{ap}}}$$

Here S_{al} is the allowable strength, σ_{ap} is the applied stress, and FS is the factor of safety. If the material properties are known *precisely* and there is no variation in them—and the same holds for the load and geometry—then the part can be designed with a factor of safety of 1, the applied stress can be equal to the allowable strength, and the resulting design will not fail (just barely). However, not only are these measures not known with precision, they are not constant from sample to sample or use to use. In a statistical sense all these measures have some variance about their mean values (see App. B for the definitions of the mean and variance).

For example, typical material properties, such as ultimate strength, even when measured from the same bar of material, show a distribution of values (a variance) around a nominal mean of about 5 percent. This distribution is due to inconsistencies in the material itself and in the instrumentation used to take the data. If strength figures are taken from handbook values based on different samples and instrumentations, the variance of the values may be 15 percent or higher. Thus, the allowable strength must be characterized as a nominal or mean value with some statistical variation about it.

Even more difficult to establish are the statistics of the applied stress. The exact magnitude of the applied stress is a factor of the loading on the part (the forces and moments on the part), the geometry of the part at the critical location, and the accuracy of the analytic method used to determine the stress at the critical point due to the load.

The accuracy of the comparison of the applied stress to the allowable strength is a function of the accuracy and applicability of the failure theory used. If the stress is steady and the failure mode yielding, then accurate failure theories exist and can be used with little error. However, if the stress state is multiaxial and fluctuating (with a nonzero mean stress), there are no directly applicable failure theories and the error incurred in using the best available theory must be taken into account.

Beyond the above mechanical considerations, the factor of safety is also a function of the desired reliability for the design. As will be shown in Section C.3, the reliability can be directly linked to the factor of safety.

There are two ways to estimate the value of an acceptable factor of safety: the classical rule-of-thumb method (presented in Section C.2) and the probabilistic, or statistical, method of relating the factor of safety to the desired reliability and to knowledge of the material, loading, and geometric properties (presented in Section C.3).

An additional note on standards. Most established design disciplines and companies have factors of safety used as standards. But often these values are based on lost or outdated material specifications and quality control procedures. At a minimum the following tools will help explore the basis of these standards; at a maximum they can be used to update them.

C.2 THE CLASSICAL RULE-OF-THUMB FACTOR OF SAFETY

The factor of safety can be quickly estimated on the basis of estimated variations of the five measures previously discussed: material properties, stress, geometry, failure analysis, and desired reliability. The better known the material properties and stress, the tighter the tolerances, the more accurate and applicable the failure theory, and the lower the required reliability, the closer the factor of safety should be to 1. The less known about the material, stress, failure analysis, and geometry and the higher the required reliability, the larger the factor of safety.

The simplest way to present this technique is to associate a value greater than 1 with each of the measures and define the factor of safety as the product of these five values:

$$FS = FS_{\text{material}} \cdot FS_{\text{stress}} \cdot FS_{\text{geometry}} \cdot FS_{\text{failure analysis}} \cdot FS_{\text{reliability}}$$

Details on how to estimate these five values are given below. These values have been developed by breaking down the rules given in textbooks and handbooks into the five measures and cross checking the values with those from the statistical method described in the next section.

ESTIMATING THE CONTRIBUTION FOR THE MATERIAL

$FS_{\text{material}} = 1.0$ If the properties for the material are well known, if they have been experimentally obtained from tests on a specimen known to be identical to the component being designed and from tests representing the loading to be applied.

$FS_{\text{material}} = 1.1$ If the material properties are known from a handbook or are manufacturer's values.

$FS_{\text{material}} = 1.2–1.4$ If the material properties are not well known.

ESTIMATING THE CONTRIBUTION FOR THE LOAD STRESS

$FS_{\text{stress}} = 1.0–1.1$ If the load is well defined as static or fluctuating, if there are no anticipated overloads or shock loads, and if an accurate method of analyzing the stress has been used.

$FS_{\text{stress}} = 1.2–1.3$ If the nature of the load is defined in an average manner, with overloads of 20–50 percent, and the stress analysis method may result in errors less than 50 percent.

$FS_{\text{stress}} = 1.4–1.7$ If the load is not well known or the stress analysis method is of doubtful accuracy.

ESTIMATING THE CONTRIBUTION FOR GEOMETRY (UNIT-TO-UNIT)

$FS_{\text{geometry}} = 1.0$ If the manufacturing tolerances are tight and held well.

$FS_{\text{geometry}} = 1.0$ If the manufacturing tolerances are average.

$FS_{\text{geometry}} = 1.1–1.2$ If the dimensions are not closely held.

ESTIMATING THE CONTRIBUTION FOR FAILURE ANALYSIS

$FS_{\text{failure theory}} = 1.0–1.1$ If the failure analysis to be used is derived for the state of stress, as for uniaxial or multiaxial static stresses, or fully reversed uniaxial fatigue stresses.

$FS_{\text{failure theory}} = 1.2$ If the failure analysis to be used is a simple extension of the above theories, such as for multiaxial, fully reversed fatigue stresses or uniaxial nonzero mean fatigue stresses.

$FS_{\text{failure theory}} = 1.3\text{–}1.5$ If the failure analysis is not well developed, as with cumulative damage or multiaxial nonzero mean fatigue stresses.

ESTIMATING THE CONTRIBUTION FOR RELIABILITY

$FS_{\text{reliability}} = 1.1$ If the reliability for the part need not be high, for instance, less than 90 percent.

$FS_{\text{reliability}} = 1.2\text{–}1.3$ If the reliability is an average of 92–98 percent.

$FS_{\text{reliability}} = 1.4\text{–}1.6$ If the reliability must be high, say, greater than 99 percent.

The above values are, at best, estimates based on a verbalization of the factors affecting the design combined and on experience with how these factors affect the design. The stress on a part is fairly insensitive to tolerance variances unless they are abnormally large. This insensitivity will be more evident in the development of the statistical factor of safety.

C.3 THE STATISTICAL, RELIABILITY-BASED, FACTOR OF SAFETY

C.3.1 Introduction

As can be appreciated, the classical approach to establishing factors of safety is not very precise and the tendency is to use it very conservatively. This results in large factors of safety and over-designed components. Consider now the approach based on statistical measures of the material properties, of the stress developed in the component, of the applicability of the failure theory, and of the reliability required. This technique gives the designer a better feel of just how conservative, or nonconservative, he or she is being.

With this technique, all measures are assumed to have normal distributions (details on normal distributions appear in App. B). This assumption is a reasonable one, though not as accurate as the Weibull distribution for representing material-fatigue properties. What makes the normal distribution an acceptable representation of all the measures is the simple fact that, for most of them, not enough data are available to warrant anything more sophisticated. In addition, the normal distribution is easy to understand and work with. In sections below, each measure is discussed in terms of the two factors needed to characterize a normal distribution—the mean and the standard deviation (or variance).

The factor of safety is defined as the ratio of the allowable strength, S_{al}, to the applied stress, σ_{ap}. The allowable strength is a measure of the material properties; the applied stress is a measure of the stress (as a function of both the applied load and the stress analysis technique used to find the stress), the geometry, and the failure theory used. Since both of these measures are distributions, the factor of safety is better defined as the ratio of their mean values: $FS = \overline{S}_{\text{al}}/\overline{\sigma}_{\text{ap}}$.

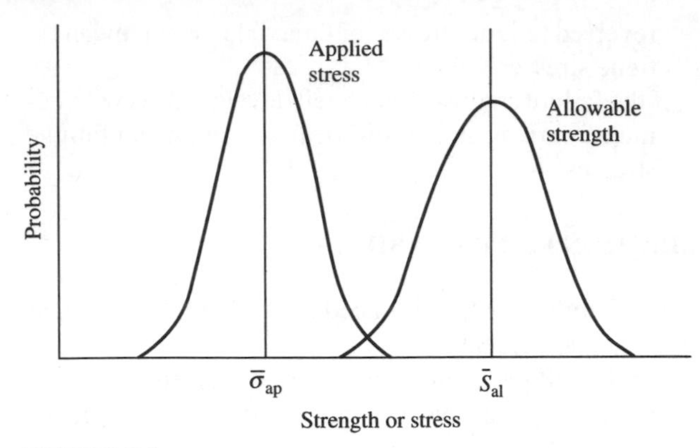

FIGURE C.1
Distribution of applied stress and allowable strength.

Figure C.1 shows the distribution about the mean for both the applied stress and the allowable strength. There is an area of overlap between these two curves no matter how large the factor of safety is and no matter how far apart the mean values are. This area of overlap is where the allowable strength has a probability of being smaller than the applied stress; the area of overlap is thus the region of potential failure. Keeping in mind that areas under normal distribution curves represent probabilities, we see that this area of overlap then is the probability of failure (PF). The reliability, the probability of no failure, is simply 1 − PF. Thus, by considering the statistical nature of these curves, the factor of safety is directly related to the reliability.

To develop this relationship more formally, we define a new variable, $z = S_{al} - \sigma_{ap}$, the difference between the allowable strength and the applied stress. If $z > 0$, then the part will not fail. But failure will occur at $z \leq 0$. The distribution of z is also normal (the difference between two normal distributions is also normal), as shown in Fig. C.2. The mean value of z is simply $\bar{z} = \bar{s}_{al} - \sigma_{ap}$. If the allowable strength and the applied stress are considered as independent variables (which is the case), then the standard deviation of z is

$$\rho_z = \sqrt{\rho_{al}^2 + \rho_{ap}^2}$$

Normalizing any value of z by subtracting the mean and dividing by the standard deviation, we can define the variable t_z as

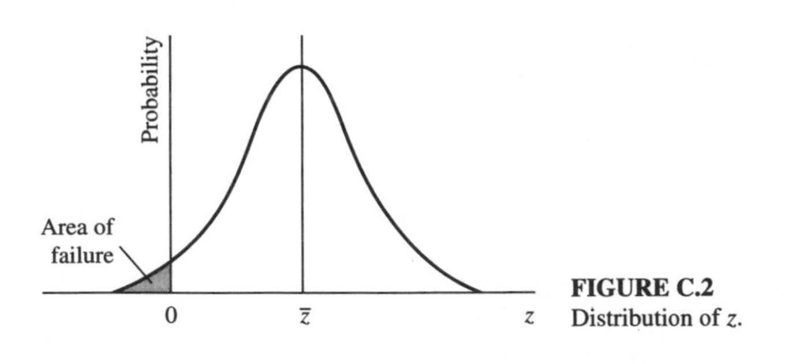

FIGURE C.2
Distribution of z.

$$t_z = \frac{(S_{\text{al}} - \sigma_{\text{ap}}) - (\overline{S}_{\text{al}} - \overline{\sigma}_{\text{ap}})}{\sqrt{\rho_{\text{al}}^2 + \rho_{\text{ap}}^2}}$$

The variable t_z has a mean value of 0 and a standard deviation of 1. Since failure will occur when the applied stress is greater than the allowable stress, a critical point to consider is when $z = 0$, $S_{\text{al}} = \sigma_{\text{ap}}$. So, for $z = 0$,

$$t_{z=0} = \frac{-(\overline{S}_{\text{al}} - \overline{\sigma}_{\text{ap}})}{\sqrt{\rho_{\text{al}}^2 + \rho_{\text{ap}}^2}}$$

Thus, any value of t that is calculated to be less than $t_{z=0}$ represents a failure situation. The probability of a failure then is $\Pr(t < t_{z=0})$, which, assuming the normal distribution, can be found directly from a normal distribution table. If the distributions of the applied stress and the allowable strength are known, $t_{z=0}$ can be found from the above equation and the probability of failure found from normal distribution tables. Finally, the reliability is 1 minus the probability of failure; thus $R = 1 - \Pr(t_z \leq t_{z=0})$. To make using normal distribution tables (App. B) easier by utilizing the symmetry of the distribution, we can drop the minus sign on the above equation and consider values of $t_z > t_{z=0}$ to represent failure. Some values showing the relation of $t_{z=0}$ to reliability are given in Table C.1.

TABLE C.1
Relation of $t_{z=0}$ to R

R	$t_{z=0}$
0.50	0.00
0.90	1.28
0.95	1.64
0.99	2.33

To reduce the equations to a usable form in which the factor of safety is the independent variable, we rewrite the previous equation, dividing by the mean value of the applied stress and using the definition of the factor of safety:

$$t_{z=0} = \frac{\text{FS} - 1}{\sqrt{\text{FS}^2(\rho_{\text{al}}/\overline{S}_{\text{al}})^2 + (\rho_{\text{ap}}/\overline{\sigma}_{\text{ap}})^2}}$$

With $t_{z=0}$ directly dependent on the reliability, there are four variables related by this equation: the reliability, the factor of safety, and the coefficients of variation (standard deviation divided by the mean) for the allowable and applied stresses. In the development here the unknown will be the factor of safety. Thus the final form of the statistical factor of safety equation is

$$\text{FS} = 1 + t_{z=0}\frac{\sqrt{(\rho_{\text{al}}/\overline{S}_{\text{al}})^2 + (\rho_{\text{ap}}/\overline{\sigma}_{\text{ap}})^2 - t_{z=0}^2(\rho_{\text{ap}}/\overline{\sigma}_{\text{ap}})^2(\rho_{\text{al}}/\overline{S}_{\text{al}})^2}}{1 - t_{z=0}^2(\rho_{\text{al}}/\overline{S}_{\text{al}})^2} \tag{C.1}$$

Before proceeding with details into the development of the applied stress and allowable strength coefficients of variation, let us look at an example of the use of the above equations. Say that the allowable strength coefficient of variation (see Sec. C.3.2) is 0.08 (the standard deviation is 8 percent of the mean value), the applied stress coefficient (see Sec. C.3.3) is 0.20 and the desired reliability is 95 percent. Using Table C.1, a 95 percent reliability gives $t_{z=0} = 1.64$. Thus, using Eq. (C.1), the design factor of safety can be computed to be 1.37. If the reliability is increased to 99 percent, the design factor of safety increases to 1.55. These design factor of safety values are not dependent on the actual values of the material properties or the stresses in the material but only on their statistics and the reliability and applicability of the failure theory. This is a very important point.

C.3.2 The Allowable Strength Coefficient of Variation

Measured material properties, such as the yield strength, the ultimate strength, the endurance strength, and the modulus of elasticity, all have distributions about their means. This is evident in Fig. B.1, which shows the result of static tests on 913 different samples of 1035 steel as hot-rolled, round bars, 1 to 9 in. diameter. Although not perfectly normal, as shown by the fit of the data on the normal distribution paper, an approximation to the straight line is not bad. Not all kinds of data fit this well. Typically, fatigue data and data on ceramic material tend not to be as evenly distributed and are better represented by a skewed distribution, such as the Weibull distribution. (Unfortunately, the four factors needed to represent the Weibull distribution have only been determined for a limited number of materials.) However, the adequacy and simplicity of the normal distribution make it the best choice for representing the material properties here. From Fig. B.2, the mean ultimate strength is 86 kpsi and its standard deviation is 4 kpsi. Note that the standard deviation is 4.6 percent of the mean, so the coefficient of variation for 1033 hot-rolled steel is 0.046. Unfortunately, it is not always simple to find the standard deviation for the material properties. In looking up the ultimate stress for 1035 HR in standard design books, the following values were found: 72, 85, 72, 82, and 67 kpsi. From this limited sample, the mean value is 75.6 kpsi with a standard deviation of 6.08 kpsi, resulting in a coefficient of variation of 0.80 (6.08/75.6). If the heat treatment on the material or the exact composition of the material are unknown, the deviation can be much higher.

The allowable stress may be based on the yield, ultimate, or endurance strengths or some combination of them, depending on the failure criteria used. In the formulation above the allowable stress appears only as a ratio of standard deviation to average value. For most materials this ratio, irrespective of which allowable stress is considered, is in the range of 0.05–0.15. It is recommended that the statistics for the strength that best represent the nature of the failure be used. For example, in cases of nonzero-mean fluctuating stresses, the allowable stress coefficient for the endurance limit should be used if the mean is small relative to the amplitude, and the coefficient for the ultimate should be used if the mean is large relative to the amplitude. Any complexity beyond this is not warranted.

C.3.3 The Applied Stress Coefficient of Variation

The applied stress coefficient of variation is somewhat more difficult to develop. The statistics of the geometry and the load obviously affect the statistics of the applied stress, as does the accuracy of the method used to find the stress. Additionally, a measure of the accuracy of the failure analysis method to be used will also be reflected in the statistics of the applied stress. (This measure could be taken into account elsewhere, but it is convenient to consider it as a correction on the applied stress.)

To see how these various factors are combined to form the applied stress coefficient of variation, consider the following example. (The statistics for the stress analysis technique and the failure theory accuracy will be included later in the example.) Consider a round, axially loaded uniform bar. In this bar the average maximum stress is given by the ratio of the average maximum force divided by the average area:

$$\overline{\sigma}_{\text{ap}} = \frac{\overline{F}}{\pi \overline{r}^2}$$

The standard deviation of the stress is a function of the independent statistics of the geometry and the load. Using standard, normal-distribution relations,

$$\frac{\rho_{\text{ap}}}{\overline{\sigma}_{\text{ap}}} = \sqrt{4\left(\frac{\rho_{\text{r}}}{r}\right)^2 + \left(\frac{\rho_F}{\overline{F}}\right)^2}$$

Thus, the applied-stress coefficient of variation is written in terms of the coefficients of variation for the geometry ($\rho_{\text{r}}/\overline{r}$) and the loading ($\rho_{\text{F}}/\overline{F}$). In general, the same form can be derived for any loading and shape. No matter whether it is normal or shear, the stress will have the form of force/area, and area always has units of length squared.

In many applications the magnitude of load forces and moments are well known through either experience or measurement. Essentially, two types of loads are considered here: static loads and fatigue, or fluctuating, loads. Regardless of which type of loading is considered, the exact magnitude of the forces and moments may have to be estimated. The determination of the statistical factor of safety takes into account the confidence in this estimation. This approach is much like that used in project planning (PERT) and requires the designer to make three estimates of the load: an optimistic estimate o; a most likely estimate m; and a pessimistic estimate p. From these three the mean $\overline{m}$, standard deviation ρ, and coefficient of variation can be found:

$$\overline{m} = \tfrac{1}{6}(o + 4m + p)$$

$$\rho = \tfrac{1}{6}(p - o)$$

$$\frac{\rho}{\overline{m}} = \frac{p - o}{o + 4m + p}$$

These equations are based on a beta distribution function rather than a normal distribution. However, if the most likely estimate is the mean load, and the optimistic and pessimistic estimates are the mean ± 3 standard deviations, then the beta distribution reduces to the normal distribution. The beauty of this is that an estimate of the important statistics can be made even if the distribution of the estimates is not symmetrical. For example, suppose the maximum load on a bracket is quoted as a force of 25,000 N. This may just be the most likely estimate. There is a possibility that the maximum load may be as low as 15,000 N or, because of light shock loading, the force may be as high as 50,000 N. Thus, from the formulas above, the expected value is 27,500 N, the standard deviation is 5833 N, and the coefficient of variation is 0.21. If the optimistic load had been 0, no load at all, the expected load would be 25,000 N and the standard deviation 8333 N. In this case the pessimistic and optimistic estimates are ± 3 standard deviations from the expected or mean value. The coefficient of variation is 0.33, reflecting the wider range of estimates. Note again that the load coefficient of variation is independent of the absolute value of the load itself and only gives information on its distribution.

The hardest factor to take into account in failure analysis is the effect of shock loads. In the example given above the potential maximum load was double the nominal value. Without dynamic modeling there is no way to find the effect of shock loads on the state of stress. The following is suggested:

- If the load is smoothly applied and released, use a ratio of optimistic: most likely: pessimistic of 1:1:2 or 1:2:4.
- If the load gives moderated shocks, use a ratio of optimistic: most likely: pessimistic of 1:1:4 or 1:4:16. Examples of moderate shock applications are blowers, cranes, reels, and calenders.
- If the load gives heavy shocks, use a ratio of optimistic: most likely: pessimistic of 1:1:10 or 1:10:100. Examples of heavy shock applications are crushers, reciprocating machinery, and mixers.

The geometry of the part is important in that, in combination with the load, the geometry determines the applied stress. Normally, the geometry is given as nominal dimensions with a bilateral tolerance (3.084 ± 0.010 inches). The nominal is the mean value, and the tolerance is usually considered to be three times the standard deviation. This implies that, assuming a normal distribution, 99.74 percent of all the samples will be within the limits of the tolerance. It is assumed that there is one dimension that is most critical to the stress, and the coefficient of variation for this dimension is used in the analysis. For the example above, the coefficient of variation is 0.0011 ($0.010/(3 \times 3.084)$), which is an order of magnitude smaller than that for the load. This is typical for most tolerances and loadings.

Using the above examples, the applied stress coefficient of variation is

$$\rho_{ap} = \sqrt{4(0.0011)^2 + 0.21^2} \doteq 0.21$$

Note the lack of sensitivity to the tolerance.

The above does not take into account the accuracy of the stress analysis technique used to find the stress state from the loading and geometry nor the adequacy of the failure analysis method. To include these factors, the allowable strength needs to be compared with the calculated applied stress corrected for the stress analysis and the failure analysis accuracy. Thus,

$$\sigma_{\text{ap}} = \sigma_{\text{calc}} \times N_{\text{sa}} \times N_{\text{fa}}$$

where N_{sa} is a correction multiplier for the accuracy of the stress analysis technique and N_{fa} is a correction multiplier for the failure analysis accuracy.

If the two corrections are assumed to have normal distributions, they can be represented as coefficients of variation. With the product of normally distributed independent standard deviations being the square root of the sum of the squares (see App. B), we have for the applied stress coefficient of variation

$$\frac{\rho_{\text{ap}}}{\overline{\sigma}_{\text{ap}}} = \sqrt{4\left(\frac{\rho_r}{\overline{r}}\right)^2 + \left(\frac{\rho_F}{\overline{F}}\right)^2 + \left(\frac{\rho_{\text{sa}}}{\overline{N}_{\text{sa}}}\right)^2 + \left(\frac{\rho_{\text{fa}}}{\overline{N}_{\text{fa}}}\right)^2} \tag{C.2}$$

This is the same as before, with the addition of the coefficient of variation s for the stress analysis method and for the failure theory.

The coefficient of variation for the stress analysis method can be estimated using the same technique as for estimating the statistics on the loading—namely, estimate an optimistic, pessimistic, and most likely value for the stress, based on the most likely load. Again, consider a load of 25,000 N (the most likely estimate of the maximum load). Assume that at the critical point the normal stress caused by this load is 40.9 kpsi (282.0 Mpa), with a stress concentration factor of 3.55. The most likely normal stress is the product of the load and the stress concentration factor, 145 kpsi. However, confidence in the method used to find the nominal stress and the stress concentration factor is not high. In fact, the maximum stress may really be as high as 160 kpsi or as low as 140 kpsi. With these two values as the pessimistic and the optimistic estimates, the coefficient of variation is calculated at 0.023. For strain gauge data or other measured results, the stress analysis method coefficient of variation will be very small and can, like the geometry statistics, be ignored.

The adequacy of the failure analysis technique, as discussed in the development of the classical factor of safety method, has a marked effect on the design factor of safety. On the basis of experience and the limited data in the references, the coefficient of variations recommended for the different types of loadings are as follows:

Static failure theories: 0.02
Fully reversed uniaxial infinite life fatigue failure theory: 0.02
Fully reversed uniaxial finite life fatigue failure theory: 0.05
Nonzero mean uniaxial fatigue failure theory: 0.10
Fully reversed multiaxial fatigue failure theory: 0.20
Nonzero mean multiaxial fatigue failure theory: 0.25
Cumulative damage load history: 0.50

These values imply that for well-defined failure analysis techniques, where the failure mode is identical to that found with the allowable strength material test, the standard deviation is small, namely, 2 percent of the mean. When the failure theory is comparing a dissimilar applied stress state to an allowable strength, the margin for error increases. The rule used in cumulative damage failure estimation can be off by as much as a factor of 2 and is therefore used with high uncertainty.

C.3.4 Steps for Finding the Reliability-Based Factor of Safety

We can summarize the method discussed in the previous two sections as an eight-step procedure:

STEP 1: SELECT RELIABILITY. From Table C.1, find the value of $t_{z=0}$ for the desired reliability.

STEP 2: FIND THE ALLOWABLE STRENGTH COEFFICIENT OF VARIATION. This can be found experimentally or by following the following rules of thumb. If the material properties are well known, use a coefficient of 0.05; if the material properties are not well known, use a coefficient of 0.01–0.15.

STEP 3: FIND THE CRITICAL DIMENSION COEFFICIENT OF VARIATION. This value is generally small and can be ignored except when the variation in the critical dimensions are large because of manufacturing, environmental, or aging effects.

STEP 4: FIND THE LOAD COEFFICIENT OF VARIATION. This is an estimate of how well the maximum loading is known. It can be estimated using the PERT method given in Sec. C.3.2.

STEP 5: FIND THE ACCURACY OF THE STRESS ANALYSIS COEFFICIENT OF VARIATION. Even though the variation of the load was taken into account in step 4, knowledge about the effect of the load on the structure is a separate issue. The stress due to a well-known load may be hard to determine because of complex geometry. Conversely, the stress caused by a poorly known load on a simple structure is no more poorly known than the load itself. This measure then takes into account how well the stress can be found for a known load.

STEP 6: FIND THE FAILURE ANALYSIS TECHNIQUE COEFFICIENT OF VARIATION. Guidance for this is given near the end of Sec. C.3.2.

STEP 7: CALCULATE THE APPLIED STRESS COEFFICIENT OF VARIATION. This is found using Eq. (C.2).

STEP 8: CALCULATE THE FACTOR OF SAFETY. This is found using Eq. (C.1).

C.4 SOURCES

Ullman, D. G.: "Less Fudging with Fudge Factors," *Machine Design*, Oct. 9, 1986, pp. 107–111.
———: *Mechanical Design Failure Analysis*, Marcel Dekker, New York, 1986.

APPENDIX D

HUMAN FACTORS IN DESIGN

D.1 INTRODUCTION

Most machines work in coordination with people. Consider the types of interactions you have with a standard gas-powered lawn mower. First, in starting and pushing the mower you *occupy a workspace* around the mower. You have to stoop or bend in this space to reach the starting mechanism, then you have to position yourself while holding your arms at a certain height to push and steer the mower. Second, you *provide a source of power* to the mower to start it and to push it. (Even if it is electrically started, you have to push a button or turn a key.) Additionally, it takes muscle power to steer the mower, whether you are walking behind it or riding on it. Third, you *act as a sensor,* listening to determine if anything is stuck in the mower, seeing where you are going so that you can guide the mower, and feeling with your hands any feedback motion through the steering that might give you information on how well you are guiding the mower. Fourth, based on the information received by the sensory inputs, you *act as a controller.* You determine how much power to provide and in what direction to keep the mower under control.

These four ways a person interacts with the product—as occupant of workspace, as power source, as sensor, and as controller—form the basis for the study of the *human factors* that play a major role in the design of a device. Beyond these four basic types of interactions between person and product, there are further human interaction issues that must be considered during design. First, even those devices that spend their operating life remote from all human interaction, at the bottom of a well or in deep space, for example, must first be assembled. The assembler must interface with the device in the same four ways as described in the lawn-mower example. Second, most devices have to be maintained, which presents yet another situation for the consideration of human interaction in the design of a product. *Human factors must be taken into account for every person who comes into contact with the product, whether during manufacture, operation, maintenance and repair, or disposal.*

Two reasons for this concern with human factors are quality and safety. In Table 1.2 we saw that in determining quality in a product, the most important factor was perceived to be that it "work as it should." This design requirement relates directly to the four components of human-product interface. Products are perceived to work as they should if they are comfortable to use (there is a good match between the device and the person in the workspace), they are easy to use (minimal power is required), their operating condition is easily sensed, and their control logic is natural, or user-friendly. Of equal importance is the concern for safety. Although not listed as one of the factors in the survey, it is readily assumed that an unsafe design will never be perceived as a quality product. Customers assume that neither they nor others will be injured, and that no property will be destroyed, when a product is in use (obvious exceptions are products that are designed to destroy or injure).

In the following sections all of these issues will be further explored, with emphasis on understanding the interactions between humans and machines in order to ensure that quality and safety are designed into the product.

D.2 THE HUMAN IN THE WORKSPACE

It is vital that a product "fit" its intended user; in other words, it must be comfortable for a person to use. The lawn-mower pull starter must be at the right height, or it

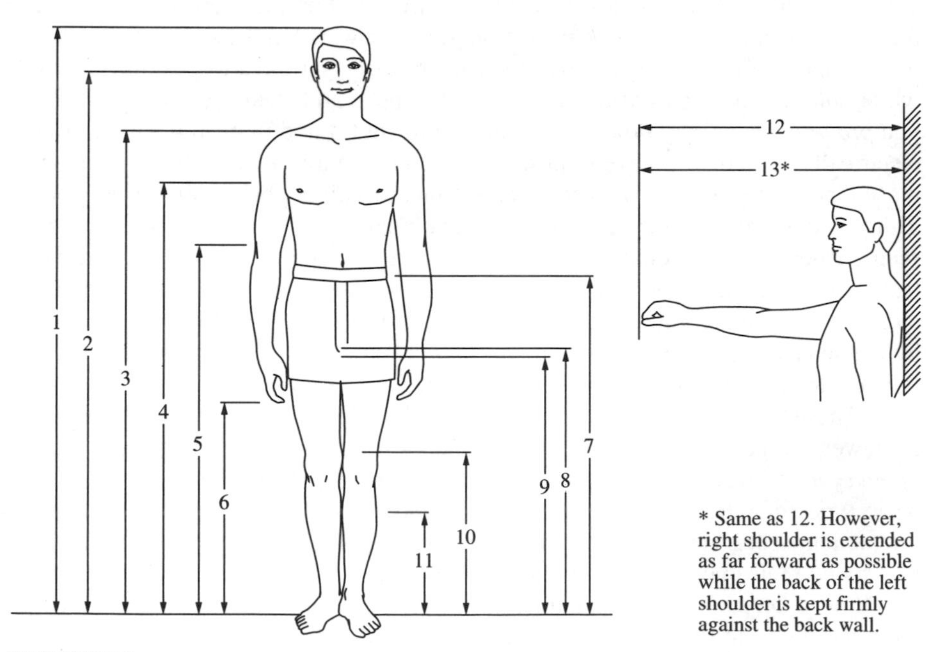
1
2
3
4
5
6
7
8
9
10
11
12
13*
* Same as 12. However, right shoulder is extended as far forward as possible while the back of the left shoulder is kept firmly against the back wall.

FIGURE D.1
Anthropometric man (from MIL-STD 1472D).

will be hard to reach and even harder to pull. Likewise, the handle on a push mower must be at a height that is comfortable for a majority of people or the mower will judged to be of poor quality. The geometric properties of humans—their height, reach, and seating requirements, the size of the holes they can fit through, etc.—are called *anthropometric data* (literally, "human-measure data"). Many such data have been collected by the armed forces because so many different people must operate military equipment on a day-to-day basis. Typical anthropometric data given in MIL-STD (Military Standard) 1472D are shown in Fig. D.1 and Table D.1.

TABLE D.1
Anthropometric data (from MIL-STD 1472D)

	Percentile values in centimeters					
	5th percentile			95th percentile		
	Ground troops	Aviators	Women	Ground troops	Aviators	Women
Weight (kg)	65.5	60.4	46.6	91.6	96.0	74.5
Standing body dimensions						
1. Stature	162.8	164.2	152.4	185.6	187.7	174.1
2. Eye height (standing)	151.1	152.1	140.9	173.3	175.2	162.2
3. Shoulder (acromial) height	133.6	133.3	123.0	154.2	154.8	143.7
4. Chest (nipple) height*	117.9	120.8	109.3	136.5	138.5	127.6
5. Elbow (radiate) height	101.0	104.8	94.9	117.8	120.0	110.7
6. Fingertip (dactylion) height		61.5			73.2	
7. Waist height	96.6	97.6	93.1	115.2	115.1	110.3
8. Crotch height	76.3	74.7	68.1	91.8	92.0	83.9
9. Gluteal furrow height	73.3	74.6	66.4	87.7	88.1	81.0
10. Kneecap height	47.5	46.8	43.8	58.6	57.8	52.5
11. Calf height	31.1	30.9	29.0	40.6	39.3	36.6
12. Functional reach	72.6	73.1	64.0	90.9	87.0	80.4
13. Functional reach, extended	84.2	82.3	73.5	101.2	97.3	92.7
	Percentile values in inches					
Weight (lb)	122.4	133.1	102.3	201.9	211.6	164.3
Standing body dimensions						
1. Stature	64.1	64.6	60.0	73.1	73.9	68.5
2. Eye height (standing)	59.5	59.9	55.5	68.2	69.0	63.9
3. Shoulder (acromial) height	52.6	52.5	48.4	60.7	60.9	56.6
4. Chest (nipple) height*	46.4	47.5	43.0	53.7	54.5	50.3
5. Elbow (radiate) height	39.8	41.3	37.4	46.4	47.2	43.6
6. Fingertip (dactylion) height		24.2			28.8	
7. Waist height	38.0	38.4	36.6	45.3	45.3	43.4
8. Crotch height	30.0	29.4	26.8	36.1	36.2	33.0
9. Gluteal furrow height	28.8	29.4	26.2	34.5	34.7	31.9
10. Kneecap height	18.7	18.4	17.2	23.1	22.8	20.7
11. Calf height	12.2	12.2	11.4	16.0	15.5	14.4
12. Functional reach	28.6	28.8	25.2	35.8	34.3	31.7
13. Functional reach, extended	33.2	32.4	28.9	39.8	38.3	36.5

*Bustpoint height for women.

Since people come in a variety of shapes and sizes, it is important that anthropometric data give a range of dimensions. The measures of humans are well represented as normal distributions. (See App. B for details on normal distributions.) Typically, these measures are given for the 5th and 95th percentile, as in Table D.1. It is safe to assume that data for civilians do not differ significantly from that for military

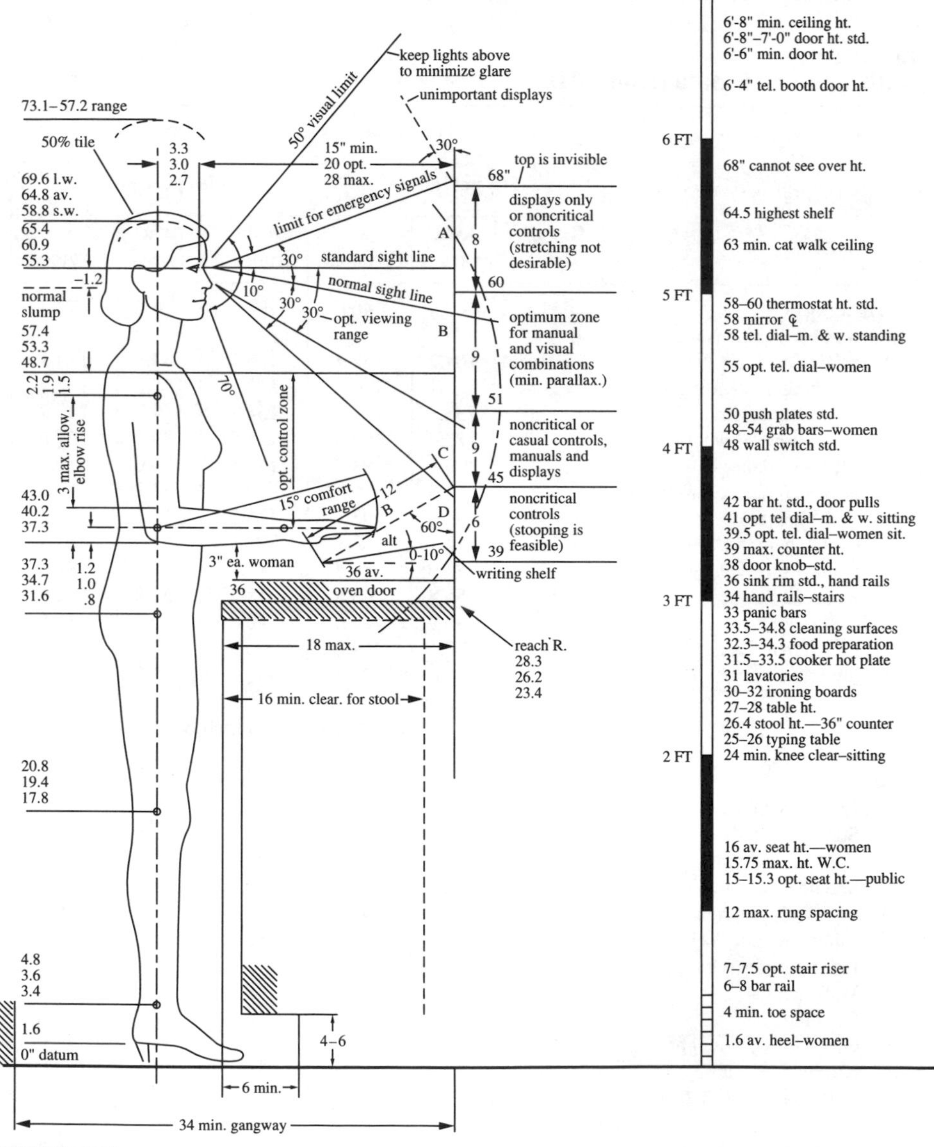

FIGURE D.2
Anthropometric woman at control panel (from H. Dreyfuss, *The Measure of Man: Human Factors in Design*, Whitney Library of Design, New York, 1967).

personnel. Similar data are available for children. (See Sources at the end of this appendix.)

Besides measures for humans standing with their arms at their sides, measures for humans performing various activities are also available. For example, in Fig. D.2, an anthropometric woman is shown standing at a control panel. This drawing

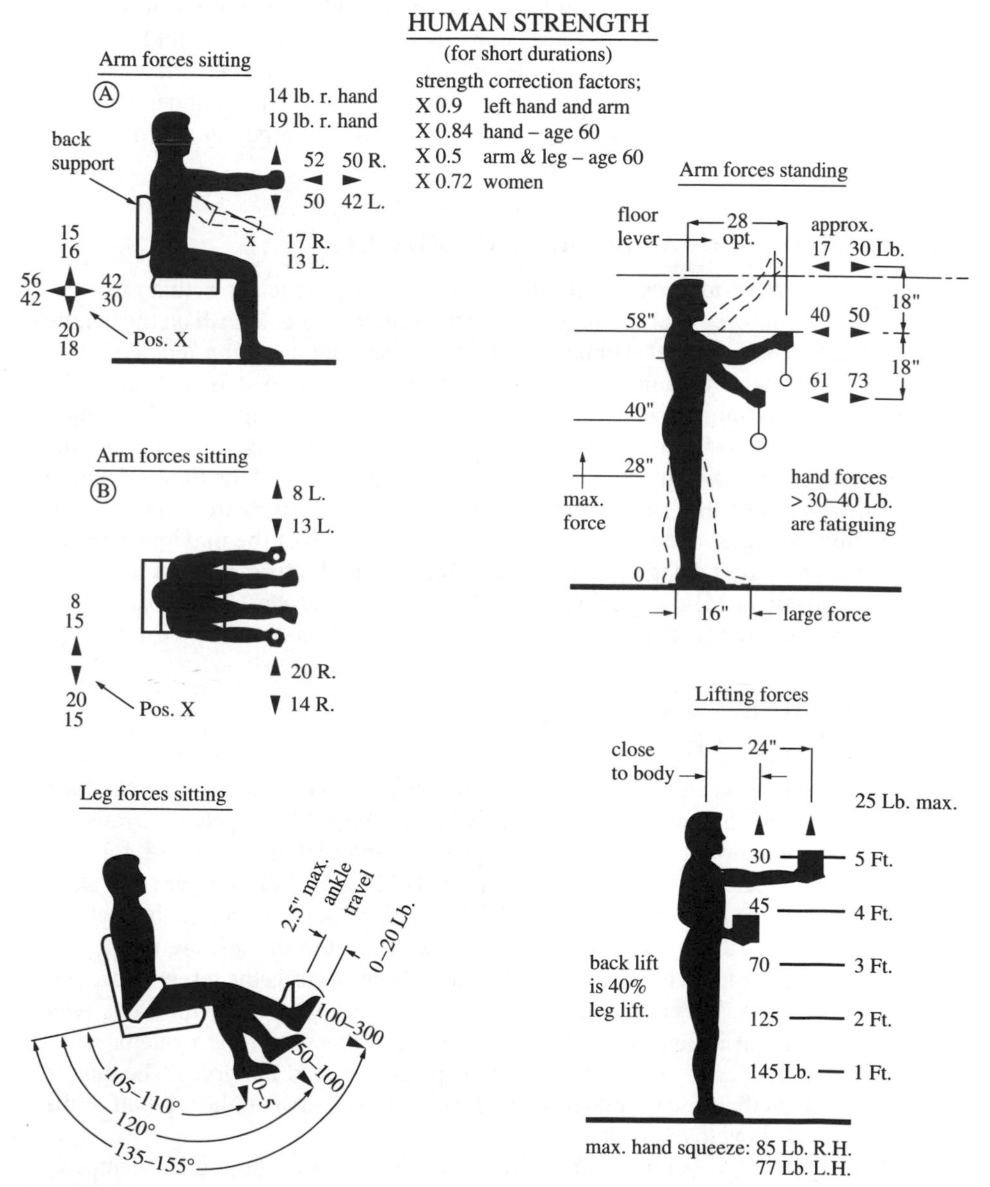

FIGURE D.3
Average human strength for different tasks (from H. Dreyfuss, *The Measure of Man: Human Factors in Design*, Whitney Library of Design, New York, 1967).

is of a 50th-percentile woman. The dimensions on the control panel are such that a majority of women will feel comfortable looking at the displays and working the controls.

Returning to our lawn mower: the handle should be at about elbow level, height 5 in Fig. D.1 and Table D.1. To fit all men and women between the 5th and 95th percentiles, the handle must be adjustable between 94.9 cm (37.4 in.) for the 5th-percentile woman and 117.8 cm (46.4 in.) for the 95th-percentile man. Anthropometric data from the references also show that the pull starter should be 69 cm (27 in.) off the ground for the average person. For this uncommon position, only an average value is given in the references. For positions even more unique, the engineer may have to develop measurements of a typical user community in order to get the data necessary for quality products.

D.3 THE HUMAN AS SOURCE OF POWER

Humans often have to supply some force to power a product or actuate its controls. The lawn-mower operator must pull on the starter cord and push on the handle or move the steering wheel. Human force-generation data are often included with anthropometric data. This information comes from the study of *biomechanics* (the mechanics of the human body). Listed in Fig. D.3 is the average human strength for differing body positions. In the data for "arm forces standing," we find that the average pushing force 40 in. off the ground (the average height of the mower handle) is 73 lb, with a note that hand forces of greater than 30 to 40 lb are fatiguing. Although only averages, these values do give some indication of the maximum forces that should be used as design requirements. More detailed information on biomechanics is available in MIL-HDBK (Military Handbook) 759A and *The Human Body in Equipment Design* (see Sources at the end of this appendix).

D.4 THE HUMAN AS SENSOR AND CONTROLLER

Most interfaces between humans and machines require that humans *sense* the state of the device and, based on the data received, *control* it. Thus products must be designed with important features readily apparent, and they must provide for easy control of these features. Consider the control panel from a clothes dryer (Fig. D.4). The panel has three controls, each of which is intended both to actuate the features and to relate the settings to the person using the dryer. On the left are two toggle switches. The top switch is a three-position switch that controls the temperature setting to either "Low," "Permanent Press," or "High." The bottom switch is a two-position switch that is automatically toggled to off at the end of the cycle or when the dryer door is opened. This switch must be pushed to start the dryer. The dial on the right controls the time for either the no-heat cycle (air dry) on the top half of the dial or the heated cycle on the bottom half.

The dryer controls must communicate two functions to the human: temperature setting and time. Unfortunately, the temperature settings on this panel are hard to sense because the "Temperature" rocker switch does not clearly indicate the status of the setting and the air-dry setting for temperature is on the dial that can override

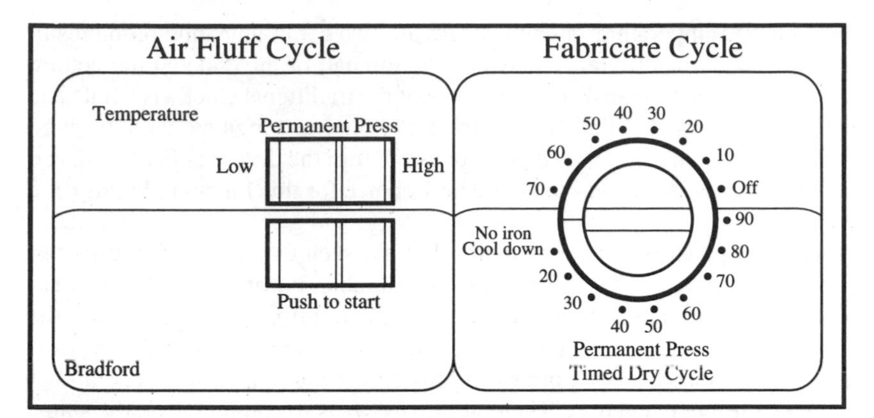

FIGURE D.4
Clothes dryer control panel (from J. H. Burgess, *Designing for Humans: The Human Factor in Engineering,* Petrocelli Books, Princeton, N.J., 1986).

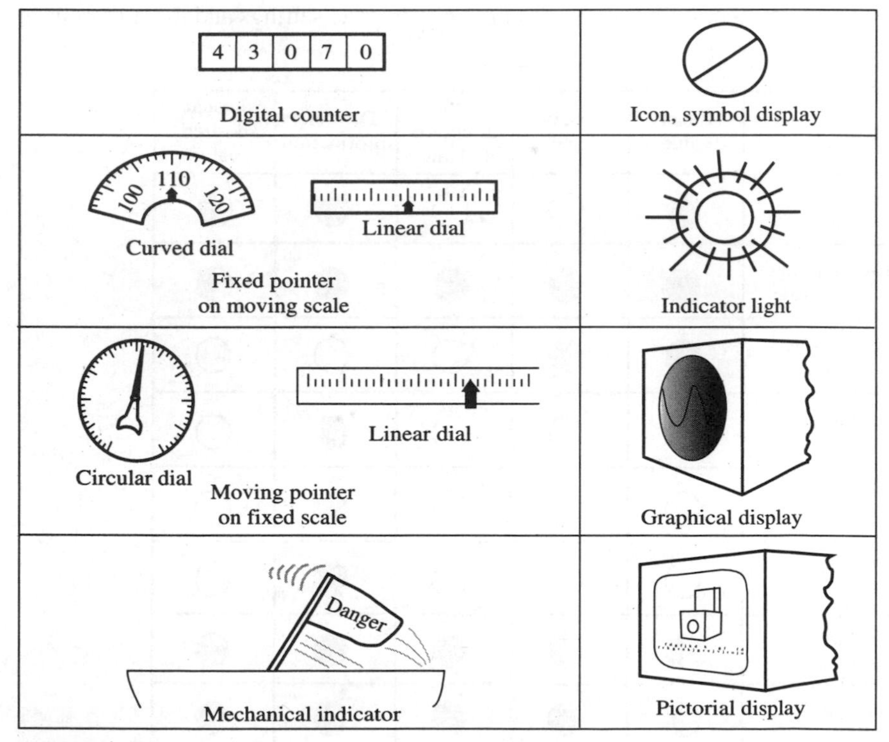

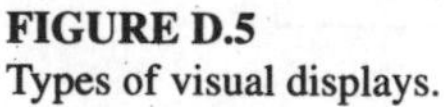
FIGURE D.5
Types of visual displays.

the setting of the "Temperature" switch. There are two communication problems in the time setting also: the difference between the top half of the dial and the bottom half is not clear and the time scale is the reverse of the traditional clockwise dial. The user must not only sense the time and temperature but must regulate them through the controls. Additionally, there must be a control to turn the dryer on. For this dryer, the rocker switch does not appear to be the best choice for this function. Finally, the labeling is confusing.

This control panel is typical of many that are seen every day. The user can figure out what to do and what information is available, but it takes some conjecturing. The more guessing required to understand the information and to control the action of the product, the lower the perceived quality of the product. If the controls and labeling were as unclear on a fire extinguisher, for example, it would be all but useless—and therefore dangerous. There are many ways to communicate the status of a product to a human. Usually the communication is visual; however, it can also be through tactile or audible signals. The basic types of visual displays are shown in Fig. D.5. When choosing which of these displays to use, it is important to consider the type of information that needs to be communicated. Figure D.6 relates five different types of information to the types of displays.

Comparing the clothes-dryer control panel of Fig. D.4 to the information of Fig. D.6, the temperature controls require only discrete settings and the time control

	Exact value	Rate of change	Trend, direction of change	Discrete information	Adjusted to desired value
Digital counter	Recommended	Not suitable	Not suitable	Recommended	Acceptable
Moving pointer on fixed scale	Recommended	Recommended	Recommended	Recommended	Acceptable
Fixed pointer on moving scale	Recommended	Recommended	Not suitable	Not suitable	Not suitable
Mechanical indicator	Not suitable	Not suitable	Not suitable	Recommended	Not suitable
Symbol display	Not suitable	Not suitable	Not suitable	Recommended	Not suitable
Indicator light	Not suitable	Not suitable	Not suitable	Recommended	Not suitable
Graphical display	Acceptable	Acceptable	Recommended	Recommended	Recommended
Pictorial display	Acceptable	Recommended	Recommended	Recommended	Recommended

○ Not suitable ◑ Acceptable ● Recommended

FIGURE D.6
Appropriate uses of common visual displays.

a continuous (but not accurate) numerical value. Since toggle switches are not very good at displaying information, the top switch on the panel of Fig. D.4 should be replaced by any of the displays recommended for discrete information. The use of the dial to communicate the time setting seems satisfactory.

To input information into the product, there must be controls that readily interface with the human. Figure D.7 shows 18 common types of controls and their

	Control	Dimension, mm	Force F, N Moment M, N · m	2 positions	>2 positions	Continuous adjustment	Precise adjustment	Quick adjustment	Large force application	Tactile feedback	Setting visible	Accidental actuation
Turning movement	Handwheel (d, D)	D: 160–800 d: 30–40	D / M: 160–800 mm / 2–40 N · m 200–250 mm / 4–60 N · m	◑	◑	●	●	◑	●	○	○	◑
Turning movement	Crank (r, l, d)	Hand (finger) r: <250 (<100) l: 100 (30) d: 32 (16)	r / M: <100 mm / 0.6–3 N · m 100–250 mm / 5–14 N · m	◑	◑	●	◑	●	◑	◑	◑	○
Turning movement	Rotary knob (h, D)	Hand (finger) D: 25–100 (15–25) h: >20 (>15)	D / M: 15–25 mm / 0.02–0.05 N · m 25–100 mm / 0.3–0.7 N · m	◕	◕	●	●	◑	○	○	○	◑
Turning movement	Rotary selector switch (b, l, h)	l: 30–70 h: >20 b: 10–25	l / M: 30 mm / 0.1–0.3 N · m 30–70 mm / 0.3–0.6 N · m	●	●	◔	◕	◔	◑	●	●	◑
Turning movement	Thumbwheel (b)	b: >8	F = 0.4–5 N	◑	◑	●	●	◑	○	○	○	◑
Turning movement	Rollball (D)	D: 60–120	F = 0.4–5 N	○	○	●	●	◕	○	○	○	◑
Linear movement	Handle (slide) (d, l, 1, 2)	d: 30–40 l: 100–120	F_1 = 10–200 N F_2 = 7–140 N	●	●	●	◕	◑	●	◑	◑	○
Linear movement	D-handle (d, b)	d: 30–40 b: 110–130	F = 10–200 N	●	●	●	◑	◑	●	◔	◔	○
Linear movement	Push button (d)	Finger: d > 15 Hand: d > 50 Foot: d > 50	Finger: F = 1–8 N Hand: F = 4–16 N Foot: F = 15–90 N	●	○	○	○	●	○ ◑ ●	○	○	●* ○
Linear movement	Slide (b, l)	l: >15 b: >15	F = 1–5 N (Touch grip)	●	◕	◑	◑	◑	○	○	◕	●
Linear movement	Slide (b, h)	b: >10 h: >15	F = 1–10 N (Thumb-finger grip)	●	◕	◕	◕	◑	◑	○	◕	◑
Linear movement	Sensor key (b, l)	l: >14 b: >14		●	○	○	○	●	○	○	○	◑

*Recessed installation

FIGURE D.7

Appropriate uses of hand- and foot-operated controls (from G. Salvendy (ed.), *Handbook of Human Factors,* Wiley, 1987).

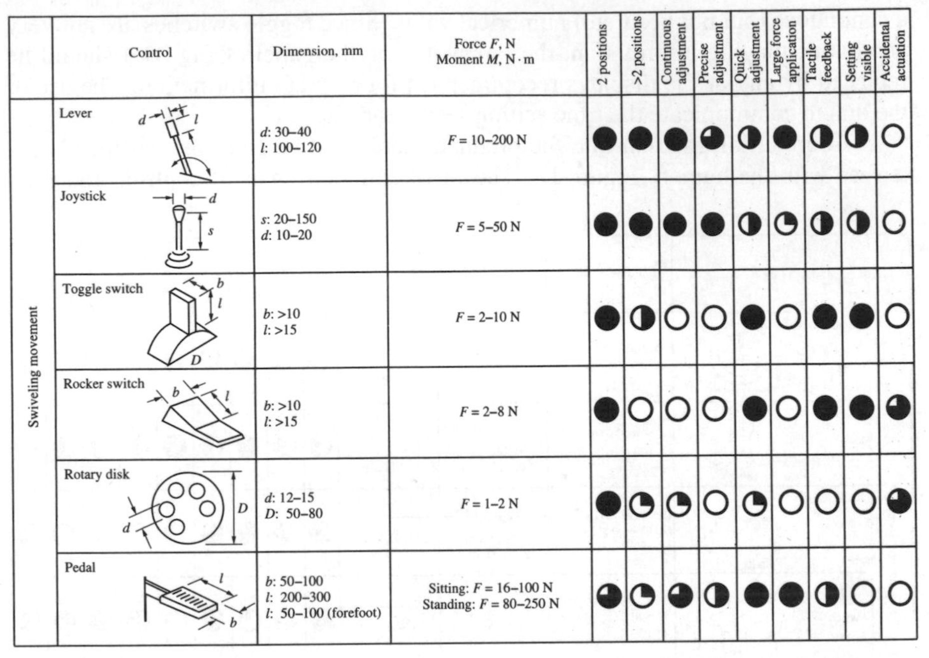

	Control	Dimension, mm	Force F, N Moment M, N · m	2 positions	>2 positions	Continuous adjustment	Precise adjustment	Quick adjustment	Large force application	Tactile feedback	Setting visible	Accidental actuation
Swiveling movement	Lever	d: 30–40 l: 100–120	F = 10–200 N	●	●	●	◕	◑	●	◑	◑	○
Swiveling movement	Joystick	s: 20–150 d: 10–20	F = 5–50 N	●	●	●	●	◑	◔	◑	◑	○
Swiveling movement	Toggle switch	b: >10 l: >15	F = 2–10 N	●	◑	○	○	●	○	●	●	○
Swiveling movement	Rocker switch	b: >10 l: >15	F = 2–8 N	●	○	○	○	●	○	●	●	◕
Swiveling movement	Rotary disk	d: 12–15 D: 50–80	F = 1–2 N	●	◔	◔	○	◔	○	○	○	◕
Swiveling movement	Pedal	b: 50–100 l: 200–300 l: 50–100 (forefoot)	Sitting: F = 16–100 N Standing: F = 80–250 N	◕	◔	◕	◑	●	●	◑	○	○

FIGURE D.7 (*continued*)
Appropriate uses of hand- and foot-operated controls (from *Handbook of Human Factors,* Wiley, 1987).

use characteristics; it also gives dimensional, force, and recommended use information. Note that the rotary selector switch is recommended for more than two positions and is rated between "acceptable" and "recommended" for precise adjustment. Thus the rotary switch is a good choice for the time control of the dryer. Also, for rotary switches with diameters between 30 and 70 mm, the torque to rotate them should be in the range from 0.3 to 0.6 N · m. This is important information when one is designing or selecting the timing switch mechanism. In addition, note that for the rocker switch no more than two positions are recommended. Thus, the top switch on the dryer, Fig. D.4, is not a good choice for the temperature setting.

An alternative design of the dryer control panel is shown in Fig. D.8. The functions of the dryer have been separated, with the temperature control on one rotary switch. The "Start" function, a discrete control action, is now a button, and the timer switch has been given a single scale and made to rotate clockwise. Additionally, the labeling is clear and the model number is displayed for easy reference in service calls.

In general, when designing controls for interface with humans, it is always best to simplify the structure of the tasks required to operate the product. Recall the characteristics of the short-term memory discussed in Chap. 3. We learned there that humans can deal with only seven unrelated items at a time. Thus, it is important not to expect the user of any product to remember more than four or five steps. One way to overcome the need for numerous steps is to give the user mental aids. Office

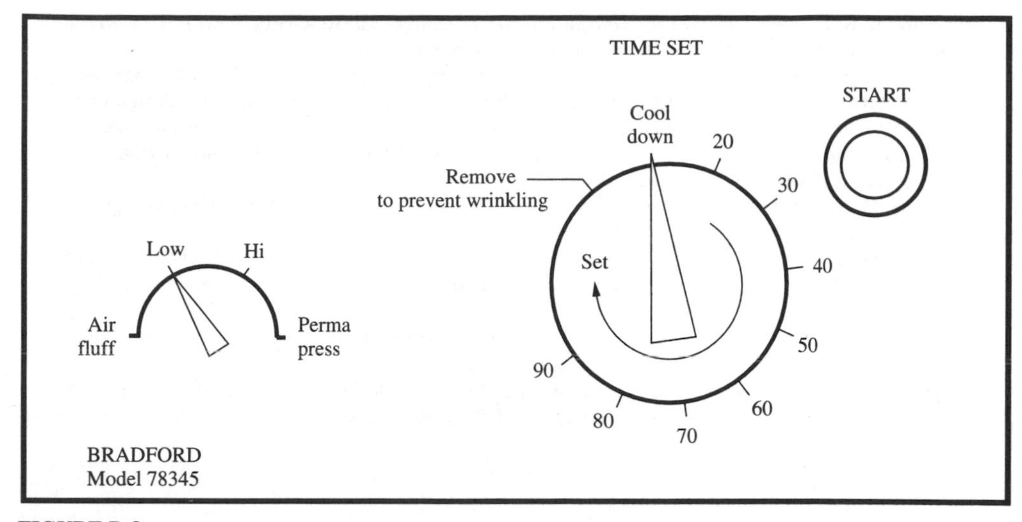

FIGURE D.8
Redesign of the clothes dryer control panel of Fig. D.4.

reproducing machines often have a clearly numbered sequence (symbol display) marked on the parts to show how to clear a paper jam, for example.

In selecting the type of controller, it is important to make the actions required by the system match the intentions of the human. An obvious example of a mismatch would be to design the steering wheel of a car so that it rotates clockwise for a left turn—opposite to the intention of the driver and inconsistent with the effect on the system. This is an extreme example; the effect of controls is not always so obvious. It is important to make sure that people can easily determine the relationship between the intention and the action and the relationship *between the action and the effect* on the system. *A product must be designed so that when a person interacts with it, there is only one obviously correct thing to do.* If the action required is ambiguous, the person might or might not do the right thing. The odds are that many people will not do what was wanted, will make an error, and, as a result, will have a low opinion of the product.

D.5 SOURCES

Burgess, J. H.: *Designing for Humans: The Human Factor in Engineering,* Petrocelli Books, Princeton, N.J., 1986. A good text on human factors written for use by engineers; the dryer example is from this book.

Damon, A., H. W. Stoudt, and R. A. McFarland: *The Human Body in Equipment Design,* Harvard University Press, Cambridge, Mass., 1966. Broad range of anthropometric and biomechanical tables.

Dreyfuss, H.: *The Measure of Man: Human Factor in Design,* Whitney Library of Design, New York, 1967. This is a loose-leaf book of 30 anthropometric and biomechanical charts suitable for mounting; two are life-size, showing a 50th-percentile man and woman. A classic.

Human Engineering Design Criteria for Military Systems, Equipment, and Facilities, MIL-STD 1472D. Four hundred pages of human factors information.

Human Factors Engineer Design for Army Material, MIL-HDBK 759A. Almost 700 pages of information on all aspects of human factors.

Hutchinson, R. D.: *New Horizons for Human Factors in Design,* McGraw-Hill, New York, 1981. A good text on human factors, organized by areas of application.

Jones, J. V.: *Engineering Design: Reliability, Maintainability and Testability,* TAB Professional and Reference books, Blue Ridge Summit, Pa., 1988. This book considers engineering design from the view of military procurement, relying strongly on military specifications and handbooks.

Norman, D.: *The Psychology of Everyday Things,* Basic Books, New York, 1988. Guidance for designing good interfaces for humans; light reading.

Salvendy, G. (ed.): *Handbook of Human Factors,* Wiley, New York, 1987. Eighteen hundred pages of information on every aspect of human factors.

Sonar, D. G.: *The Expert Witness Handbook: A Guide for Engineers,* Professional Publications, San Carlos, Calif., 1985. A paperback, with details on being an expert witness for products liability litigation.

System Safety Program Requirements, MIL-STD 882B. U.S. Government Printing Office, Washington, D.C. The hazard assessment is from this standard.

Tilly, A. R.: *The Measure of Man and Woman,* Whitney Library of Design, New York, 1993. An updated version of the above classic rewritten by one of Dreyfuss's associates.

INDEX

abstraction, level of, 33, 154
accuracy, 215
action verb, 126
aging effects, 218
allowable strength, 314
analytical modeling, 84, 220
ANSI standards, 110
anthropometric data, 327
applied stress, 314
assemblies, 19
assembly, design for, 255–275
assembly drawings, 177, 288
assembly manager, 81
assembly sequence evaluation, 265
ASTM, 110

behavior, human problem solving, 47
behavior, product, 21, 212
benchmarking, 113, 148
best practices, 66
bicycle Splashgard, 75, 92, 201, 238
bill of materials (BOM), 180
biomechanics, 330
brainstorming, 145
brainwriting, 147

codes of practice, 111
cognitive psychology, 38
communication, 14, 72, 169
comparison, in evaluation, 153
competition benchmarking, 113
component(s), 19
component design, 189
computer-aided design (CAD), 83, 175
concept development, 65
concept evaluation, 152–172
concept generation, 120–151
conceptual design, 120–172
concurrent engineering, 8
configuration design, 24
configurations, component, 187
connections, 188
constraints, 31, 186
control, design process, 43, 47
control factors, 236
cost of design, 2
cost estimating, 243
  machined components, 248
  plastic components, 251
cost requirements, 110
CPM, 87
creativity, 51–54
critical dimensions, 188
critical functions, 188
critical parameters, 110, 157, 219
customer:
  definition, 102
  identification, 102
  satisfaction, 104
customers' requirements:
  collection of, 104
  importance of, 112
  quality in, 104
  types of, 108

Da Vinci, Leonardo, 140
decisions, 31, 32
decision making, 153

decision matrix, 160
decomposition:
  functional, 125
  mechanical systems, 18
design:
  cost of, 2
  decisions, 31
  effect on cost, 2
engineer, 79
  freedom, 15
  phases, 9
  problems, 12
    configuration, 24
    original, 27
    parametric, 26
    selection, 23
    redesign, 27
  project types, 78
  refinement, 33
design-build-test cycle, 154
design-test-build cycle, 154
design for assembly, 255
design of experiments, 215, 234
design for manufacture, 255–274
design for test and maintenance, 279
design for the environment, 279–282
design for reliability, 274–278
design life, 9
design notebook, 73
design process:
  definition, 7, 60
  measures, 2
  over-the-wall, 7
  concurrent, 8
  simultaneous, 8
  integrated product and process design, 8
design records, 73, 284
design review, 65
detail drawings, 177
detailer, 80
deterministic, models, 216
dimensioning, 179
dimensions, sensitive, 180
disassembly, 281
documentation, 73
domain knowledge, 40
drafter, 80
drawings:
  assembly, 177
  detail, 177
  importance in the design process, 175
  layout, 176

electromechanical design, 20
energy flow, 122
engineering change notice, 289
engineering changes, 4
engineering specifications, 64, 115
engineering targets, 116
environmental concerns, 280
evaluation, 47, 153
experiments, 215, 236
  design of, 215, 234

factor of safety, 314–324
failure modes and effects analysis, (FMEA), 276
Fastex clips, 204–210
feasibility judgment, 155
feature, 20
flow of energy, information, and materials, 122
focus groups, 104
force flow, 193
form generation, 186–199
form-function relation, 184
function, importance of, 21, 120, 184
functional performance, 108
functional changes, 212
functional decomposition, 122–133

Gantt chart, 87
generation, 120–151, 184–211
goals, 31
go/no-go screening, 159
graphical language, 30
green design, 279
group technology, 195

hazard assessment, 169, 275
house of quality, 99–119
human:
  as controller, 330
  as sensor, 330
  as source of power, 330
  as work space occupant, 326
human factors, 108, 325–336
human problem solving, 38–51

ill-defined problem, 13
industrial designer, 81
information:
  flow, 122
  value of, 32
information-centered design, 8, 85
information processing system, human, 38–45
injection-molding costs, 251
installation instructions, 288
integrated product and process design, 8
interface design, 188

judgment, 32

Kano model, 104
knowledge, types of, 40
knowledge gain during design, 15, 32

languages of design, 30, 154
layout drawings, 176
learning during design, 15
level of abstraction, 33
liability, 167
life-cycle requirements, 109
life of product, 10
line fallout, 4
long-term memory, 42–43

maintainability, 279
maintenance manual, 288
make or buy decision, 248
manufacturing engineer, 80
manufacturing variations, 218
Mariner IV satellite, 98
market-driven projects, 62
materials:
  commonly used, 292
  flow of, 122
  properties, 294–305
  selection, 199
materials specialist, 80
mature design, 28
mean square deviation, 235
mean time between failures (MTBF), 277
mechatronics, 19
memory, human:
  controller 43
  long-term, 42–43
  short-term, 40–42
milestone chart, 87
models, 32, 83, 218
morphology, 133

nail clippers, 261
noise, 215, 236
  types of, 218
normal distribution, 216, 309–313

observation of customers, 104
operating instructions, 288
original design, 27
over-the-wall design process, 7
overhead, 245

P-diagram, 214
parametric design, 26
parameter design, 222, 232
parameters, critical, 110, 157, 219
part fallout, 4
parts list, 181

patching, 195
patent:
  application, 286–288
  searching, 141–144
performance, 21, 108
  evaluation importance, 212
  evaluation goals, 213
personnel requirements, 86
PERT, 87
physical modeling, 83
planning, 64
  definition, 77
  steps, 85
probability, normal, 309–313
problem:
  ill-defined, 13
  understanding of, 45
problem solving:
  actions, 14
  behavior, 47
  human, 38–51
process selection, 199
product design, 173–291
product design engineer, 79
product development, 65, 173–283
  plan, 77
product:
  cost, 244
  design phase, 174
  evaluation, 212–283
  generation, 184–211
  liability, 167
  life cycle, 10
  models, 84
  safety, 165
product data management, 181
product manager, 80
production, 185
project types, 78
prototypes, 83–85
Pugh's method, 160–164

quality, 4
  basic, performance, and excitement 104
quality assurance, 288
quality control, 224, 288
quality control/quality assurance specialist, 80
quality function deployment, 99–119
quality product, 2

rapid prototyping, 221
recycling, 280
redesign, 27
  project initiation by need for, 62
refinement, 33, 195
reliability, 109, 277

requirements:
customer, 108
engineering, 115
retirement, product, 280, 288
reuse, 280
robust design, 179, 221, 234
routine design, 28

safety, 165
selection design, 23
semantic language, 30
sensing, human, 330
sensitivity analysis, 225–231
short-term memory, 40–42
signal-to-noise ratio, 235
simulation, 83
simultaneous engineering, 8
6–3–5 method, 147
space shuttle, 62, 70, 132
spatial constraint, 186
specification development, 98–119
Sperry Gyroscope Co., 140
Splashgard example, 75, 92, 202, 238
standard deviation, 310
standards, 110
stochastic analytic methods, 218
strong shapes, 191
supplier's representation, 81
surveys, 104
system, 19

Taguchi's method, 225
targets for engineering requirements, 116
tasks, 85
teams:
goals, 55
members, 79–81
performance building, 56
roles, 55–56
structure, 81
technician, 80
technology-driven design, 62
technology readiness, 157
testability, 279
time requirements, 110
time to design, 2, 86
tolerance design, 223, 232
tolerance stack-up, 227–229
tolerances, 179, 218
trade journals, 145

understanding the problem, 45
Underwriters Laboratories (UL), 110
Unilever, 140

value engineering, 253
variance, 311
variation, 215
vendor development, 201, 289
vendor's representative, 81

worst-case analysis, 227